W0230107

Mohammad Aslam Khan Khalil   (Ed.)

Atmospheric Methane

Springer

Berlin
Heidelberg
New York
Barcelona
Hong Kong
London
Milan
Paris
Singapore
Tokyo

Mohammad Aslam Khan Khalil (Ed.)

Atmospheric Methane

Its Role in the Global Environment

With 61 Figures and 49 Tables

 Springer

Professor Dr. MOHAMMAD ASLAM KHAN KHALIL
Department of Physics
Portland State University
P.O. Box 751
Portland, OR 97207-0751
USA

ISBN 3-540-65099-7 Springer-Verlag Berlin Heidelberg New York

Library of Congress Cataloging-in-Publication Data

Khalil, M. A. K. (Mohammad Aslam Khan), 1950-
    Atmospheric methane : its role in the global environment / M.A.K. Khalil.
        p. cm.
    Includes bibliographical references and index.
    ISBN 3540650997
    1. Atmospheric methane. I. Title

    QC879.85 .K43 2000
551.51'12--dc21

© Springer-Verlag Berlin Heidelberg 2000
Printed in Germany

Cover design:
Typesetting: Camera ready by editors
SPIN 10555439   31/3136 - 5 4 3 2 1 0 - Printed on acid-free paper

Preface

Atmospheric methane is an important trace gas involved in man-made climate change. It may be second only to carbon dioxide in causing global warming. Methane affects also the oxidizing capacity of the atmosphere by controlling tropospheric OH radicals and creating O_3, and it affects the ozone layer in the stratosphere by contributing water vapor and removing chlorine atoms. In the long term, methane is a natural product of life on earth, reaching high concentrations during warm and biologically productive epochs. Yet the scientific understanding of atmospheric methane has evolved mostly during the past two decade after it was shown that concentrations were rapidly rising.

The first edition of this book was commissioned by North Atlantic Treaty Organization's Scientific and Environmental Affairs Division as a product of an Advanced Research Workshop (NATO-ARW). The conference was held during the week of October 6, 1991, at Timberline Lodge on Mount Hood near Portland, Oregon. About 100 scientists participated.

This is the second edition of the book that has been revised, trimmed, and brought up to date. The book is not a collection of specialized research articles, but rather it consists of review articles that form a coordinated whole. The logic of the organization is to answer four questions that follow each other. How are methane concentrations measured in the atmosphere? What do the measurements tell us about atmospheric distributions and trends? How do we explain the observed concentrations in terms of the specific sources and sinks? And finally, what are the implications of these observations for the environment? We go from the most experimental to the most theoretical. Results of original research, I believe, belong in peer-reviewed journals instead of a book such as this. Accordingly, some 55 technical papers, arising from the NATO-ARW were published in *Chemosphere, 26* #s 1-4 (1993) written by 125 authors from all over the world. This publication remains a companion volume for the present version of the book.

In putting together this book I was fortunate to have the support of my family, of many friends and colleagues, and of generous sponsors. I especially want to thank Martha J. Shearer, Edie Taylor, and Francis Moraes. I received much valuable advice, encouragement, and major contributions from the organizing committee for the Advanced Research Workshop, which consisted of Paul Crutzen, Robert Harriss, Rei Rasmussen, Dominique Raynaud, and Wolfgang Seiler. Major financial support from NATO's Scientific and Environmental Affairs Division was the foundation for this work and is gratefully acknowledged. The critical involvement of scientists from many non-NATO countries and the excess of U.S. scientists was supported by supplementary grants from the National Science Foundation (ATM-9120070) and the United States Environmental protection agency (Order No. 1-W-1085-NASA). Both these grants were given to Andarz Company, which provided additional support.

M.A.K. Khalil, Professor   July 1999
Department of Physics
Portland State University
P.O. Box 751
Portland, Oregon 97207-0751, USA

Contents

1. Atmospheric Methane: An Introduction

M.A.K. Khalil
Department of Physics, Portland State University
P.O. Box 751, Portland, Oregon 97207, USA

Methane is a greenhouse gas thought to be second only to CO_2 as an agent of future global warming. The increasing trend is the single most important reason for the current interest in methane. It is the foundation for much of the research on methane during the past decade, particularly on the global budget, which is tied directly to explaining why it is increasing. Figure 1 (a-f) shows the concentration and trends of methane over the recent decades and back to a thousand years. I want to describe briefly the progress towards the understanding of global budgets and trends, and to provide a guide to the book.

The existence of methane in the Earth's atmosphere has been known since 1948 when Migeotte published the first measurements showing concentrations of about 2 ppmv. It was listed under "non-variable components of atmospheric air" by Glueckauf in 1951 and similarly in many textbooks since. In his 1951 paper Glueckauf also pointed out that, from the isotopic analysis by F.W. Libby, atmospheric methane appeared to be of mostly biogenic origin, and there was evidence that the concentration was likely to be around 1200 ppbv rather than the more commonly accepted 2 ppmv. From the 1950s to the early part of the 1980s, there was a prevalent belief that methane was a stable gas in the Earth's atmosphere. In retrospect, by 1975, there were three important published results that could have established that methane had more than doubled during the last century and was rapidly increasing. These facts were not discovered at the time because methane was thought to be among the stable and non-variable components of the Earth's atmosphere.

1. During the 1960s and the 1970s, sporadic measurements of methane had been reported by many scientists as gas chromatographic techniques came into common use. These measurements taken together contained evidence of increasing concentrations, but no trend analyses were reported until much later when we analyzed these data to support the results of our systematic time series of atmospheric measurements (Rasmussen and Khalil, 1981; Khalil et al., 1989).

2. Early global budgets of methane, such as by Ehhalt (1974), attributed 100-220 Tg/yr to animals (mostly cattle) and 280 Tg/yr to rice agriculture out of total emissions of 590-1060 Tg/yr. Accordingly, at least half the global source of methane was from human activities, although in the papers these emissions were reported as biogenic and not considered to be anthropogenic. With so much of the source being controlled by human activities it was reasonable to believe that the methane concentrations should be increasing as Singer suggested in 1971.

Mohammad Aslam Khan Khalil (Ed.)
Atmospheric Methane
© *Springer-Verlag Berlin Heidelberg 2000*

3. In 1973, Robbins et al. showed that methane concentrations were only 560 ppbv in bubbles or air extracted from polar ice. Because high CO concentrations were observed in the same samples, the authors attributed the low methane levels to chemical processes in ice that transformed methane ultimately to CO. Later experiments showed that even freshly fallen snow can contain large amounts of CO, suggesting that the CO in ice cores may not come from methane. Besides, there was no clear mechanism that would cause methane to transform to CO in ice deep below the Earth's surface. The obvious alternative explanation of the Robbins experiments was that the low concentrations in polar ice represented atmospheric levels of the past (Rasmussen et al., 1982; Khalil and Rasmussen, 1982).

Around 1980, the early work on the increasing trend of methane met with considerable skepticism and encountered much opposition. Soon, however, the atmospheric record became long enough that the increasing trend was undeniable, especially because the trend was so fast that it was easily seen even in measurements lasting only a year or two. During the intervening years, there has been much argument as to whether the increase of methane is caused by increasing emissions or a slowdown in the removal rates (sinks). For those who want to control methane concentrations, this is a most important issue (Rasmussen and Khalil, 1981; Khalil and Rasmussen, 1982, 1985; Madronich and Granier, 1992; Prinn et al., 1992; Krol et al., 1998)

In the 1980s there was a bias towards the sink explanation, perhaps because earlier papers emphasized such a mechanism to predict an increasing methane trend. The idea was that an increase of CO would lead to a decrease of OH because the two react with each other. Since OH also reacts with CH_4 and is in fact the principal mechanism for removing it from the atmosphere, less OH would lead to less methane being removed annually, causing an increasing trend in its concentration (Sze, 1977; Chameides et al., 1977). This was only a hypothetical mechanism since there was no observational evidence that CO had increased in the global atmosphere, and even if it had, there are other atmospheric processes that could counteract the effect of CO on OH levels in the atmosphere (Pinto and Khalil, 1991) . On the other hand, an evaluation of the methane sources clearly shows that they have increased significantly during the past 100-300 years. When these changes are taken into account, the trends of methane can be explained entirely by the increases in emissions (Khalil and Rasmussen, 1985). But there are enough uncertainties that changes in OH may also have affected methane trends but, most likely, to a much lesser extent than the sources. The present slowdown in trends is also probably caused by the lack of increase of emissions from the major anthropogenic sources such as rice agriculture and cattle populations (Khalil et al., 1996).

The causes of the trends relate fundamentally to the mass balance of methane, which, for any infinitesimal or finite region or box, can be expressed as the balance between processes that add or remove methane concentration in the box. Processes that add methane are direct emissions from sources inside and transport of methane from outside the box, and processes that reduce concentrations in the box are transport outward and removal by chemical reactions or deposition. Trends arise when these competing processes are out of balance. The elements of the mass balance are thus seen to be sources, sinks, transport, and concentrations. Although all these elements are discussed in the book, there is a special emphasis on estimates of methane emissions from its various sources and the factors that control these emissions; this subject is more than half of the book. Although this is a complex and difficult aspect of the global methane cycle, it is amenable to direct experimental observations and is valued because it provides the opportunity to control global warming by reducing methane emissions. Current estimates, based on the results presented in this book, of the global emissions from various sources are given in Table 1

On the global scale, the present understanding of sources, sinks, atmospheric burden, and the trends are quantitatively balanced. The fraction of anthropogenic emissions is constrained by the existing atmospheric and ice core data and is likely to be quite accurate. The contribution of individual sources to the present atmospheric concentrations is substantially more uncertain and is an active area of research. On yet smaller scales of time and space, such as emissions from countries, small regions, or over seasons, there are practically no constraints, leading to mass balances that are not unique and may be completely unreliable in many cases. At these small scales current technology for the assessment of methane creates irreducible uncertainties that prevent accurate estimates of emissions.

In the early measurements, around 1980, the trends were almost 2%/yr , giving it a doubling time of about 40 years. These rates were not sustained, as can be seen in Figure 1. At present, methane concentrations are increasing at a slow rate of about 5 ppbv/yr or 0.3%/yr. What does this mean? Will methane stabilize without any effort on our part or need for inter-governmental agreements? Or are we in a cusp when one class of sources is reaching stability because of limits of growth while another class may be on the rise? For instance, sources related to food production such as rice and cattle are reaching their limits, while sources related to energy and waste disposal are not. If so, methane trends may revert to faster increases in the future and rise to levels we cannot predict. Now that we are beginning to understand past changes, the past is getting uncoupled from the future. This book provides the information and tools necessary to address the issue of how methane is likely to affect the future climate and atmospheric chemistry.

References

Chameides, W.L., S.C. Liu, R.J. Cicerone. 1977. Possible variations in atmospheric methane. *J. Geophys. Res., 82*:1,795-1,798.

Ehhalt, D.H. 1974. The atmospheric cycle of methane. *Tellus, 26*:58-70.

Glueckauf, E. 1951. The composition of atmospheric air. In: *Compendium of Meteorology* (Thomas F. Malone, ed.), American Meteorological Society, Boston, U.S.A., pp 3-9.

Khalil, M.A.K., R.A. Rasmussen. 1982. Secular trend of atmospheric methane. *Chemosphere, 11*:877-883.

Khalil, M.A.K., R.A. Rasmussen. 1985. Causes of increasing methane: Depletion of hydroxyl radicals and the rise of emissions. *Atmos. Environ., 19*:397-407.

Khalil, M.A.K., R.A. Rasmussen. 1990. Atmospheric methane: Recent global trends. *Environ. Sci. Technol., 24*:549-553.

Khalil, M.A.K., R.A. Rasmussen, M.J. Shearer. 1989. Trends of atmospheric methane during the 1960s and 1970s. *J. Geophys. Res., 94*:18,279-18,288.

Khalil, M.A.K., M.J. Shearer, and R.A. Rasmussen. 1996. Atmospheric Methane over the Last Century. *World Resource Review, 8*: 481-492.

Krol, M., P.J. van Leeuwen and J.Leilieveld. 1998. Global OH trend inferred from methylchloroform measurements. *J. Geophys. Res. 103:* 10697-10712.

Madronich, S., C. Granier. 1992. Impact of recent total ozone changes on tropospheric ozone photodissociation, hydroxyl radicals and methane trends. *Geophys. Res. Lett., 19:*465-467.

Migeotte, M.V. 1948a. Spectroscopic evidence of methane in the Earth's atmosphere. *Phys. Rev., 73:*519-520.

Migeotte, M.V. 1948b. Methane in the Earth's atmosphere. *J. Astrophys., 107:*400-403.

Pinto, J.P., M.A.K. Khalil. 1991. The stability of tropospheric OH during ice ages, inter-glacial epochs and modern times. *Tellus, 43B:*347-352.

Prinn, R.G., et al. 1992. Global average concentration and trend of hydroxyl radicals deduced from ALE/GAGE trichloroethane (methyl chloroform) data for 1978-1990. *J. Geophys. Res., 97:*2,445-2,461.

Rasmussen, R.A., M.A.K. Khalil. 1981. Atmospheric methane: Trends and seasonal cycles. *J. Geophys. Res., 86:*9,826-9,832.

Rasmussen, R.A., M.A.K. Khalil, S.D. Hoyt. 1982. Methane and carbon monoxide in snow. *JAPCA, 32:*176-178.

Robbins, R.C., L.A. Cavanagh, L.J. Salas. 1973. Analysis of ancient atmospheres. *J. Geophys. Res., 78:*5,341-5,344.

Singer, S.F. 1971. Stratospheric water vapor increase due to human activities. *Nature, 233:*543-545.

Steele, L.P., et al. 1992. Slowing down of the global accumulation of atmospheric methane during the 1980s. *Nature, 358:*313-316.

Sze, N.D. 1977. Anthropogenic CO emissions: Implications for the atmospheric CO-OH-CH_4 cycle. *Nature, 195:*673-675.

Table 1. Current estimates of global emissions from various sources.

	Tg/yr	Range (if given)	Chapter
Natural Sources			
Wetlands	100		Matthews
Termites	20	15-35	Judd et al., 1993 (first book)
Open ocean	4		Judd et al., 1993 (first book) Lambert, 1993 (first book)
Marine sediments	5	0.4-12.2	Judd (Appendix)
Geological	14	12-36	Judd, after Lacroix, 1993
Wild fire	2		Levine et al.: Biomass burning in temperature and boreal forests is assumed to be wild fire
Total Natural	145		
Anthropogenic Sources			
Rice	60	40-90	Shearer and Khalil (this volume)
Animals	81		Johnson et al. (this volume)
Manure	14		Johnson et al. (this volume)
Landfills	22		Thorneloe et al. (this volume)
Wastewater treatment	25		Thorneloe et al. (this volume)
Biomass burning	50	27-80	Levine et al. (this volume)
Coal mining	46		Kirchgessner (this volume)
Natural gas	30*	7-70	In Kirchgessner, from a summary of estimates; new estimate was not made
Other anthropogenic	13	7-30	Judd, after Lacroix (1993)
Low temperature fuels	17		Khalil et al. (1993): Methane from coal burning (Chemosphere)
Total Anthropogenic	358		
TOTAL	503		

* mid-range

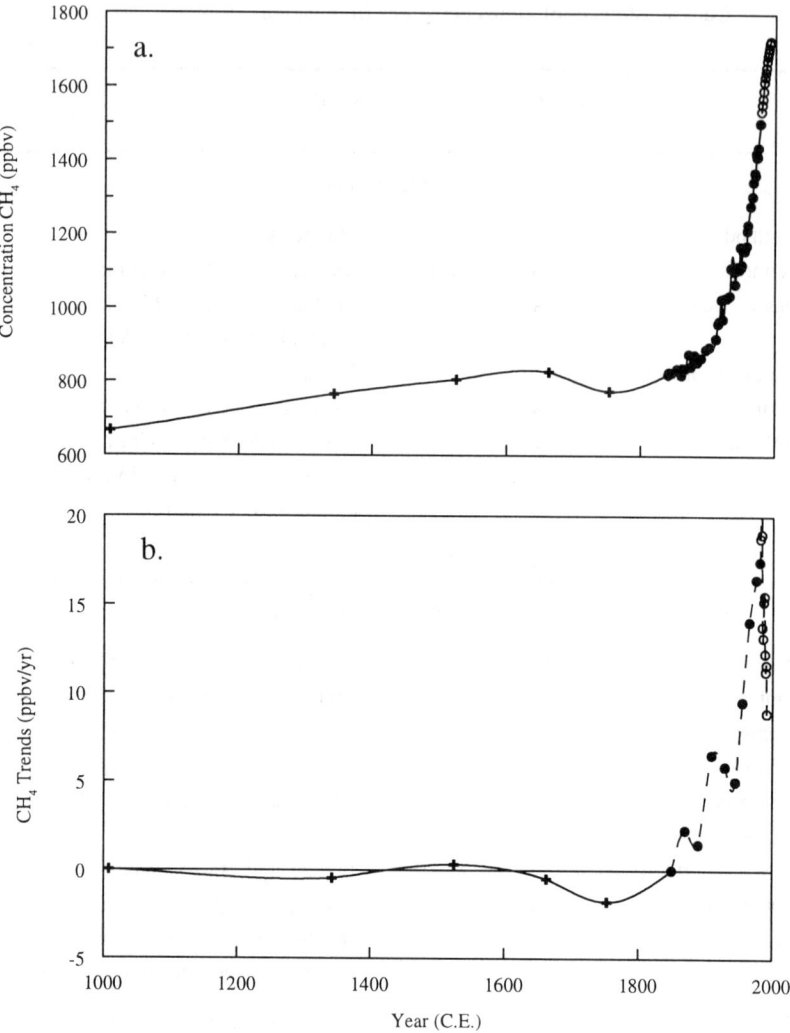

Fig. 1. Concentrations (a) and trends (b) of methane over the last 1000 years. Data of Rasmussen and Khalil (1984) (⊣) and Etheridge et al. (1992) (●) are from ice core samples. Data of Khalil and Rasmusen (○) are global averages from weekly flask samples collected at various latitudes. Trends of methane concentrations during the last 1000 years are calculated from the composite concentration data set. Rapid increases in methane started only about 200 years ago. These are linear regression estimates of trends over various (non-overlapping) periods of time between about 1000 c.e. and the present. For data between 1840 and 1940, trends were calculated over 20-year periods; for 1940-1980, over 10-year periods; and for 1980-1998, over 2-year periods. The calculated trends were placed at the middle of the time span in each calculation. The earlier data are more sparse. Trends for the period between 1000 c.e. and 1800 c.e. were calculated for every 10 data points, and the trends are placed at the average time spanned by the 10 data.

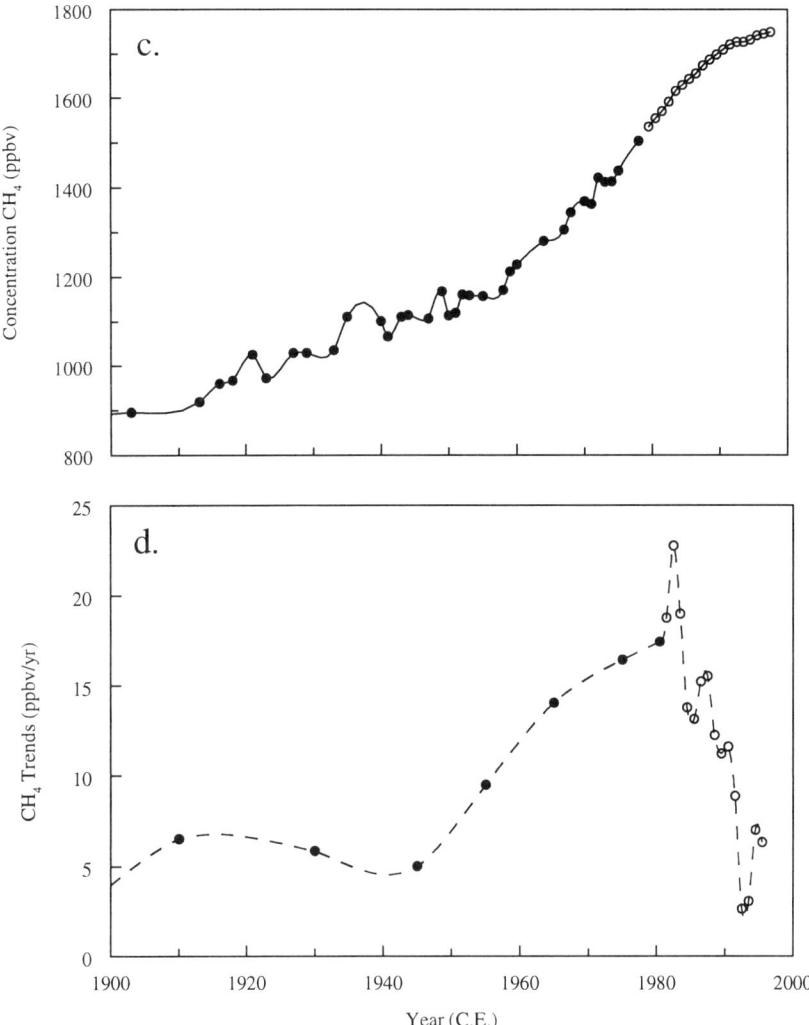

Fig. 1. Concentrations (c) and trends (d) of methane over the last century. Data of Etheridge et al. (●) are from ice core samples, and data of Khalil and Rasmussen (○) are global averages from weekly flask samples collected at various latitudes. Trends of methane concentrations were calculated from the concentration data. These are linear regression estimates of trend over various (non-overlapping) periods of time: for data between 1900 and 1940, trends were calculated over 20-year periods; for 1940-1980, over 10-year periods; and for 1980-1998, over 2-year periods. The calculated trends were placed at the middle of the time span in each calculation.

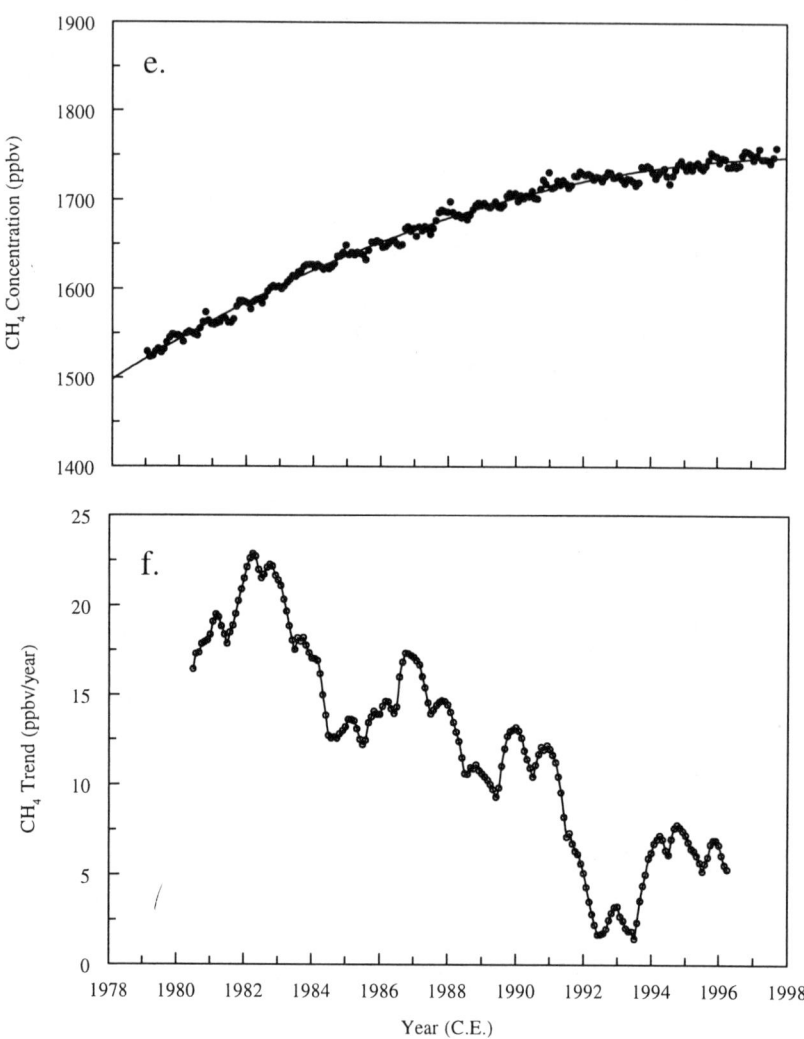

Fig. 1. Global average concentration (e) of methane from weekly flask samples collected from six different sites, and the trend (f) of methane calculated from the global average by linear regression of 3-year overlapping periods of time, plotted at the center point of the time period.

2. The Ice Core Record of Atmospheric Methane

J. Chappellaz and D. Raynaud
CNRS Laboratoire de Glaciologie et Géophysique de l'Environnement
54, rue Molière, BP 96
38402 St Martin d'Hères Cedex France

T. Blunier and B. Stauffer
Physikalisches Institut, Sidlerstrasse 5
Bern 3200, Switzerland

1. Introduction

The global evolution of atmospheric CH_4 has been documented by sporadic direct measurements in the atmosphere during the 1960s and 1970s and by systematic survey only since 1979. The data from this time up to 1983 indicate an increasing trend at a rate of about 1% per year (Rasmussen and Khalil, 1986; Steele et al., 1987; Blake and Rowland, 1988). The most recent measurements indicate a decrease of the global accumulation of atmospheric CH_4 during the years 1991 and 1992, and a return to a 1%/yr increase afterward (Steele et al., 1992; Dlugokencky et al., 1994; Lowe et al., 1994). The analysis of infrared solar absorption spectra (Rinsland et al., 1985; Zander et al., 1989) provides additional data of global concentrations for a few specific years (1951, 1975, 1981, 1984-87) and shows a mean increase of about 30% over the past 40 years. This long-term accumulation of CH_4 in the atmosphere is related to human activities, particularly from agriculture.

The analysis of gases trapped in ice makes possible the documentation of the anthropogenic influence back in time and the extension of the CH_4 record over periods such as the little ice age, the Holocene, and the glacial-interglacial cycles, encompassing climatic change with little or no human disturbance.

In this chapter we will first discuss the method itself. In particular, we will show how an atmospheric record of CH_4 from ice core analysis is obtained and what kind of accuracy can be expected from such reconstruction. We will then review the existing ice records for CH_4 and discuss what can be learnt in terms of the CH_4 cycle and its relation to the climate.

2. Process of Atmospheric CH_4 Recording in Ice

The air trapping process in natural ice is illustrated in Figure 1. Snow deposited on ice sheets in areas free of summer melting is compressed and sintered under the weight of subsequently fallen snow (Schwander, 1989). During this stage, the

Mohammad Aslam Khan Khalil   (Ed.)
Atmospheric Methane
© *Springer-Verlag Berlin Heidelberg 2000*

interstitial air in the open pores of the snow layers diffuses slowly between the atmosphere and the deepest snow layers of the firn column. Its composition is slightly changed under the combined effect of gravitation (heavier molecules are enriched in the bottom compared with the atmosphere) and of molecular diffusion (mixing ratios are controlled by the diffusion coefficient of each molecule in air and by the gradient of mixing ratio). During the last step of the sintering process, at the bottom of the firn column, ice starts to form, and the air found in the open pores becomes trapped in bubbles. When all the pores are finally closed, the resulting material - ice - contains in volume about 90% of solid water and 10% of air. With increasing depth, the bubble volume gets smaller due to the hydrostatic pressure and below about 1000 m, the trapped gases begin to associate with water molecules under the effect of increased pressure, creating hydrate structures in the ice, and the bubbles progressively disappear. There are no visible bubbles when the ice layers reach about 1000 m of depth in the ice sheet.

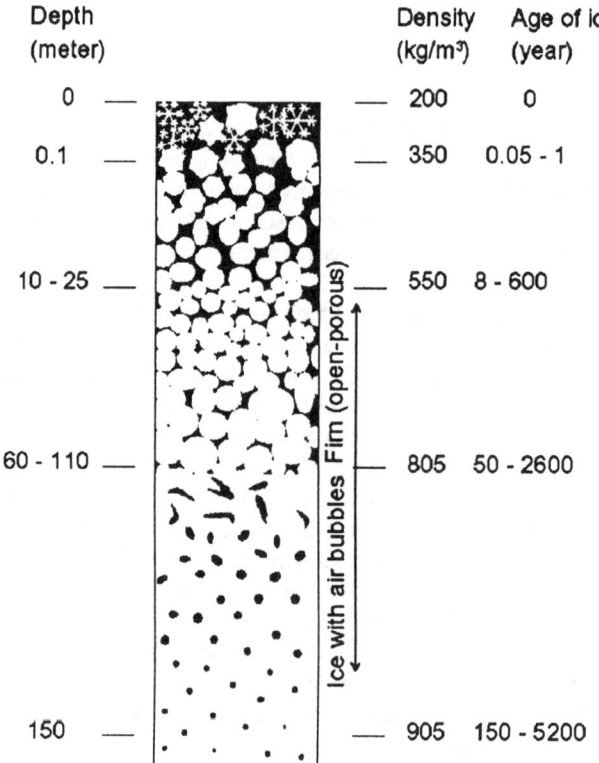

Fig. 1. Diagram of the snow-ice transformation, indicating the typical depth, density and age ranges for polar ice sheets.

2.1 Dating the Ice Record of CH₄

In order to date the air trapped in a given ice core, there are two major steps:
   (1) To date the ice itself. This can be obtained by various techniques with different

orders of precision (annual layer counting, observation of well-dated horizons, radioactive dating, curve matching with another record, ice-flow modeling, Legrand and Mayewski, 1997, and references therein).

(2) To determine the difference of age between the trapped air and the surrounding ice. As the air is isolated from the atmosphere at the snow/ice transition, the air is younger than the ice. The air/ice age difference depends on the isolation depth (mainly function of the local temperature and accumulation rate), on the mean accumulation rate from the surface to the close-off depth, and on the time required for the air to diffuse through the firn column. Through the use of models of various complexity, it is possible to calculate the difference of age for any given site (Schwander et al., 1997).

Also, the bubble close-off is a progressive phenomenon implying that the composition of the trapped air in a given ice sample (including several thousands of bubbles) integrates the atmospheric composition over a time interval ranging from a few years to several hundreds of years.

The accuracy of the air age estimate depends on the site and on the age. It ranges from a few years for young ice in high accumulation rate sites, to several thousands of years for old ice in low accumulation rate sites (Barnola et al., 1991).

2.2 Potential Artifacts Affecting the CH_4 Record in Ice

Potentially, several processes could make the CH_4 record measured in ice samples to be different from the real atmospheric CH_4 record.

(1) Physi- and chemi-sorption of CH_4 on the surface of snow and ice could enrich its mixing ratio compared to the atmosphere. But experiments have shown that, although potentially important for some trace gases, this phenomenon is negligible for CH_4 (Rasmussen et al., 1982; Chaix et al., 1996).

(2) CH_4 could react on long time scales with other trace species in the ice. However, the good agreement between CH_4 profiles obtained from ice cores with different chemical composition argues against such hypothesis (Brook et al., 1996).

(3) Gravitational fractionation in the firn column is a well-known phenomenon, which can be quantified both theoretically and based on the measurement of an inert tracer such as the isotopic composition of molecular nitrogen. The maximum gravitational effect observed today leads to CH_4 mixing ratios at the base of the firn, which are 0.7% smaller than the ones prevailing at the top of the ice sheet (Sowers et al., 1989).

(4) The bubble/hydrate transition deep in the ice sheets could change the CH_4 mixing ratio in each of these structures. Indeed, such phenomenon has been observed indirectly on ice samples from the transition zone in Greenland. In this zone, hydrates appear to be depleted in CH_4 whereas the remaining bubbles are enriched (Blunier et al., 1995). If the air extraction technique used to measure the trapped CH_4 favors one of the two air structures (for instance by grinding or shaving crudely the ice under vacuum), the measured CH_4 mixing ratio can be biased. Below the transition zone, where air is only present as hydrates in the ice, the phenomenon disappears.

(5) Fractures in the ice cores can bring contaminants from the outside to the inside, or can let some air escape from the inner part of the core, with the risk of fractionation. CH_4 measurements on ice of poor quality have generally revealed

higher CH_4 levels than expected, as well as unusual scattering of results (Chappellaz et al., 1990, 1997). Thus, ice samples for CH_4 analysis are as often as possible selected to be free of fractures or thermal cracks.

(6) If the ice experiences partial melting during its history, gas solubility and liquid-state chemical reactions can alter the mixing ratio of trapped CH_4. This was observed on a Chinese ice core, where air content measurements showed the loss of nearly all the trapped air on some samples (Chappellaz, 1990). The associated CH_4 levels were up to three times higher than the expected levels.

Overall, by selecting the right sites (no melting) and the right gas extraction technique, the resulting CH_4 records reflect very closely (within a few per cent) the true CH_4 mixing ratio prevailing in the atmosphere at the time of bubble formation. Confidence in the mean results comes from the direct comparison of ice core measurements with contemporaneous atmospheric measurements. This was achieved by the Australians, using the Antarctic site of Law Dome, with exceptionally high accumulation rate (Etheridge et al., 1992). The overlap of the two records over a few years indicates a remarkable connection. Furthermore, the good agreement obtained over the last 110, 000 years between CH_4 records from both Greenland and Antarctic ice cores support the accuracy of the atmospheric CH_4 record based on ice core data (Brook et al., 1996).

2.3 Measuring CH_4 in Ice Cores

Experimental procedures may constitute the major source of uncertainty concerning the reliability of the atmospheric CH_4 record from polar ice. The first step is to extract the trapped gases from the ice lattice. This is performed either by crushing or melting the ice. The former technique requires to avoid as much as possible frictions between the metal container, the crushing system and the ice, generating CH_4 out gassing from metal surfaces. Such systems exist in three laboratories measuring CH_4 in ice cores (Etheridge et al., 1992; Fuchs et al., 1993; Nakazawa et al., 1993a). In each case, blanks of the extraction systems show a contamination of the order of 10-20 ppbv. Melting the ice, as done in four other laboratories (Brook et al., 1996; Chappellaz et al., 1997; Sowers et al., 1997) is the easiest way to extract CH_4 from ice cores. It has the advantage on the crushing technique to be insensitive to the bubble/clathrate distribution in the ice (Blunier et al., 1995). This method also shows a contamination of the order of 15-20 ppbv, which origin is currently unknown. For both methods, the contamination is subtracted systematically from the raw CH_4 mixing ratios obtained on ice samples.

For the CH_4 measurement itself, all laboratories working on ice cores use a gas chromatograph equipped with a flame ionization detector. A recent comparison on standard gases and ice samples between four laboratories measuring CH_4 in ice cores has shown that all lead to comparable results within the experimental uncertainties (Sowers et al., 1997). For most laboratories, this experimental uncertainty amounts to +/- 20 ppbv.

3. The Industrial and Pre-industrial Ice Record of CH₄

Apart from the historical work of Robbins et al. (1973), and mainly since the beginning of the 80s, several laboratories have measured the CH₄ concentration in industrial and pre-industrial ice (Craig and Chou, 1982; Khalil and Rasmussen, 1982; Rasmussen and Khalil, 1984; Stauffer et al., 1985; Pearman et al., 1986; Etheridge et al., 1988; Raynaud et al., 1988; Etheridge et al., 1992; Blunier et al., 1993; Nakazawa et al., 1993b; Dibb et al., 1993). Figure 2 shows a compilation of the most recent studies covering the last 1000 years.

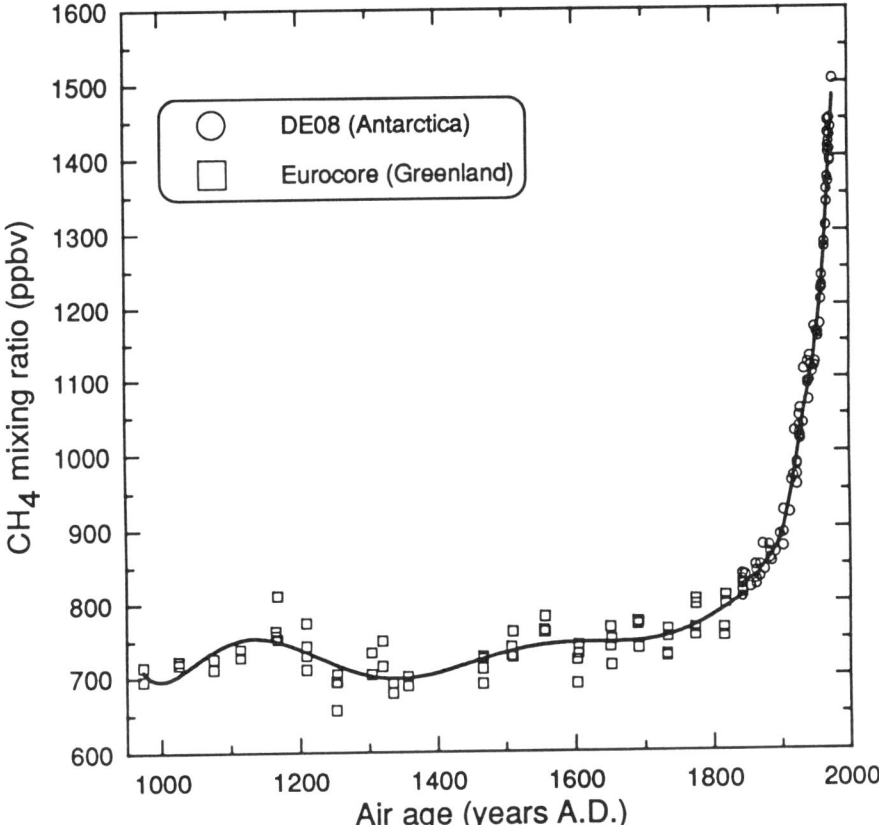

Fig. 2. Atmospheric methane variations over the last 1000 years from recent ice core data: DE08 in Antarctica (Etheridge et al., 1992), Eurocore in Greenland (Blunier et al., 1993). The two continuous lines are spline functions running in each data set.

The most peculiar and well-known characteristic observed from this figure is the considerable increase of CH_4 levels over the last 200 years. Starting from a mean pre-industrial level around 700 ppbv, the CH_4 mixing ratio then increases progressively during the 19th and 20th centuries until reaching its present-day level around 1700 ppbv. The start of the increase remains quite difficult to date precisely from published data. Results from the Greenland ice core of Eurocore suggest a significant increase from 1800 A.D. (Blunier et al., 1993).

The CH_4 increase over the last 200 years is highly correlated with the evolution of human population (Rasmussen and Khalil, 1984). It results primarily from the addition to the natural CH_4 budget of anthropogenic contributions such as domestic ruminants, rice paddies, human-induced fires, landfills, and fossil fuel exploitation (Fung et al., 1991 and references therein). A secondary contribution, estimated to account for ~20% of the CH_4 increase, lies in a decreased CH_4 sink due to increased atmospheric levels of CH_4, CO, non-methane hydrocarbons and nitrogen oxides (Thompson, 1992).

With its unique time resolution, the recent CH_4 record from the Antarctic ice core of DE08 provides the only estimate available of changes in the growth rate of atmospheric CH_4 since the onset of the industrial revolution (Etheridge et al., 1992). The growth rate seems to increase regularly between 1841 and 1978, but gets stabilized between 1920 and 1945. A reduction or stabilization of fossil fuel emissions during this period is the most plausible explanation provided by Etheridge et al. (1992), although other explanations cannot be ruled out.

The more ancient part of the CH_4 record shown in Figure 2 allows to characterize the variability of CH_4 mixing ratios on a secular time scale. It indicates changes of about 70 ppbv, or 10% of the pre-industrial level, with maximum levels during the medieval warm period followed by minimum levels during the 14th and 15th centuries, in phase with the beginning of the little ice age. However, the raise of CH_4 since the 16th century complicates the relationship between well-known climatic events of the last millennium and CH_4 variations. As a first guess, these secular variations are supposed to be driven by changes in the extent of wetlands, the main source of CH_4 before the industrial revolution, with a small amplification from associated fluctuations of the atmospheric CH_4 sink (Blunier et al., 1993). Also the role of the pre-industrial anthropogenic sources may be significant but probably cannot explain the entire variations.

Another piece of information potentially available from ice core measurements during the pre-industrial and industrial periods arises from the isotopic composition of CH_4 ($^{13}C/^{12}C$ and D/H ratios). Only one exploratory study has been carried out so far (Craig et al., 1988), as such measurements in ice cores remain a technical challenge. It gives a mean pre-industrial $^{13}C/^{12}C$ ratio of -49.6±0.7‰, significantly lighter than the mean present-day ratio of -47.2 ‰ (Stevens, 1993). Taking into account the air/ice age difference, the study of Craig et al. (1988) suggest that this increase took place in a rather short period, between the 1960s and 1980s. The authors interpreted it as the result of a massive increase in CH_4 emissions from global biomass burning. However, a recent study of the $^{13}C/^{12}C$ ratio of CH_4 in firn air indicates that the signal in ice sheets can be significantly biased by processes such as gravitational fractionation and diffusion (Trudinger et al., 1997). Thus, a more thorough consideration of such phenomenons must take place before attempting to interpret the existing ice core $^{13}C/^{12}C$ of CH_4 data as direct atmospheric signals.

4. The Holocene Record

The GRIP ice core recently drilled in Central Greenland has given access to the first detailed record of atmospheric CH_4 variability over the present interglacial period, the Holocene (Blunier et al., 1995). This record is particularly valuable due to the precision of the ice dating (Johnsen et al., 1992), ±50 years over the last 11,500 years. A refined version of this record (Chappellaz et al., 1997), with a mean sampling resolution of 85 years, is shown in Figure 3. It shows a clear long-term trend with two periods of maxima during the early Holocene (11,500 - 9,000 yr BP) and the last millennium, with mean mixing ratios of 720 ppbv, surrounding a minimum of 570 ppbv centered around 5,000 yr BP. Superimposed on the trend, sharp drops of nearly 70 ppbv and with a duration of 200 years are observed at 11,300, 9700 and 8200 yr B.P. The last one is synchronous with a well-defined climatic event observed in other proxies (Alley et al., 1997).

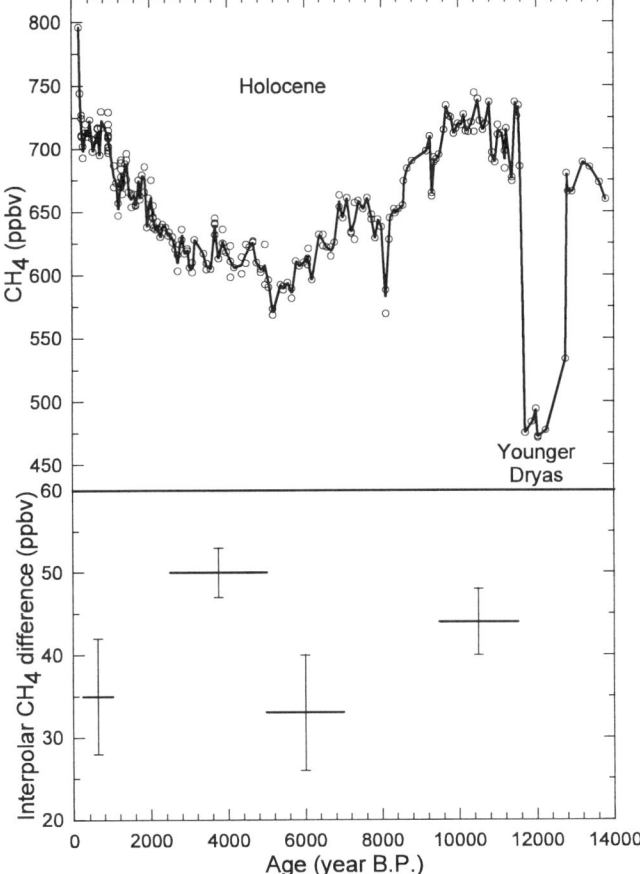

Fig. 3. Changes of the CH_4 mixing ratio and CH_4 interpolar difference over the Holocene, from the GRIP (Greenland), D47 and Byrd (Antarctica) ice cores (Blunier et al., 1995; Chappellaz et al., 1997). The vertical bar on the interpolar differences is one-sigma error.

The analyses of the companion Antarctic ice cores of D47 and Byrd have allowed to measure the interpolar difference of CH_4 mixing ratio over 75% of the Holocene (Chappellaz et al., 1997). This difference is a function of the latitudinal distribution of CH_4 sources and sinks, and of the inter-hemispheric exchange time. Significant changes of this difference are observed, around a mean value of 45±3 ppbv (Figure 3), a number to compare with the present-day difference of 140 ppbv (Dlugokencky et al., 1994). This brings an additional constraint on the causes of natural CH_4 variability at the scale of an interglacial period, as discussed in section 6.

5. The Glacial-Interglacial Cycles

Measurements on Greenland and Antarctic ice during the 80s showed that during the last two glacial-interglacial climatic transitions CH_4 levels nearly doubled, from 350 to 700 ppbv (Stauffer et al., 1988; Raynaud et al., 1988). Subsequently the CH_4 record has been documented over the complete sequence of the last climatic cycle (Chappellaz et al., 1990). Overall the record showed a strong correlation with the Antarctic climate, and it is characterized by a strong periodicity in the precession domain (~ 20,000 yr).

The picture of CH_4 variability during glacial-interglacial cycles has recently become more accurate with the analysis of the Greenland ice cores of GRIP and GISP2. Figure 4 shows a compilation of these records compared to the climatic variability recorded in the isotopic composition of the Greenland ice. Superimposed on the precessional CH_4 variability, sharp and important CH_4 changes are associated with nearly each climatic event (called Dansgaard/Oeschger event) recorded in the Greenland ice, including the well-known cooling event of the Younger Dryas during the last deglaciation. These climatic events are characterized by temperature changes of more than 10°C on time scales of decades to centuries. Two notable exceptions in this CH_4/climate association are the Dansgaard/Oeschger events # 19 and 20 which have a negligible counterpart in the CH_4 signal. The general concomitance between the CH_4 changes and Greenland climate clearly indicates that these climatic features were not restricted to the North Atlantic region but indeed involved a large part of the globe. It also reveals a very tight response of the natural CH_4 cycle to climate changes.

These millennial scale CH_4 variations are now used to put ice cores from Greenland and Antarctica on a common time scale and to quantify the phase relationship between the climate of the two hemispheres during the past (Blunier et al., 1997). Discussion of these studies on past CH_4 changes is beyond the scope of this chapter.

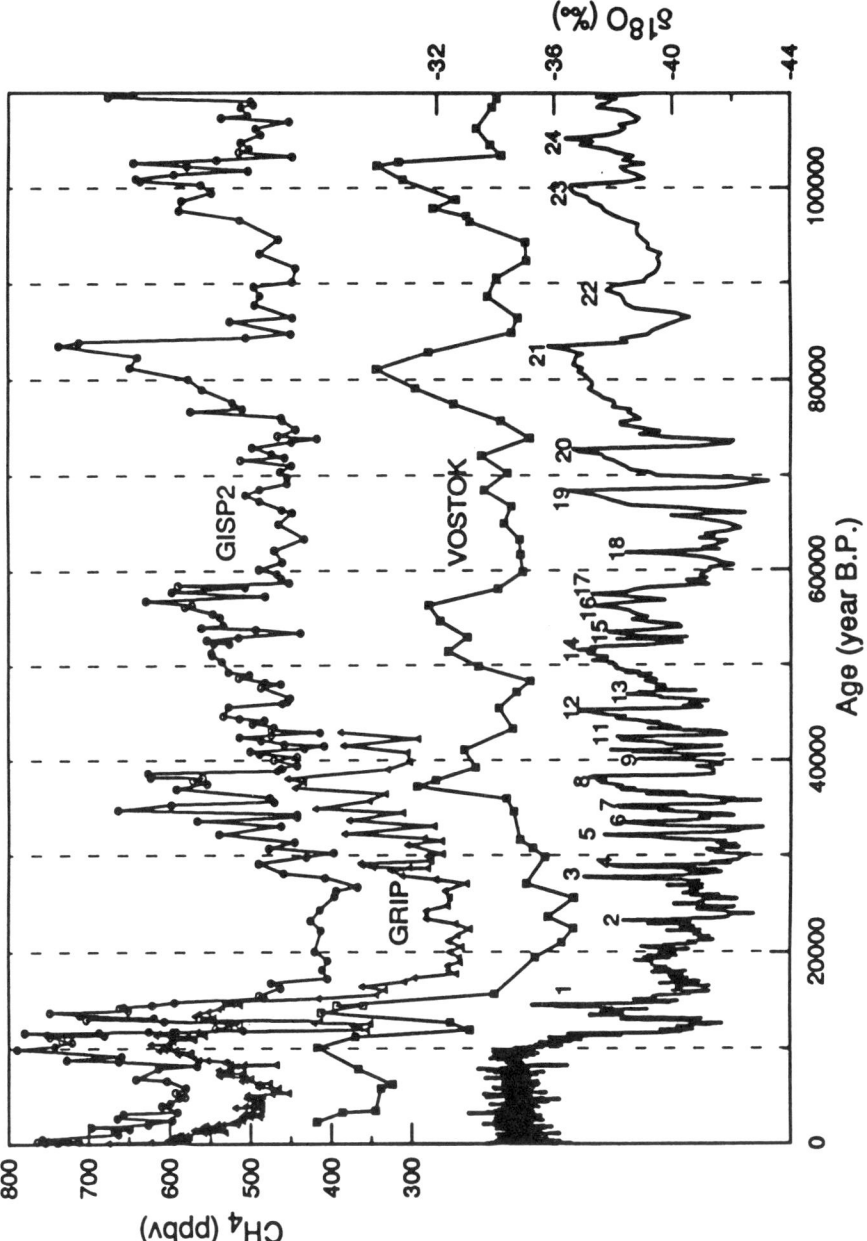

Fig. 4. Rapid CH_4 variations during the last 110,000 years from the GISP2 (Brook et al., 1996), GRIP (Chappellaz et al., 1993) and Vostok (Chappellaz et al., 1990) ice cores. The time scale for all three CH_4 profiles is from Brook et al. (1996). The GRIP and Vostok profiles are decreased by 120 and 250 ppbv, for clarity. The GISP2 isotopic profile is shown for comparison (Grootes et al., 1993). Numbers 1 to 24 indicate the Dansgaard/Oeschger events.

6. Climatic Impact of the CH_4 Changes Over the Climatic Cycle

CH_4 changes can affect the Earth's surface temperature in three ways: the direct radiative effect, the chemical feedbacks of CH_4 oxidation, and the climatic feedback.

The direct radiative effect of the glacial-interglacial increase of about 300 ppbv results in a global surface equilibrium warming of 0.08°C (Chappellaz et al., 1990).

The chemical feedbacks involve the tropospheric ozone and stratospheric water vapor produced through CH_4 oxidation in the atmosphere, the former taking place only in regions of the atmosphere with high NO and NO_2 levels. Both ozone and water vapor are strong greenhouse gases. These two feedbacks are estimated in present-day conditions to amplify the CH_4 radiative forcing by 35-40% (Lelieveld and Crutzen, 1992). They could thus have produced an additional warming of 0.03°C for glacial-interglacial transitions, leading to a CH_4 forcing of 0.11°C. Although small compared to the 4-5°C mean global warming of the deglaciation, this forcing represents about 20% of the concomitant CO_2 forcing (0.50°C).

The climatic feedbacks involve changes in the climate system, such as tropospheric water vapor content, sea ice extent, cloud cover, vegetation type, potentially amplifying an initial radiative forcing. As the CH_4 changes observed on glacial-interglacial time scales result themselves from the dynamic of the climate system, the pure effect of climatic feedbacks on a given CH_4 forcing is difficult to quantify. Some estimates suggest that they could have amplified the CH_4 greenhouse forcing by a factor 3-4 (Broccoli and Manabe, 1987; Rind et al., 1989; Lorius et al., 1990).

7. Causes of CH_4 Changes Over the Climatic Cycle

As indicated in previous sections, ice core studies of atmospheric CH_4 reveal a close link between the CH_4 cycle and climate on time scales from a few decades to a few tens of thousands of years. Based on the present-day knowledge of the CH_4 budget, qualitative explanations for such relationship have essentially involved the CH_4 emission from natural wetlands and the CH_4 oxidation with OH radicals in the atmosphere, the main constituents of the natural CH_4 budget (Chappellaz et al., 1990). Other sources, including termites, wild animals, wildfires, oceans and CH_4 hydrates in permafrost and continental shelves, would have weakly contributed to past CH_4 changes, due to their small contribution to the natural CH_4 budget even today.

Although field studies have generally indicated a correlation between CH_4 flux and soil temperature in wetlands, it is believed that the main cause of potential changes in the CH_4 emission from wetlands is a change in global wetland area. As a large part of the present-day wetlands lie between 50° and 70°N, a direct impact of the growth and decay of continental ice sheets at these latitudes was suspected. On the other hand, the main part of CH_4 emission from today's wetlands is located in the Tropics, between 20°N and 30°S (Fung et al., 1991). Thus, past changes in monsoon intensity and the hydrological cycle at low latitudes could also be responsible for the CH_4 variability observed from ice cores. Based on a comparison with proxy data at high and low latitudes over the last climatic cycle, Chappellaz et al. (1990) proposed a principal role of the monsoon in controlling past CH_4 concentrations. Later on, Chappellaz et al. (1993) used a parameterization of wetlands and other CH_4 sources on vegetation and topography distribution to propose a quantification of CH_4 sources during the Last Glacial Maximum and the pre-industrial Holocene. They concluded that changes in the wetland CH_4 source were the major factor driving the CH_4 increase during the last deglaciation, with a dominant contribution from the northern mid and high latitudes. Finally, Brook et al. (1996) interpreted the lack of CH_4 increase during Dansgaard/Oeschger events #19 and 20 as a result of a substantial increase in continental ice volume, reducing the CH_4 emissions from boreal wetlands. They consequently favored a boreal control on the CH_4 budget on short and long time scales.

Concerning the CH_4 sink through oxidation by OH, potential changes in its strength have been determined with several photochemical models of the past atmosphere (e.g., Thompson, 1992; Thompson et al., 1993; Crutzen and Brühl, 1993; Martinerie et al., 1995). All models suggest that OH concentrations were higher during glacial conditions than today, but not much greater than during pre-industrial time. This means that a small part (10 to 30%) of the glacial-interglacial CH_4 increases are due to a decrease in the sink strength. Still, the reliability of model estimates is strongly restricted by the lack of other chemical constraints from ice cores and by uncertain assessments of the fluxes of other trace species (NO, NO_2, CO, O_3, NMHC) involved in the atmospheric oxidative capacity (Thompson, 1996).

As indicated in section 4, an additional constraint on the cause of past CH_4 changes comes from the quantification of the interpolar CH_4 difference. Using a three-box model of the atmosphere, Chappellaz et al. (1997) translated the measured differences during the Holocene into quantitative contributions of CH_4 sources in the tropics and the middle to high latitudes of the northern hemisphere. The model assumed that the inter-box exchange time, the OH latitudinal distribution, and the CH_4 source between 30° and 90°S remained similar as today. Source distributions among the model boxes are shown in Figure 5. They show that the tropics had a major contribution to the CH_4 budget over the last 20,000 years, modulated by significant changes in the middle to high northern latitude sources. Such exercise remains to be performed over the last glacial period.

Another theory on the cause of the fast deglacial CH_4 increase involves the sudden release of vast quantities of methane hydrates from marine sediments or permafrost deposits (Paull et al., 1991; Nisbet, 1992). Due to the quick removal of CH_4 through atmospheric oxidation and to the smoothing of atmospheric variations created by the air trapping process in ice sheets, the CH_4 record from ice cores over the deglaciation would be, according to this theory, an underestimate of the true CH_4 changes which happened due to the hydrate degassing. Thus, CH_4 change would be a trigger, instead of a response, of past climate change. Thorpe et al. (1996) have modeled the CH_4 concentration in the atmosphere and in ice cores, as well as the radiative forcing, resulting from such sudden methane burst. They concluded that a sampling interval of 50 yr in the ice core record is required to detect such event. Still, growing evidences from the GRIP and GISP2 ice cores disprove such catastrophic scenario to have happened during the second part of the last deglaciation, in particular at the end of the Younger Dryas and during the early Holocene (Chappellaz et al., 1997; Severinghaus et al., 1997). Further ice-core analyzes over the first half of the deglaciation should provide a full test of the hydrate theory.

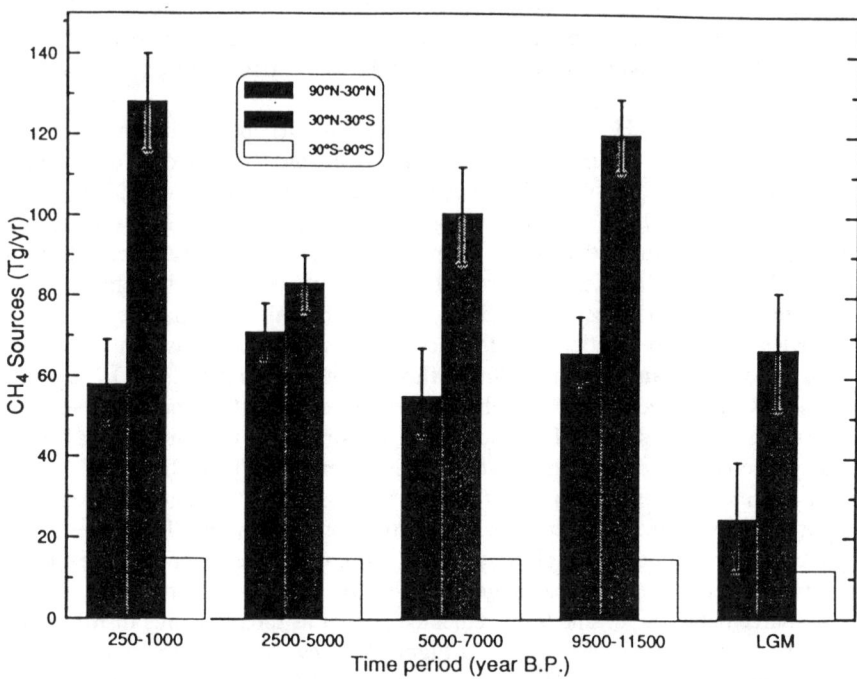

Fig. 5. CH$_4$ source distribution among three boxes of the atmosphere, for selected time intervals of the Holocene and for the Last Glacial Maximum (LGM), as deduced from changes in the interpolar CH$_4$ difference shown in Figure 3 (from Chappellaz et al., 1997). Vertical bars are one-sigma confidence intervals.

8. Conclusions

Ice cores are the only means available to reconstruct atmospheric mixing ratios of methane over long time periods. Several arguments support their overall good reliability for such purposes.

Numerous ice-core studies over the past 15 years have provided a very detailed picture of past CH$_4$ changes. The main features revealed by these records are: (1) a large increase from 700 to 1700 ppbv during the last 200 years due to human activities; (2) a natural variability of up to 150 ppbv during the present interglacial, coupled with significant changes in the interpolar CH_4 difference, probably related to the wetland dynamic in tropical and boreal regions; (3) rapid variations in the 350-700 ppbv range during the last glacial-interglacial cycle, remarkably correlated with climate changes recorded in Greenland, their cause still being subject to debate.

Detailed CH_4 profiles are now performed on various ice cores to use this signal as a stratigraphic marker, allowing detailed time-correlation between the climate of the two hemispheres. Past CH_4 budget studies are still in progress: future work includes the interpolar CH_4 difference during glacial times, the CH_4 variability during previous climatic cycles (down to 400,000 years), and the isotopic composition of CH_4 ($^{13}C/^{12}C$ and D/H) in the atmosphere during the pre-industrial and glacial times.

Acknowledgments. This work was supported by the European Community, by the Fondation de France and the Institut National des Sciences de l'Univers in France, and by the Swiss National Science Foundation in Switzerland.

References

Alley, R. B., P.A. Mayewski, T. Sowers, M. Stuiver, K.C. Taylor, P. U. Clark. 1997. Holocene climate instability: A prominent, widespread event 8200 yr ago. *Geology, 25*:483-486.

Barnola, J.-M., P. Pimienta, D. Raynaud, Y. S. Korotkevich. 1991. CO_2-climate relationship as deduced from the Vostok ice core: A re-examination based on new measurements and on a re-evaluation of the air dating. *Tellus, 43B*:83-90.

Blake, D. R., F. S. Rowland. 1988. Continuing worldwide increase in tropospheric methane, 1978 to 1987. *Science, 239*:1129-1131.

Blunier, T., J. Chappellaz, J. Schwander, J. M. Barnola, T. Desperts, B. Stauffer, D. Raynaud. 1993. Atmospheric methane, record from a Greenland ice core over the last 1,000 years. *Geophys. Res. Lett., 20*:2219-2222.

Blunier, T., J. Chappellaz, J. Schwander, B. Stauffer, D. Raynaud. 1995. Variations in atmospheric methane concentration during the Holocene epoch. *Nature, 374*:46-49.

Blunier, T., J. Schwander, B. Stauffer, T. Stocker, A. Dällenbach, A. Indermühle, J. Tschumi, J. Chappellaz, D. Raynaud, J.-M. Barnola. 1997. Timing of the Antarctic Cold Reversal and the atmospheric CO_2 increase with respect to the Younger Dryas event. *Geophys. Res. Lett., 24*:2683.

Broccoli, A. J., S. Manabe. 1987. The influence of continental ice, atmospheric CO_2, land albedo on the climate of the Last Glacial Maximum. *Clim. Dyn., 1*:87-99.

Brook, E. J., T. Sowers, J. Orchardo. 1996. Rapid variations in atmospheric methane concentrations during the past 110,000 years. *Science, 273*:1087-1091.

Chaix, L., J. Ocampo, F. Dominé. 1996. Adsorption of CH_4 on laboratory-made crushed ice and on natural snow at 77 K. Atmospheric implications. *C.R. Acad. Sci. Paris, 322*:609-616.

Chappellaz, J. 1990. Etude du méthane atmosphérique au cours du dernier cycle climatique à partir de l'analyse de l'air piégé dans la glace antarctique. *Ph-D thesis*, Grenoble University, France, 214 pp.

Chappellaz, J., J. M. Barnola, D. Raynaud, Y. S. Korotkevich, C. Lorius. 1990. Ice-core record of atmospheric methane over the past 160,000 years, *Nature, 345*:127-131.

Chappellaz, J., I. Y Fung, A.M. Thompson. 1993. The atmospheric CH_4 increase since the Last Glacial Maximum: 1. Source estimates. *Tellus, 45B*:228-241.

Chappellaz, J., T. Blunier, D. Raynaud, J. M. Barnola, J. Schwander, B. Stauffer. 1993. Synchronous changes in atmospheric CH_4 and Greenland climate between 40 and 8 kyr BP.

Nature, 366:443-445.

Chappellaz, J., T. Blunier, S. Kints, A. Dällenbach, J. M. Barnola, J. Schwander, D. Raynaud, B. Stauffer. 1997. Changes in the atmospheric CH_4 gradient between Greenland and Antarctica during the Holocene. *J. Geophys. Res., 102*:15,987-15,997.

Craig, H., C. C. Chou. 1982. Methane: the record in polar ice cores. *Geophys. Res. Lett., 9*:1221-1224.

Craig, H., C. C. Chou, J. A. Welhan, C.M. Stevens, A. Engelkemeir. 1988. The isotopic composition of methane in polar ice cores. *Science, 242*:1535-1539.

Crutzen, P. J., C. Brühl. 1993. A model study of atmospheric temperatures and the concentrations of ozone, hydroxyl, and some other photochemically active gases during the glacial, the pre-industrial Holocene and the present. *Geophys. Res. Lett., 20*:1047-1050.

Dibb, J. E., R.A. Rasmussen, P.A. Mayewski, G. Holdsworth. 1993. Northern Hemisphere concentrations of methane and nitrous oxide since 1800: results from the Mt Logan and 20D ice cores. *Chemosphere, 27*:2413-2423.

Dlugokencky, E. J., L.P. Steele, P.M. Lang, K. A. Masarie. 1994. The growth rate and distribution of atmospheric methane. *J. Geophys. Res., 99*:17,021-17,043.

Etheridge, D. M., G. I. Pearman, F. De Silva. 1988. Atmospheric trace-gas variations as revealed by air trapped in an ice core from Law Dome, Antarctica. *Ann. Glaciol., 10*:28-33.

Etheridge, D. M., G. I. Pearman, P. J. Fraser. 1992. Changes in tropospheric methane between 1841 and 1978 from a high accumulation-rate Antarctic ice core. *Tellus, 44B*: 282-294.

Fuchs, A., J. Schwander, B. Stauffer. 1993. A new ice mill allows precise concentration determination of methane and most probably also other trace gases in the bubble air of very small ice samples. *J. Glaciol., 39*:199-203.

Fung, I., J. John, J. Lerner, E. Matthews, M. Prather, L.P. Steele, P. J. Fraser. 1991. Three-dimensional model synthesis of the global methane cycle. *J. Geophys. Res., 96*:13,033-13,065.

Grootes, P., M. Stuiver, J. W. C. White, S. Johnsen, J. Jouzel. 1993. Comparison of oxygen isotope records from the GISP2 and GRIP ice cores. *Nature, 366*:552-554.

Johnsen, S. J., H. B. Clausen, W. Dansgaard, K. Fuhrer, N. Gundestrup, C. U. Hammer, P. Iversen, J. Jouzel, B. Stauffer, J.P. Steffensen. 1992. Irregular glacial interstadials recorded in a new Greenland ice core. *Nature, 359*:311-313.

Khalil, M.A.K., R.A. Rasmussen. 1982. Secular trends of atmospheric methane (CH_4). *Chemosphere, 11*:877-883.

Legrand, M., P. Mayewski. 1997. Glaciochemistry of polar ice cores: a review. *Rev. Geophys., 35*:219-243.

Lelieveld, J., P. J. Crutzen. 1992. Indirect chemical effects of methane on climate warming. *Nature, 355*:339-342.

Lorius, C., J. Jouzel, D. Raynaud, J. Hansen, H. Le Treut. 1990. The ice-core record: climate sensitivity and future greenhouse warming. *Nature, 347*:139-145.

Lowe, D.C., C. A. M. Brenninkmeijer, G. W. Brailsford, K. R. Lassey, A. J. Gomez. 1994. Concentration and ^{13}C records of atmospheric methane in New Zealand and Antarctica: evidence for changes in methane sources. *J. Geophys. Res., 99*:16,913-16,925.

Martinerie, P., G. P. Brasseur, C. Granier. 1995. The chemical composition of ancient atmospheres: A model study constrained by ice core data. *J. Geophys. Res., 100*:14,291-14,304.

Nakazawa, T., T. Machida, K. Esumi, M. Tanaka, Y. Fujii, S. Aoki, O. Watanabe. 1993a. Measurements of CO_2 and CH_4 concentrations in air in a polar ice core. *J. Glaciol., 39*:209-215.

Nakazawa, T., T. Machida, M. Tanaka, Y. Fujii, S. Aoki, O. Watanabe. 1993b. Differences of the atmospheric CH_4 concentration between the Arctic and Antarctic regions in pre-industrial/pre-agricultural era. *Geophys. Res. Lett., 20*:943-946.

Nisbet, E.G. 1992. Sources of atmospheric CH_4 in early postglacial time. *J. Geophys. Res., 97*:12,859-12,867.

Paull, C. K., W. Ussler, W. P. Dillon. 1991. Is the extent of glaciation limited by marine gas hydrates? *Geophys. Res. Lett., 18*:432-434.

Pearman, G.I., D. Etheridge, F. De Silva, P. J. Fraser. 1986. Evidence of changing concentrations of atmospheric CO_2, N_2O and CH_4 from air bubbles in Antarctic ice. *Nature, 320*:248-250.

Rasmussen, R.A., M.A.K. Khalil. 1984. Atmospheric methane in the recent and ancient atmospheres: concentrations, trends and interhemispheric gradient. *J. Geophys. Res., 89*:11,599-11,605.

Rasmussen, R.A., M.A.K. Khalil. 1986. Atmospheric trace gases: trends and distributions over the last decade. *Science, 232*:1623-1624.

Rasmussen, R.A., M.A.K. Khalil, S.D. Hoyt. 1982. Methane and carbon monoxide in snow. *J. Air Poll. Cont. Asso., 32*:176-178.

Raynaud, D., J. Chappellaz, J. M. Barnola, Y. S. Korotkevich, C. Lorius. 1988. Climatic and CH_4-cycle implications of glacial-interglacial CH_4 change in the Vostok ice core. *Nature, 333*:655-657.

Rind, D., D. Peteet, G. Kukla. 1989. Can Milankovitch orbital variation initiate the growth of ice sheets in a General Circulation Model ? *J. Geophys. Res., 41*:12,851-12,871.

Rinsland, C.P., J. S. Levine, T. Miles. 1985. Concentration of methane in the troposphere deduced from 1951 infrared solar spectra. *Nature, 330*:245-249.

Robbins, R. C., L.A. Cavanagh, L. J. Salas, E. Robinson. 1973. Analysis of ancient atmosphere. *J. Geophys. Res., 78*:5341-5344.

Schwander, J. 1989. The transformation of snow to ice and the occlusion of gases. In: *The Environmental Record in Glaciers and Ice Sheets*, Report of the Dahlem Workshop held in Berlin 1988, March 13-18, John Wiley and Sons, Chichester, 51-67.

Schwander, J., T. Sowers, J.-M. Barnola, T. Blunier, A. Fuchs, B. Malaizé. 1997. Age scale of the air in the summit ice: Implications for glacial-interglacial temperature change. *J. Geophys. Res., 102*:19,483-19,493.

Severinghaus, J.P., T. Sowers, E.J. Brook, R. B. Alley, M.L. Bender. 1997. Timing of abrupt climate change at the end of the Younger Dryas from thermally fractionated gases in polar ice. *Nature*, in press.

Sowers, T., M. Bender, D. Raynaud. 1989. Elemental and isotopic composition of occluded O_2 and N_2 in polar ice. *J. Geophys. Res., 94*:5137-5150.

Sowers, T., E. Brook, D. Etheridge, T. Blunier, A. Fuchs, M. Leuenberger, J. Chappellaz, J. M. Barnola, M. Wahlen, B. Deck, C. Weyhenmeyer. 1997. An interlaboratory comparison of techniques for extracting and analyzing trapped gases in ice cores. *J. Geophys. Res., 102*:26,527-26,538.

Stauffer, B., G. Fischer, A. Neftel, H. Oeschger. 1985. Increase of atmospheric methane in Antarctic ice core. *Science, 229*:1386-1388.

Stauffer, B., E. Lochbronner, H. Oeschger, J. Schwander. 1988. Methane concentrations in the glacial atmosphere was only half that of the preindustrial Holocene. *Nature, 332*:812-814.

Steele, L.P., P. J. Fraser, R.A. Rasmussen, M.A.K. Khalil, T. J. Conway, A. J. Crawford, R.H. Gammon, K. A. Masarie, K. W. Thoning. 1987. The global distribution of methane in the troposphere. *J. Atmos. Chem., 5*:125-171.

Steele, L.P., E. J. Dlugokencky, P.M. Lang, P.P. Tans, R. C. Martin, K. A. Masarie. 1992. Slowing down of the global accumulation of atmospheric methane during the 1980s. *Nature, 358*:313-316.

Stevens, C.M. 1993. Isotopic abundances in the atmosphere and sources. In: *Atmospheric Methane: Sources, Sinks, and Role in Global Change*, NATO ASI Series I13, Springer-Verlag, New York, 62-88.

Thompson, A.M. 1992. The oxidizing capacity of the Earth's atmosphere: probable past and future changes. *Science, 256*:1157-1168.

Thompson, A.M., J. A. Chappellaz, I. Y. Fung, T. L. Kucsera. 1993. Atmospheric methane increase since the Last Glacial Maximum. 2. Interactions with oxidants. *Tellus, 45B*:242-

257.

Thompson, A.M. 1996. Modeling framework for atmospheric trace gas measurements at the air-snow interface. In: *Chemical Exchange Between the Atmosphere and Polar Snow*, NATO ASI Series I43, Springer-Verlag, New York, 225-248.

Thorpe, R. B., K. S. Law, S. Bekki, J. A. Pyle, E.G. Nisbet. 1996. Is methane-driven deglaciation consistent with the ice core record ? *J. Geophys. Res., 101*:28,627-28,635.

Trudinger, C. M., I.G. Enting, D. M. Etheridge, R. J. Francey, V. A. Levchenko, L. P. Steele, D. Raynaud, L. Arnaud. 1997. Modeling air movement and bubble trapping in firn. *J. Geophys. Res., 102*:6747-6763.

Zander, R., P. Demoulin, D. H. Ehhalt, U. Schmidt. 1989. Secular increases of the vertical abundance of methane derived from IR solar spectra recorded at the Jungfraujoch Station. *J. Geophys. Res., 94*:1129-1139.

3. The Isotopic Composition of Atmospheric Methane and Its Sources

C.M. Stevens [1] and M. Wahlen [2]
[1] Chemical Technology Division, Argonne National Laboratory, Argonne, IL 60439
[2] Scripps Institution of Oceanography, University of California, San Diego, La Jolla, CA 92093

1. Introduction

It has been almost 20 years since major studies of atmospheric methane, including isotopic measurements, were undertaken. The NATO conference in 1991 summarized these efforts with the publication of the book "Atmospheric Methane: Sources, Sinks, and Role in Global Change."

For this update of the field we present all the new isotopic measurements carried out since the 1991 NATO book. An in-depth analysis of the interpretations by the respective authors of the isotopic data will not be attempted, but a brief review of the conclusions of some applications of isotopic budgets and trends will be presented. Since 1991 the progress related to isotopic studies consists of the following:

1. Measurements of the trend of the isotopic composition of atmospheric methane have been continued by Lowe et al. (1994), Gupta et al. (1996), and Quay et al. (1991). All these groups have done detailed studies that show seasonal variations as well as recent secular trends in both hemispheres (see Figure 1). Also there has been a cooperative effort between laboratories to cross check analytical procedures and standards.

2. Modeling calculations applied to the recent trends in the isotopic composition of atmospheric methane by Gupta et al. (1996) and Lowe et al. (1994) have become more sophisticated, with higher resolution and taking into account the effect of other sink mechanisms for the first time, namely, oxidation by soil bacteria and reaction with stratospheric Cl radicals. The latter reaction is important in the higher stratosphere and has the large fractionation effect of 1.066 at 297 °K (Saueressig et al., 1995).

3. Studies have been done by Whiting and Chanton (1993) and Hornibrooke et al. (1997) to elucidate and quantify methanogenic processes in salt and fresh water environments. New measurements of the isotopic composition of mene from various wetlands environments are listed in an updated summary of these sources in Table 2.

4. A comprehensive set of measurements of the isotopic methane from numerous landfills has been carried out by Hackley et al. (1996) with an average result about 3 per mil more depleted in carbon-13 than the previously measured values.

Mohammad Aslam Khan Khalil (Ed.)
Atmospheric Methane
© *Springer-Verlag Berlin Heidelberg 2000*

5. There have not been any additional isotopic measurements of methane from cattle, rice, or biomass burning since the last tabulation in 1993.

New measurements on the deuterium and carbon-14 isotopic compositions of atmospheric methane and its sources are in the unpublished work of Quay (1997).

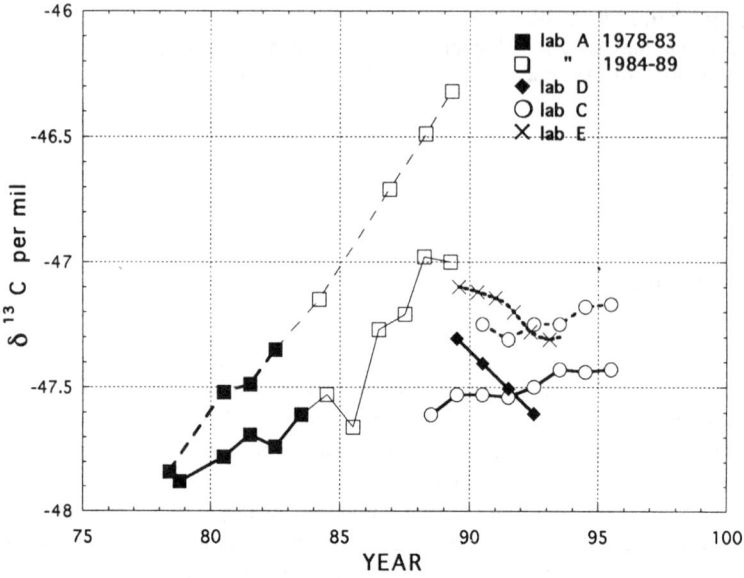

Fig. 1. Comparison of the trend of the carbon-13 isotopic composition of atmospheric methane by different laboratories. See Table 3 for laboratory identification. Northern hemisphere (solid lines), southern hemisphere (dotted lines). The results for the northern and southern hemispheres for the years 1984 to 1989 by lab A (open squares) are shown as lighter lines because of a large uncertainty (see discussion).

2. Isotopic Composition of the Sources

Table 1 lists the averages of the measured values of the $\delta^{13}C$ for all the major anthropogenic sources of atmospheric methane as reported by several laboratories in the literature. The only additions to this list are the new values for methane from a number of landfills by Hackley et al. (1996) and Bergamaschi and Harris (1995). The goal of the latter study was to determine the microbiological processes occurring in the landfill, sampling the gases in a closed system that prevented mixing with atmospheric oxygen and aerobic oxidation of the methane. Because these do not represent the usual conditions for methane escaping a landfill, these results are not included in determining an average landfill value. Bergamaschi and Harris also measured the deuterium isotopic compositions and found an inverse correlation with the $\delta^{13}C$ values for the situation of no evidence of oxidation from admixture with air, which they attribute to slightly varying conditions from the two methanogenic pathways, CO_2 reduction and acetate fermentation

Table 1. The isotopic composition of methane from the interstitial water and sediments of wetland and marine waters.   Listings are in order of latitude north to south within a continent.

Wetlands	Sampling method	Mean δ13C (per mil). Range and number of samples in brackets.	Reference
		Asia	
USSR marshes	Not specified	-64 (-52 to -69) [12]	Ovsyannikov & Lebedev, 1967
		Africa	
Kenya river	Flux chamber	-54 [2]	Tyler, 1986
Kenya lake	Flux chamber	-48 (-44 to -50) [3]	Tyler, 1986
Kenya papyrus swamp	Flux chamber	-51 (-31 to -62) [3]	Tyler, 1986
		Europe	
Heidelberg, Germany			
Natural swamp	Flux chamber	-57.4 ± 2.1 [7]	Levin et al., 1993
Artificial swamp, not oxidized	Flux chamber	-58.5 ± 6.8 [14]	Levin et al., 1993
Artificial swamp, oxidized	Flux chamber	-50.8 ± 3.4 [5]	Levin et al., 1993
		Pacific	
Checker Reef, Kaneohe Bay, Hawaii	Interstitial porewater	-54.5 to -51.0 for 8 samples collected between surface and 100 cm. depth.	Popp et al., 1995
		North America	

Wetlands	Sampling method	Mean δ13C (per mil). Range and number of samples in brackets.	Reference
Alaska tundra	Flux chamber	-66 (-55 to -73) [15]	Quay et al., 1988
Alaska tundra	Bubbles	-62 (-57 to -72) [4]	Quay et al., 1988
Canadian tundra	Surface	-63 (-61 to -65) [na]	Wahlen et al., 1989
Ontario, Point Melee Marsh	Porewater collected by piezometer	-48 to -71.6 for 17 samples collected between surface and 180 cm. depth, Sept. 1995.	Hornibrooke et al., 1997
Ontario, Sifton Bay, London	Porewater collected by piezometer	-49.6 to -63.4 for 8 samples collected between surface and 90 cm. depth, Aug. 1995	Hornibrooke et al., 1997
Ontario, Sifton Bay, London	Porewater collected by piezometer	-52 to -67.5 for 13 samples collected between surface and 180 cm. depth, Oct. 1995	Hornibrooke et al., 1997
Minnesota peat bog	Inversion	-67 (-64 to -71) [3]	Stevens & Engelkemeir, 1988
Minnesota peat bog	Flux chamber	-66 (-57 to -77) [17]	Quay et al., 1988
Minnesota peat bog	Bubbles	-66 (-50 to -86) [9]	Quay et al., 1988
Washington King peat bog	Static flux chamber	-74 (-61 to -85 for 42 samples)	Lansdown et al., 1992
King peat bog	Bubbles	-72 ± 4 [23]	Lansdown et al., 1992
King peat bog	Inversion	-57 [1]	Lansdown et al., 1992
King peat bog	Soil water	-69 to -84 for 24 samples collected at depths between 70 cm. and the surface	Lansdown et al., 1992
New England lakes	Mud gases	-64 (-57 to -80) [13]	Oone & Deevey, 1960
New York wetlands	Surface	-58 (± 2.4) [na]	Wahlen et al., 1989
New York wetlands	Inversion	-52 (-49 to -56) [7]	Wahlen et al., 1989
Illinois slough	Flux chamber	-50 (-49 to -51) [4]	Stevens & Engelkemeir, 1988
Illinois slough	Sediment gases	-56 (-55 to -58) [4]	Stevens & Engelkemeir, 1988

Wetlands	Sampling method	Mean $\delta13C$ (per mil). Range and number of samples in brackets.	Reference
Colorado pond	Sediment gases	-53.1 ±1.1 [4]	Tyler, 1986
Colorado marsh	Sediment gases	-48.3 [1]	Tyler, 1986
North Carolina marine basin, Upper Fork	Bubbles	-66.3 ± 0.4 [32]	Chanton & Martens, 1988
North Carolina marine basin, Goldhaber Island	Bubbles	-69.5 ± 0.6 [15]	Chanton & Martens, 1988
North Carolina marine basin	Bubbles	-58 to -68 for 9 samples collected between Jan. and Aug. 1983	Martens et al., 1986
North Carolina marine basin	Bubbles	-58 to -69 for 8 samples collected during Jan., Aug., and Oct. 1984	Martens et al., 1986
North Carolina marine basin	Flux chamber	-56 to -62 for 8 samples collected between Apr. and Dec. 1986	Boehme et al., 1996
North Carolina marine basin	Porewater samples from closed box cores	-50.5 (-50.16 to -50.79)	Boehme et al., 1996
North Carolina marine basin	Headspace gas from sediment cores	-47 to -63 for 67 samples collected between 1986 and 1990 at depths between surface and 65 cm.	Boehme et al., 1996
South Carolina pond	Headspace gas from sediment cores	-53 (-51 to -55) [3]	Burke & Sackett, 1986
California, Tamales Bay	Sediment porewater	-57 to -92 for 26 samples collected between surface and 330 cm depth	Burke & Sackett, 1986

Wetlands	Sampling method	Mean δ13C (per mil). Range and number of samples in brackets.	Reference
Florida Everglades	Inversion	-55 (-53 to -58) [6]	Stevens & Engelkemeir, 1988
Florida swamp forest	Flux chamber, dissolved in surface H$_2$O bubbles	-52.7 ± 6 [28]	Hoppell et al., 1994
Florida Everglades	Sediment bubbles	-65 (-63 to -70) [23]	Chanton et al, 1988
Florida, Crescent Lake	Flux chamber	-64 (-71 to -79) [7]	Burke & Sackett, 1986
Florida, Mirror Lake	Flux chamber	-55 (-52 to -56) [4]	Burke & Saclett. 1986
Florida, Lake Dias	Flux chamber	-63 (-60 to -67) [3]	Burke & Sackett, 1986
Florida, Tampa Bay estuary	Flux chamber	-66 (-63 to -71) [47]	Burke & Sackett, 1986
Mississippi River delta	Flux chamber	-60 (-59 to -60) [3]	Burke & Sackett, 1986
		South America	
Amazon River	Surface	-64 [2]	Tyler, 1986
Amazon flood plain	Bubbles, 1985	-62 (-47 to -73) [16]	Quay et al., 1988
Amazon flood plain	Bubbles, 1987	-52 (-42 to -60) [10]	Quay et al., 1988
Amazon flood plain	Dissolved in lakes	-56 (-41 to -66) [16]	Quay et al., 1988
Amazon flood plain	Flux chamber	-51 (-42 to -73)]10]	Quay et al., 1988

The natural sources of methane consist of natural wetlands, natural fires (forest and savanna fires), termites, wild animals, and the oceans. Table 1 lists the carbon isotopic measurements for methane emitted or contained in the interstitial water or sediments of fresh and marine wetlands. The new measurements listed here are those of Hornibrooke et al. (1997), Popp et al. (1995), and Boehme et al. (1996). These studies consist of systematic measurements of the variation of the isotopic composition of methane with depth in the interstitial water column and sediments of shallow wetlands.

Table 2. Carbon isotopic composition of the anthropogenic sources

Source	Method and number of samples	Mean $\delta^{13}C$ and range (‰)	Reference
Rice paddies			
California	Inversion (4)	-67 (-66 to -68)	Stevens and Engelkemeir, 1988
Louisiana	Flux chamber (8)	-63.2 ± 2.9	Wahlen et al., 1989
Kenya	Flux chamber (10)	-59.4 (-57 to -63)	Tyler et al., 1988
Vercelli, Italy	Flux chamber (7)	-65.4 ± 1.6	Levin et al., 1993
Weighted average: -63.8 ± 1.5			
Ruminants C3 diet			
Cattle	Fistula (5)	-63.7 (-61 to -76)	Rust, 1981
Cattle	Barn (1)	-61.1	Rust, 1981
Sheep	Fistula (2)	-68.6 (-67 to -70)	Rust, 1981
Cattle	Barn (4)	-71.3 ± 4	Wahlen et al., 1989
Cattle	Bag (1)	-70.6	Wahlen et al., 1989
Goat	Bag (1)	-65.2	Wahlen et al., 1989
Weighted average: -63 ± 1.0			
Ruminants C4 diet			
Cattle	Fistula (3)	-50.3 (-47 to -52)	Rust, 1981
Cattle	Barn (1)	-45.4	Rust, 1981
Cattle (60-80% C4 diet)	Bag (3)	-55.6 ± 1.4	Levin et al., 1993
Weighted average: -50 ± 3			

Source	Method and number of samples	Mean $\delta^{13}C$ and range (‰)	Reference
Landfills			
Indiana		-50 (-48 to -52)	Games and Hayes, 1976
Colorado	Flux chamber (2)	-53 (-51 to -55)	Tyler, 1986
Heidelberg, Germany	Upper layers (1)	-55.6	Levin et al., 1993
Mainz, Germany	Closed collection (23)	-60.8 (-55 to -62)*	Bergamaschi and Harris, 1995
Eight U.S.A. landfills	Vent wells (22)	-53.7 (-48 to -60)	Hackley et al., 1996

Weighted average: -53.6 ± 0.3   (* not included in average)

Natural gas			
Thermogenic 80%		-38 (-25 to -52)	Schoell, 1980
Biogenic 20%		-65 (-60 to -70)	Rice and Claypool, 1981

Weighted average: -43 ± 4

Coal mining		-37 ± 4 (-14 to -60)	Deines, 1980

Biomass burning			
Wood fire	Plume (3)	-27 (-24 to -32)	Stevens and Engelkemeir, 1988
Grass fire	Plume (1)	-32	Stevens and Engelkemeir, 1988
Brush fire	Plume (2)	-26.6	Wahlen et al., 1989
Wood fire	Plume (1)	-26.4	Levin et al., 1993

Weighted average: -28

3. The $\delta^{13}C$ Ratio and Its Temporal Trend in Atmospheric Methane

Table 3 lists the published measurements of the ^{13}C isotopic ratios measured by seven laboratories from 1978 to 1995. New published values since the NATO report are those of Lowe et al. (1994), Gupta et al. (1996), and Sugawara et al. (1996). Other unpublished results include some of Paul Quay's group at the University of Washington and Roger Francey's group at CSIRO, Aspendale, Australia. Quay's group have carried out very extensive measurements on approximately one thousand samples as a function of latitude, season, and annual trends from 1985 to 1995.

Table 3. Comparison of measurements of the carbon-13 isotopic composition of atmospheric methane. Errors are 1 SD with numbers of samples in parentheses.   A: Stevens, 1988. B: Wahlen et al., 1989. C: Quay et al., 1991, and Quay, 1997 (unpublished). D: Gupta et al., 1996. E: Lowe et al., 1994. F: Sugawara et al., 1996. Francey et al., 1998 (in preparation).

Labora-tory	Year	Latitude	$\delta^{13}C$ (per mil)	
			Southern hemisphere	Northern hemisphere
A	1978.4	13-57 S	-47.84 ± 0.5 (4)	
	1978.8	45 N		-47.88 ± .14 (7)
	1980.5	45 N	-47.52 ± .09 (3)	-47.78 ± .08 (5)
	1981.5	19-45 N	-47.49 ± .09 (3)	-47.69 ± .02 (7)
	1982.5	8-50 N	-47.35 ± .04 (4)	-47.74 ± .26 (4)
A	1983.25	42 N		-47.60 ± .05 (20)
	1983.5	45 N		-47.61 ± .05 (7)
	1983.75	42 N		-47.61 ± .04 (20)
	1984.2	90 S	-47.15 ± .15 (1)	
	1984.5	42 N		-47.53 ± .10 (5)
A	1985.5	42 N		-47.66 ± .08 (5)
	1986.25	42 N		-47.28 ± .06 (11)
	1986.75	42 N		-47.25 ± .06 (10)
	1986.9	15 S	-46.71 ± .08 (6)	
	1987.5	42 N		-47.21 ± .10 (6)
A	1988.25	42 N		-46.98 ± .05 (15)
	1988.3	15 S	-46.49 ± .09 (6)	
	1989.25	42 N		-47.00 ± .10 (7)
	1989.3	35 S	-46.32 ± .10 (2)	
B	1976.5	43 N		-47.3 ± .2 (1)
	1977.5	42 N		-46.9 ± .2 (1)
	1978.5	43 N		-46.5 ± .2 (1)
	1986.5	5 to 9 S	-46.0 ± .2 (5)	
	1987.5	10 to 32 S	-45.2 ± .3 (2)	
B	1986 to 1988	9 to 65 N		-46.7 ± .2 (12)
	1988.0	90 S	-46.6 ± .3 (4)	
C	1987 to 1989	4-60 S	-47.04 ± .2 (10)[a]	
	1987 to 1989	70 N		-47.6 (70)[a]
	1987 to 1989	47 N		-47.4 (63)[a]
	1987 to 1989	19 N		-47.3 (50)[a]
	1988.5	47 N		-47.61 (~100)[d]

Labora-tory	Year	Latitude	$\delta^{13}C$ (per mil) Southern hemisphere	$\delta^{13}C$ (per mil) Northern hemisphere
C	1989.5	47 N		-47.53 (~100)[d]
	1990.5	15 S & 47 N	-47.25[d]	-47.53 (~100)[d]
	1991.5	15 S & 47 N	-47.31[d]	-47.54 (~100)[d]
	1992.5	15 S & 47 N	-47.25[d]	-47.50 (~100)[d]
	1993.5	15 S & 47 N	-47.25[d]	-47.43 (~100)[d]
C	1994.5	15 S & 47 N	-47.18[d]	-47.44 (~100)[d]
	1995.5	15 S & 47 N	-47.17[d]	-47.43 (~100)[d]
D	1989.5	40 N		-47.3[b]
	1990.5	40 N		-47.4[b]
	1991.5	40 N		-47.5[b]
	1992.5	40 N		-47.6 [b]
E	1989.6	41 S	-47.1 ± .07[c]	
	1990.0	41 S	-47.11 ± .04[c]	
	1990.5	41 S	-47.13 ± .04[c]	
	1991.0	41 S	-47.15 ± .04[c]	
	1991.5	41 S	-47.18 ± .04[c]	
E	1992.0	41 S	-47.23± .04[c]	
	1992.5	41 S	-47.29 ± .04[c]	
	1993.0	41 S	-47.31 ± .04[c]	
	1993.5	41 S	-47.30 ± .06[c]	
F	1993.5	55-62 N		-47.9 ± .3
	1994.5	55-62 N		-47.8 ± .2 (30)
G	1978.5	41 S	-47.78	
	1979.0	41 S	-47.65	
	1980.7	41 S	-47.64	
	1981.8	41 S	-47.79	
	1984.4	41 S	-47.52	
G	1986.2	41 S	-47.50	
	1987.2	41 S	-47.47	
	1988.6	41 S	-47.38	
	1989.6	41 S	-47.27	
	1990.5	41 S	-27.35	
G	1991.8	41 S	-47.33	
	1992.1	41 S	-47.14	
	1992.8	41 S	-47.28	
	1993.4	41 S	-47.16	
	1994.1	41 S	-47.16	
	1995.0	41 S	-47.09	

[a] These values are the annual mean of samples collected regularly in all seasons (Quay et al., 1991).
[b] Averaged annual values determined from $\delta^{13}C$ = -47.26 – [2.76 × 10^{-4}] T + 0.0219 sin(ωT) – 0.1269 cos(ωT) (43 measurements) where T = days relative to January 1, 1989, and ω = 2π/365.
[c] Semi-annual values determined from the deseasonalized trend (procedure STL: Seasonal Trend Decomposition Procedure based on Loess) from data for weekly samples analyzed between 1989.6 and 1993.5 (Lowe et al., 1994).
[d] Quay, 1997.

Most of these are unpublished except for 208 measurements for the period 1985 to 1987, which have been published (Quay et al., 1991) and were included in the annual averages in the NATO report and in Table 3. With permission of Quay (1997), the annual mean isotopic values for the years 1988 to 1995 for samples from the Olympic Peninsula and Samoa are listed in Table 3; Quay's measurements at other latitudes in both hemispheres are not listed but are in the process of being published. They show significant variations with latitude and season. From Point Barrow (70° N) compared to Hawaii (19° N), the results of Quay et al. (1991) show as much as 0.6 per mil depletion in the winter, decreasing to 0.2 per mil in the early summer.

The measurements of Francey et al. (1998), shown in Table 3, were made on samples from Cape Grim collected and archived in high pressure metal containers. This group has done other measurements on Antarctic firn air and air samples collected at Cape Grim in 5 l glass flasks, which agree well with the archived samples but are not reported here.

Results not shown listed here of analyses by laboratories C, D, and E all show a seasonal variation in both hemispheres with an amplitude of 0.2 to 0.3 per mil. The seasonal variation in the southern hemisphere is attributed by Lowe et al. (1994) as due mainly to the seasonal variation in the OH oxidation sink with an additional effect from seasonal source fluxes and transport. In the northern hemisphere the amplitudes of the seasonal variation are explained as a combination of both seasonal variations in the sink processes, mainly due to OH oxidation, and seasonal variations in the fluxes of isotopically different sources (Gupta et al., 1996, and Quay, 1997). Results of detailed model calculations are presented by the above referenced investigators.

Figure 1 shows plots of the trends of the mean annual values measured by five laboratories and listed in Table 3. All the plots of the trends in the northern hemisphere are for mid temperate latitudes 40° to 47° N in order to compare the results as nearly as possible to the same latitude. The comparison of the long term temporal trends shown in Figure 1 shows major disagreements in the southern hemisphere values for lab A compared to lab G for the years 1984 to 1989, and lab A compared to labs E and G for year 1989, and in the northern hemisphere for lab A compared to labs C and D. These large disagreements were caused by errors in the analyses for samples done between 1984 and 1989 by Stevens (1988) (lab). There was no problem with the results for the years 1978 to 1983. For this period of 1978 to 1983 all their samples were supplied from the storage bank maintained by Rasmussen and Khalil at the Oregon Graduate Center and analyzed at Argonne National Laboratory in random order, not chronologically, and using a working laboratory isotope standard with a value of -24.8 per mil. The atmospheric methane samples were generally 23 per mil more depleted in carbon-13 than this standard; because of this large difference, these analyses have an uncertainty due to memory effects between the sample and the standard estimated at ± 0.1 per mil. This standard was available in large quantity and was frequently replenished to ensure that no change occurred in its isotopic composition due to fractionation during use. Thus, the slope of the trend in both hemispheres between 1978 and 1983 is not in doubt although there may be an uncertainty of ± 0.1 per mil in the absolute value. On the other hand, the procedure between 1984 and 1989 was changed in that a different isotopic reference standard was used, one that was made up to have a value close to that of the atmospheric samples of ~ -46 per mil in order to reduce the uncertainty due to memory between samples and the standard. However, because the supply of this standard was limited and the amount of it was small, it was the practice to not replenish it often. Furthermore, during the same period the samples from the Oregon sample bank were no longer available or needed, so fresh samples of background air were obtained in rural northern Illinois as a northern hemisphere site, while southern hemisphere samples were collected by a NOAA observatory in Samoa. Tests were carried out on the samples for 1983 to ensure (a) that the samples from the rural northern Illinois site gave the same value as those from the sample bank collected on the Oregon coast at Cape Meares, which they did, and (b) that the samples collected in Samoa did not change due to contamination from the cylinder walls during the month-long round trip transit time, which they did not. The upshot of this change in procedure meant that the samples were analyzed chronologically but within a few months of the same time for samples from both hemispheres so the results gave the *appearance* of a uniform changing isotopic trend even though, as we believe now, the isotopic standard fractionated during this period, becoming more enriched due to preferential loss of ^{12}C during flow into the mass spectrometer in the analytical procedure. In 1989 the mass spectrometer used for all these analyses was destroyed. However, an analysis of the isotopic standard used for the analyses in the period 1984 to 1989 was checked on a new isotope ratio mass spectrometer and found to be 0.6 per mil heavier than the original isotopic composition. This is the same as the amount of the discrepancy between the results of the samples for 1989 by lab A and lab C. This confirms that the standard underwent a uniform enrichment between 1984 and 1989, leading to progressively isotopically heavier and incorrect values for the atmospheric samples for both hemispheres during this period.

The most significant difference between lab C and lab B in the northern hemisphere is the opposite trend for the slope between 1989 and 1993; except for that, the differences may be due to slightly different latitudes or locations. For the southern hemisphere lab C and lab D have a small difference in the slope of the trend between 1989 and 1993, which is similar to the difference in slope for the northern hemisphere case, namely, that the lab C values are more depleted than those of B or D in 1989, but, with the different trend slopes, become more enriched in 1993. These disagreements may not be significant because of the statistical uncertainty of the analyses. The publication by Quay of the large data sets for other latitudes may help explain some of these small discrepancies.

All hemispheric values show a difference between hemispheres of 0.2 to 0.3 per mil for the years 1989 to 1995, while lab A showed no difference in 1978 indicating a trend of increasing enrichment of the methane in the southern hemisphere compared to that for the northern hemisphere during the years 1978 to 1984. This is consistent with the general consensus that increasing biomass burning in the southern hemisphere was a main contributor to the increasing concentration in the 1970's and early 1980's, and being isotopically much heavier than all other sources caused the increasing difference in the values between the hemispheres. This is also supported by the enrichment increasing more rapidly in the southern hemisphere than in the northern hemisphere between 1978 and 1984 as shown by the results from lab A.

4. Deuterium Isotopic Composition

The isotopic abundance of deuterium in atmospheric methane and its sources provide an additional constraint for a global methane budget as shown by Wahlen et al. (1990). Their measurements show the global average δD for atmospheric methane is -82 per mil, while δD for the biogenic methane from the major wet environments sources is -290 to -360 per mil (see Figure 2). The deuterium data provide additional constraints for calculating a budget of fluxes from the different isotopic source types: first, the δD of the methane from wet environments correlates with the δD of the local precipitation, which is latitude-dependent. This makes it possible to distinguish methane from tropical and Arctic wetlands; secondly, the δD of fossil fuel from wetlands has a different relative composition compared to the value for methane from wetlands than in the case for the carbon isotopes. The fractionation factor of $(\alpha-1)$ = -0.330 measured by Gordon and Mulac (1975) for the oxidation of CH_3D/CH_4 by OH radicals seems too large, leading to the average value for the sources more depleted than the value of the most depleted source. Wahlen et al. (1990) derive a value of -0.150 to -0.170, based on the enrichment of CH_3D in the stratosphere where the concentration decreases with altitude because of OH oxidation. This is in good agreement with the isotopic distributions in Figure 2. Saueressig et al. (1996) have measured the kinetic isotope effect for the reaction of CH_4 with Cl for deuterium, $(KIE[CH_4/CH_3D] = 1.508 \pm 0.04$ at $297°K)$, which is important in the higher stratosphere. The unpublished data of Quay (1997) contain the results for the δD of over 60 samples of atmospheric methane for the years 1989 to 1993 for latitudes from 70° S to 55° N. These results will hopefully refine the budget calculations.

5. The ^{14}C Abundance of Atmospheric Methane

Carbon-14 can be used to access the contributions by fossil methane to the atmospheric inventory (Wahlen, 1993). Fossil CH_4 from natural gas exploration and distribution, from coal mining, and possibly from seepage of natural gas reservoirs on land and near shore is free of ^{14}C, while biogenic methane contains more or less contemporary ^{14}C of about 120 pMC (percent modern carbon). Via ^{14}C analyses, the biogenic contribution has been determined to be $21 \pm 3\%$ of the annual input to the atmosphere by Wahlen et al. (1989), 17-25 % by Manning et al. (1990), and 16 ± 12

% by Quay et al. (1991). There is some complication due to a substantial anthropogenic contribution to the atmospheric $^{14}CH_4$ inventory from emissions from pressurized light water reactors (Kunz, 1985), which adds to the uncertainty. Statistical analyses of fossil methane releases (Cicerone and Oremland, 1988; Barns and Edmonds, 1990; Fung et al., 1991) are lower than those observed by the ^{14}C data. The discrepancy may be explained by releases of ^{14}C-depleted methane from ecosystems where old carbon is being processed. It is also possible that additional fossil CH_4 is released through seepage from natural gas reservoirs on land and near shore.

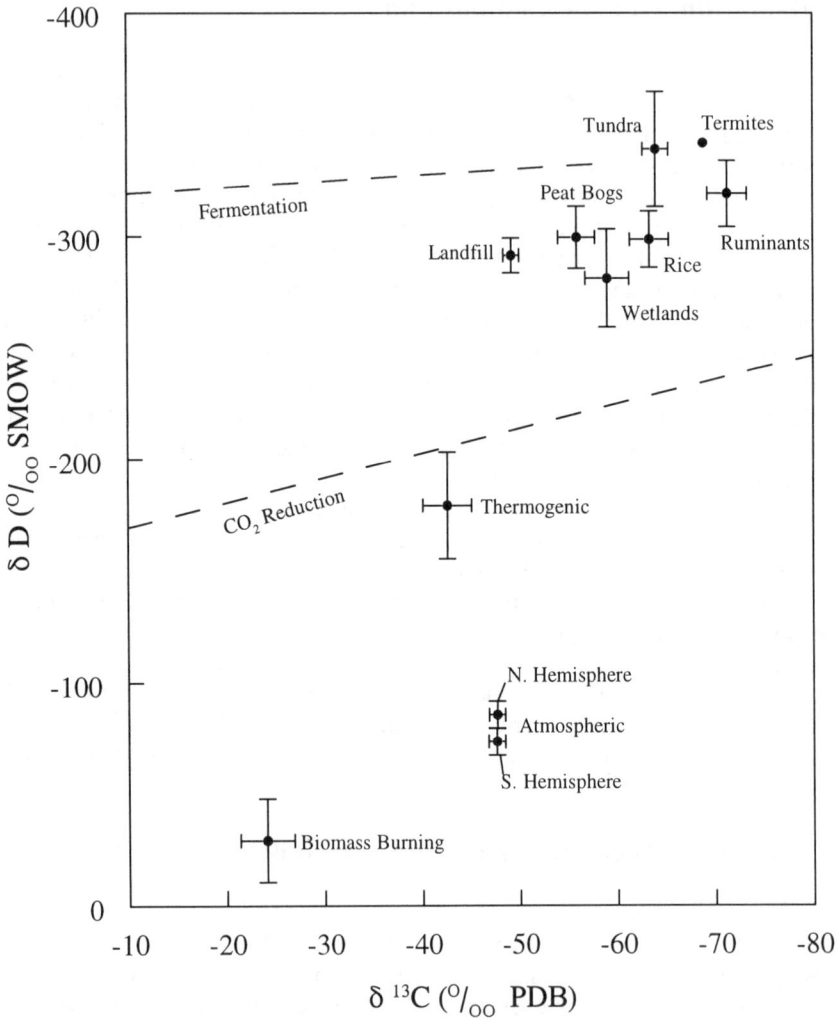

Fig. 2. Systematics among the stable isotope composition of methane from various sources and from atmospheric methane (Wahlen et al., unpublished data). Boxes show the mean values for sets of samples for ^{13}C and D in methane.

Acknowledgments. Work performed under the Office of Basic Energy Sciences, Division of Mathematical and Geosciences, U.S. Department of Energy, under Contract No. W-31-109-Eng-38.

References

Barns, D.W., J. A. Edmonds. 1990. An evaluation of the relationship between the production and use of energy and atmospheric methane emissions. *Rep.* TR047 Carbon Dioxide Res. Prog., U.S. Department of Energy, Washington, D.C. (Available as DOE/NBB-0088P, Natl. Tech. Inf. Serv., Springfield, Va., 1990.)

Bergamaschi, P., G.W. Harris. 1995. Measurements of stable isotope ratios ($^{13}CH_4/^{12}CH_4$; $^{12}CH_3D/^{12}CH_4$) in landfill methane using a tunabvle diode laser absorption spectrometer. *Global Biogeochem. Cycles*, 9:439-447.

Boehme,S.E., N.E. Blair, J.P. Chanton, C.S. Martens. 1996. A mass balance of ^{13}C and ^{12}C in an organic-rich methane-producing marine sediment. *Geoch. et Cosmoch. Acta, 20*: 3835-3848.

Burke, R.A., W.M. Sackett. 1986. Stable hydrogen and carbon isotopic compositions of biogenic methane from several shallow aquatic environments. *Organic Marine Geochemistry* (M.L. Sohn, ed.,) American Chemical Society, Washington, D.C. p.297.

Chanton, J.P., C.S. Martens. 1988. Seasonal variations in ebullitive flux and carbon isotopic composition of methane in a tidal freshwater estuary. 1988. *Global Biogeochem. Cycles.* 2:289.

Chanton, J.P., G.G. Pauly, C.S. Martens, N.E. Blair. 1988 Carbon isotopic composition of methane in Florida Everglades soils and fractionation during transport to the troposphere. *Global Biogeochem. Cycles.* 2:245.

Cicerone, R.J., R.S. Oremland. !988. Biogeochemical aspects of atmospheric methane. *Global Biogeochem. Cycles.* 2:299-327.

Deines, P. 1980 The isotopic composition of reduced organic carbon. In: *Handbook of Environmental Isotope Geochemistry, Vol. 1* (P. Fritz and J.C. Fontes, eds.), Elsevier Scientific, Chapter 9, pp.329-406.

Francey, R.J., M.R. Manning, C.E. Allison, S.A. Coram, D.M. Etheridge, R.L. Langenfelds, D.C. Lowe, and L.P. Steele. 1998. A history of $\delta^{13}C$ in atmospheric CH_4 from the Cape Grim Air Archive and Antarctic firn air. *J. Geophys. Res.,* in preparation.

Fung, I., J. John, J. Lerner, E. Matthews, M. Prather. 1991. Three-dimensional model synthesis of the global methane cycle. *J. Geophys. Res. 96:* 13,033-65.

Games, L.M., J.M. Hayes. 1976. On the mechanisms of CO_2 and CH_4 production in natural anaerobic environments. In: *Pro. of the 2nd International Conference on Environments Biogeochemistry, Vol. 1* (J.O. Nriaague, ed.), Butterworth, Stoneham, Mass., p.51.

Gordon, S., W.A. Mulac. 1975. Reactions of the OH (X^2II radical) produced by the pulse radiolysis of water vapor. *Int. J. Chem. Kinet.,* 7:289.

Gupta, M., S. Tyler, R. Cicerone. 1996. Modeling atmospheric d$^{13}CH_4$ and the causes of recent changes in atmospheric CH_4 amounts. *J. Geophys. Res,* 101: 22,923-22,932.

Hackley, K.C., C.L. Liu, D.D. Coleman. 1996. Landfill leachates and Gases. *Ground Water:* 34:827-836.

Happell, J.D., J.P. Chanton, W.S. Showers. 1994. The influence of methane oxidation on the stable isotope composition of methane emitted from Florida swamp forests. *Geoch. et Cosmoch. Acta.* 58: 4377-4388.

Hornibrook, E.R.C., F.J. Longstaffe, W.S. Fyfe. 1997. Special Distribution of microbial methane production pathways in temperate zone wetland soils: Stable carbon hydrogen isotope evidence. *Geoch. et Cosmoch. Acta.* 61(4):745-753.

Kunz, C. 1985. Carbon-14 discharges at three light-water reactors. *Health Phys.* 49:25- 35

Lansdown, J.M., P.D. Quay, S.L. King. 1992. CH_4 production via CO_2 reduction in a temperate bog; a source of ^{13}C-depleted CH_4. *Geoch. et Cosmoch. Acta.* 56:3493-3503.

Levin, I., P. Bergamaschi, H. Dorr, D. Trapp. 1993. Stable isotope signature of methane from different sources in western Europe. *Chemosphere.* 26 (1-4):161-178.

Lowe, D.C., C.A.M. Brenninkmeijer, G.W. Brailsford, K.R. Lassey, A.J. Gomez. 1994. Concentration and ^{13}C records of atmospheric methane in New Zealand and Antarctica: Evidence for changes in methane sources. *J Geophys. Res.* 99: 16,913-16,925.

Manning, M.R., D.C. Lowe, W.H. Melhuish, R.J. Sparks, C.A.M. Brenninkmeijer, et al., 1990. The use of radio-carbon measurements in atmospheric studies. *Radiocarbon,* 32:37-58.

Martens, C.S., N.E. Blair, C.D. Green, D.J. Des Marais. 1986. Seasonal variations in the stable carbon isotopic signature of biogenic methane in a coastal sediment. *Science.* 233:1300-1303.

Oona, S., E.S. Deevey. 1960. Carbon-13 in lake waters and its possible bearing on paleolimnology. *Am. J. Sci.,* 258A:253.

Ovsyannikov, V.M., V.S. Lebedev. 1967. Isotopic composition of carbon in gases of biogenic origin. *Geochem. Int.,* 4:453.

Popp, B.N., F.J. Sansone, T.M. Rust. 1995. Determination of concentration and carbon isotopic composition of dissolved methane in sediments and near shore. *Analytical Chem.* 67:405-411.

Quay, P.D., S.L. King, J. Stutsman, D.O. Wilbur, L.P. Steele, I.Fung, R.H. Gammon, T.A. Brown, G.W. Farwell, P.M. Grootes, F.H. Schmidt. 1991. Carbon isotopic composition of atmospheric CH_4: Fossil and biomass burning source strengths. *Global Biogeochem. Cycles.* 4:25-47.

Quay, P.D., S.L. King, J.M. Lansdown, D.O. Wilbur. 1988. Isotopic composition of methane released from wetlands: Implications for the increase in atmospheric methane. *Global Biogeochem. Cycles.* 2:385-397.

Quay, P.D. 1997. Personal communication.

Rice, F.E., G.E. Claypool. 1981. Generation, accumulation and resource potential of biogenic gas. *Bull. Am. Assoc. Pet. Geol.,* 65:5.

Rust, F.E. 1981. Ruminant methane d($^{13}C/^{12}C$) values: Relationship to atmospheric methane. *Science,* 211:1,044-1,046.

Schoell, M. 1980. The hydrogen and carbon isotopic composition of methane from natural gases of various origins. *Geochem. Cosmochem. Acta,* 44:649.

Stevens, C. 1988. Atmospheric methane. *Chem. Geol.,* 71:11-21.

Stevens, C., A. Engelkemeir. 1988. Stable carbon isotopic composition of methane from some natural and anthropogenic sources. *J. Geophys. Res.,* 93:725-733.

Saueressig, G., P. Bergamaschi, J.N. Crowley, H. Fischer. 1995. Carbon kinetic isotope effect in the reaction of CH_4 with Cl atoms. *Geophys. Res. Lett.,* 22:1225-1228.

Saueressig, G., P. Bergamaschi, J.N. Crowley, H. Fischer, G.W. Harris. 1996. D/H kinetic isotope effect in the reaction CH_4 + Cl. *Geophys. Res. Lett.,* 23:3619-3622.

Sugawara, S., T. Nakazawa, G. Inoue, T. Machida, H. Mukai, N.K. Vinnichenko, V. U. Khattatov. 1996. Aircraft measurements of the stable carbon isotopic ratio of atmospheric methane over Siberia. *Global Biogeochem.* Cycles. 10:223-231.

Tyler, S.C. 1986. Stable carbon isotopic ratios in atmospheric methane and some of its sources. *J. Geophys. Res.,* 91:13,232.

Tyler, S.C., P.R. Zimmerman, C. Cumberbatch, J. Greenberg, C. West berg, J.P.E.C. Arlington. 1988. Measurements and interpretation of d^{13}C of methane from termites, rice paddies, and wetlands in Kenya. *Global Biogeochem. Cycles,* 2:341.

Wahlen, M., N. Tanaka, R. Henry, B. Deck, J. Zeglen, J.S. Vogel, J. Southon, A. Shemesh, R. Fairbanks, W. Broecker. 1989. Carbon-14 in methane sources and in atmospheric methane: The contribution from fossil carbon. *Science.* 245:286-290.

Wahlen, M., B. Deck, R. Henry, N. Tanaka, A. Shemesh. 1990. Profiles of $\delta^{13}C$ and δD of

CH_4 from the lower stratosphere. *Eos, Trans. Am. Geophys. Union.* 70(43): 1017.

Wahlen, M. 1993. The global methane cycle. *Annu. Rev. Earth Planet. Sci.* 121:407-426.

Whiting, G.J., J.P. Chanton, 1993. Primary production control of methane emission from wetlands. *Nature*, 364:794-795.

4. Biological Formation and Consumption of Methane

David R. Boone
Department of Environmental Biology, Portland State University, P.O. Box 751, Portland, Oregon 97207, USA

1. Introduction

Methane is an important product formed during the bacterial degradation of organic matter in environments such as flooded soils, wetlands, estuaries, marine and freshwater sediments, and the gastrointestinal tract of animals (Whitman et al., 1992). This chapter describes the conditions that lead to biogenic methane formation in natural environments, the metabolic pathways and interactions that lead to methanogenesis, and the implications of these factors on the biogeochemistry of methane.

Methane-producing bacteria (methanogens) are a specialized group of microbes that catabolize a small number of molecules and produce methane as the major catabolic product. They are the only known life-forms that produce a hydrocarbon as a major catabolic product. The known substrates that methanogens can catabolize is very limited: $H_2 + CO_2$, formate, acetate, methanol, methylamines, and methylsulfides. In addition to these, a small number of alcohols can be oxidized by some strains. Because the substrate range of methanogens is so limited, these microbes depend on other bacteria to provide them from a wide range of more complex molecules. Methanogenic decomposition of complex organic matter is possible because many non-methanogenic bacteria convert organic compounds into the few molecules that methanogens can use.

2. Ecology of Methanogenic Decomposition

The decomposition process in anoxic environments may be divided into three major steps, fermentation (or acidogenesis), syntrophic acetigenesis, and methanogenesis (Figure 1). Each step is catalyzed by a separate group of bacteria, but all the reactions occur simultaneously so the concentrations of extracellular intermediates often remain small. For instance, H_2 concentration in such ecosystems is typically less than 100 nM (with a partial pressure of about 5 to 10 Pa, or <0.01% of the gas phase), yet its rate of production and consumption is very rapid (Bryant et al., 1967; Boone, 1982; Wolin, 1982; Conrad et al., 1986; Dwyer et al., 1988).

Mohammad Aslam Khan Khalil   (Ed.)
Atmospheric Methane
© *Springer-Verlag Berlin Heidelberg 2000*

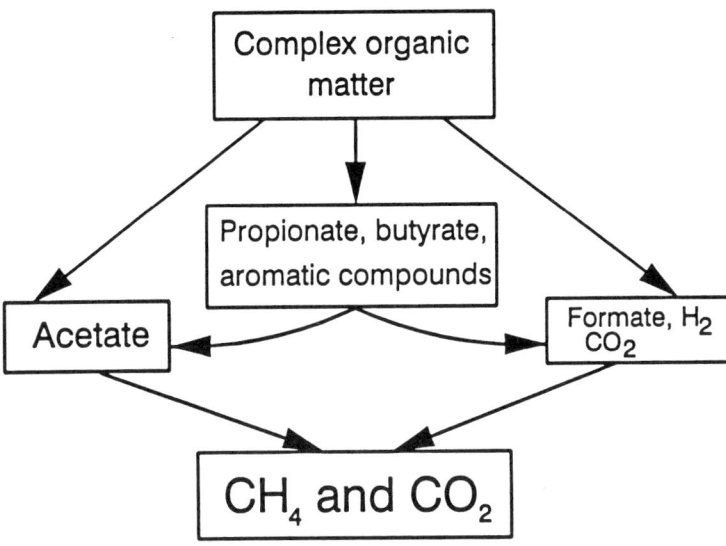

Fig. 1. Organic matter decomposition in methanogenic ecosystems

The fermentative step converts a wide array of complex organic matter in to a relatively small number of products, mainly volatile fatty acids, carbon dioxide, and hydrogen (Figure 1). The conversion of most of the components of biomass (such as proteins, carbohydrates, nucleic acids, and fats) is typically rapid. Other organic molecules such as highly polymerized lignin, hydrocarbons, and some man-made chemicals are degraded slowly or not at all in anaerobic environments (Leisinger and Brunner, 1986).

The methanogens convert some of the products of the fermentative step to methane and carbon dioxide, the terminal carbon-containing products formed during the complete degradation of organic matter. One group of molecules produced during fermentation, the volatile fatty acids longer than acetic acid, cannot be broken down by the methanogens. These acids are degraded by a specialized group of bacteria called "obligate proton-reducing acetigens," which produce acetic acid by β-oxidation of the fatty acid (McInerney, 1986; McInerney et al., 1979, 1981). The acetigenic oxidation of odd numbered fatty acids produces, in addition to acetic acid, propionic acid (a 3-carbon monocarboxylic acid). Propionate is oxidized to acetic acid and carbon dioxide, probably by the succinate pathway (Houwen et al., 1987). The oxidation of volatile fatty acids is coupled to the production of H_2 (by proton reduction) or formate (by bicarbonate reduction), and is thermodynamically favorable only when H_2 and formate concentrations are very low (Bryant, 1979; Boone et al., 1989). Low concentrations are maintained by the rapid uptake of H_2 and formate by methanogens, which requires a tight coupling of fatty acid oxidation with methane formation. This process has been called "interspecies hydrogen transfer" (McInerney et al., 1981; Wolin, 1982), but the term "interspecies electron transfer" may be more appropriate because, in addition to H_2, formate may act as an extracellular electron transporter from fatty acid oxidizers to methanogens (Thiele and Zeikus, 1988; Boone et al., 1989).

Organic matter such as lignin monomers and pectin contain methoxyl groups that may be degraded in anoxic environments by a newly discovered pathway. The removal of methoxyl groups by acetigenic bacteria was documented during the 1980s (Barik et al., 1985). These bacteria convert the methoxyl groups and carbon dioxide to acetic acid, which is their catabolic product (Ljungdahl, 1986). Recently, other bacteria were shown instead to transfer successively the methyl groups to sulfide to form methane thiol and then dimethyl sulfide. The newly discovered ability of some methanogens to catabolize dimethyl sulfide (Oremland et al., 1989; Ni and Boone, 1991, 1993; Bak and Finster, 1993) and methane thiol (Bak and Finster, 1993; Ni and Boone, 1991, 1993) could allow the methoxyl groups to be converted ultimately to methane, regardless whether they were first combined with carbon dioxide (to form acetic acid) or with sulfide (to form methane thiol or dimethyl sulfide).

3. Partial Methanogenic Fermentations, Including the Rumen Fermentation

In some methanogenic ecosystems all three decomposition steps (Figure 1) do not occur. For instance, when geothermal H_2 enters sediments, it may be converted directly to methane, so that only the methanogenic step is required (Bryant, 1979). In some environments, bacteria may be lost more rapidly than they are replaced by growth. This may be one of the factors that limit colonization of digestive tracts of man and animals by slow-growing methanogens (Bryant, 1979; Wolin, 1982; Miller, 1991). About half of the human population harbors substantial numbers of methanogens in their colons. Ungulates, or ruminants (such as cattle, sheep, deer, moose, elk), have a special forestomach (the rumen) that harbors an intense microbial fermentation. The function of this fermentation, from the standpoint of the animal, is to convert organic matter in the diet into a form that can be used by the animal. Cellulose is an important part of the diet of most of these animals, yet the animal itself cannot decompose cellulose. However, the rumen is colonized by anaerobic cellulose-fermenting microbes (Hungate, 1966), so cellulose and other biomass are fermented to compounds (mainly volatile fatty acids) which provide energy to the animal. This fermentation is not complete, however, because the slow-growing microbes responsible for volatile fatty acid degradation are not present. The methanogens which degrade acetate and the obligate proton-reducing acetigens are not active in the rumen, probably because they cannot grow fast enough to maintain large populations. The methanogens, which use H_2 and formate, do occur in the rumen, so the products of this fermentation are mainly volatile fatty acids longer than formate, methane and carbon dioxide. The gases, which are eructated, are an important source of global methane (see, for example, Miller, 1991). The volatile fatty acids are absorbed by the animal as its major source of carbon and energy. An important by-product of the rumen fermentation (again, from the standpoint of the animal) is microbial cells, which serve as a source of protein.

4. Alternate Electron Acceptors

In many environments conditions prevail that allow other catabolic pathways than methanogenesis to occur. When organic matter is completely oxidized, carbon dioxide is always the product. Thus, carbon dioxide is always available as an electron acceptor for methanogens, which can use the reducing equivalents derived from that oxidation to reduce a portion of the carbon dioxide to methane. When waters contain electron acceptors other than carbon dioxide, such as O_2, nitrate, sulfate, manganese(-IV), or iron(III), the electrons generated from organic matter oxidation are preferentially passed to one of these electron acceptors. For instance, when SO_4^{2-} is present, sulfate-reducing bacteria may pass the electrons stripped from organic matter to SO_4^{2-}, making H_2S. Methanogenesis is thought to be inhibited by the presence of alternate electron acceptors because bacteria using those electron acceptors out-compete methanogens for biodegradable molecules.

The partial pressure of H_2 in sediments or groundwater can be used as an indicator of the dominant electron acceptor present, because it reflects the degree of competition for reduced substrates (Lovley and Goodwin, 1988). H_2 partial pressure is much lower when electron acceptors other than carbon dioxide are present, presumably because bacteria using those alternate electron acceptors have more energy available to scavenge reduced substrates at lower concentrations. This ability may derive from the extra energy that is released when electrons are passed to acceptors other than carbon dioxide. For instance, aerobic bacteria release approximately 10-fold more energy by using O_2 as electron acceptor than can anaerobic bacteria, which carry out the methanogenic decomposition of organic matter. This allows aerobes in the presence of O_2 to maintain concentrations of catabolic substrates at concentrations so low that methanogens cannot compete. In addition to this indirect inhibition, O_2 itself inhibits methanogens. In general, when organic matter is degraded in the presence of alternate electron acceptors, the electron acceptors whose reduction yield the most energy disappear first, and the others are used sequentially (Figure 2 shows the energy released by the reduction of various electron acceptors by H_2). Thus, methanogenesis generally begins only after all the alternate electron acceptors are depleted.

There are at least two exceptions to the general rule of sequential use of electron acceptors. The first is that heterogeneities in concentrations may lead to microenvironments in which different reactions predominate. For instance, in flocs or biofilms the outer surface may be exposed to sulfate-containing water, but the activities of sulfate-reducing bacteria on the surface may lower sulfate concentrations so it is not available deeper in the floc. Likewise, the use of insoluble electron acceptors may be kinetically limited by their availability. Fe(III) reduction may be limited by its solubilization. A second exception to the sequential use of electron acceptors is that specific organic molecules may be used by methanogens in preference to sulfate-reducing bacteria regardless of whether sulfate is present. These molecules are methyl amines and perhaps methyl sulfides and methanol, which can be used by methanogens to make methane without complete oxidation to carbon dioxide. Such compounds, which may be converted to methane even in the presence of excess sulfate, are called "non-competitive" substrates (Oremland et al., 1982).

These "non-competitive substrates" can also be catabolized by sulfate reducers, and it has not been established what environmental conditions determine whether they are degraded by methanogens or sulfate reducers.

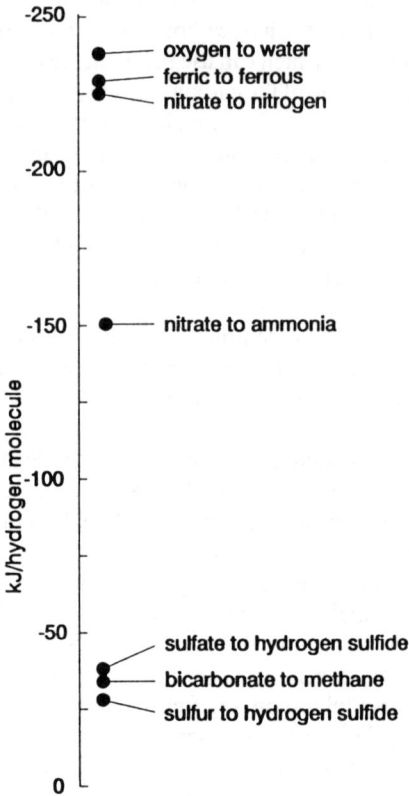

Fig. 2. Energy released by coupling H_2 oxidation to the reduction of various catabolic electron acceptors

5. Stoichiometry of Methanogenesis

When microbial cells extract all the available free energy from organic matter, decomposition is carried to its ultimate conclusion and the organic matter is in its most stable form. For organic matter composed of carbon, hydrogen, and oxygen, the most stable form (at standard temperature and pressure) is a mixture of carbon dioxide (or bicarbonate), methane, and water, in relative proportions dictated by stoichiometry (Buswell and Hatfield, 1939):

$$C_aH_bO_c + \left(a - \frac{1}{4}b - \frac{1}{2}c\right)H_2O \rightarrow \left(\frac{1}{2}a - \frac{1}{8}b + \frac{1}{4}c\right)CO_2$$
$$+ \left(\frac{1}{2}a + \frac{1}{8}b - \frac{1}{4}c\right)CH_4 \tag{1}$$

Other atoms, which may also be present in organic matter, are usually present in smaller concentrations and do not greatly affect stoichiometry. Their fates can be predicted by thermodynamics: transition metals are generally in a reduced state (Fe^{2+}, Ni^{1+}, Mn^{2+}), nitrogen, which may be found (NH_3 or N_2), sulfur as H_2S, and phosphorus as PO_4^{2-}. Their effect on the stoichiometry of methanogenesis can be calculated (McCarty, 1964), for instance for sulfur and nitrogen:

$$C_aH_bO_cS_dN_e + \left(a - \frac{1}{4}b - \frac{1}{2}c + \frac{1}{2}d + \frac{3}{4}e\right)H_2O \rightarrow$$
$$\left(\frac{1}{2}a - \frac{1}{8}b + \frac{1}{4}c + \frac{1}{4}d + \frac{3}{8}e\right)CO_2 + \left(\frac{1}{2}a + \frac{1}{8}b - \frac{1}{4}c - \frac{1}{4}d - \frac{3}{8}e\right)CH_4 \tag{2}$$
$$+ dH_2S + eNH_3$$

These theoretical calculations reflect the products of anaerobic decomposition fairly accurately in environments where organic matter is the major energy source available for the microbial community, where the degradation of organic matter is complete, and where electron acceptors (such as O_2, NO_3^-, SO_4^{2-} Fe(III), and Mn(IV)) are not present in substantial concentrations. Adding such electron acceptors reduces the yield of methane and increases that of carbon dioxide. When excess quantities of electron acceptors are added ($\frac{1}{2}a + \frac{1}{8}b - \frac{1}{4}c - \frac{1}{4}d - \frac{3}{8}e \leq 0$), methane production ceases (except perhaps for non-competitive substrates), and essentially all organic carbon is released as carbon dioxide.

6. The Methanogenic Bacteria

Reviews of the biology, ecology, and biochemistry of methanogens (Whitman, 1985; Vogels et al., 1988; DiMarco et al., 1990) indicate that all methanogens are very strictly anaerobic bacteria that all form methane as a major terminal catabolic product. However, methanogens may be found in environments that are grossly oxic, either because they inhabit microenvironments that are free of O_2 or because they can grow in the presence of O_2 when the inhibitory reaction products are removed by other bacteria within their environment.

Beginning in the late 1970s, evolutionary studies of ribosomal RNA indicated that methanogens are an evolution-coherent group of microbes that are distinct from most known bacteria (Jones et al., 1987; Woese, 1987). Together with some sulfur-dependent thermophiles and extremely halophilic bacteria, methanogens make up the "urkingdom," *Archaeobacteria*, separate from the urkingdoms of "normal" bacteria (prokaryotes) and of eukaryotes (plants, animals, fungi, and protozoa) in a number of important ways. The name "urkingdom" was coined to refer to a taxonomic grouping higher than the kingdom (such as *Planta, Animalia*). The three urkingdoms are distinct cell lines, which have evolved along different pathways since a time early in the evolution of life on Earth. Thus, the characteristics that methanogens share with prokaryotes and eukaryotes likely were developed early in the evolution of life on Earth, so studies of methanogen biochemistry and comparisons with that of prokaryote and eukaryotes may help us to understand how early life evolved. This is one of the reasons for the keen interest in the biochemistry of methanogens, including their physiology, genetics, biosynthetic pathways, structure of cell walls (Kandler and König, 1985) and membranes (Langworthy, 1985), and evolution. Table 1 lists the genera of methanogenic bacteria and their habitats.

7. Biochemistry of Methanogenesis

The biochemistry of methanogenesis summarized here has been reviewed elsewhere (Whitman, 1985; Vogels et al., 1988; DiMarco et al., 1990; Ferry, 1993). The ultimate biochemical step in the formation of methane by methanogens is a reductive demethylation of methyl carrier coenzyme M (2-mercaptoethanesulfonate [Taylor and Wolfe, 1974]) (Figure 3). This reduction is coupled to ATP synthesis, but the small amount of free energy available under in situ conditions (Boone et al., 1989) implies that the stoichiometry must be less than one ATP per mole of methane formed. Barker (1956) correctly predicted that the mechanism of methane formation from carbon dioxide reduction was a step-wise 2-electron reduction of one or more C_1 carriers, with other methyl-group donors (e.g., methanol, acetate, methyl sulfides) feeding into the pathway at various points.

The first step of methanogenesis from H_2 plus carbon dioxide is perhaps the least-well understood (Figure 3). Carbon dioxide is reductively condensed with the unique coenzyme methanofuran in a reaction that is driven in some way by energy released in by later reactions. Of all of the reductive steps in methanogenesis (Figure 3) only the final cleavage of methane from methyl-coenzyme M can yield substantial energy. After the carbon dioxide is fixed as a C_1 group on methanofuran at the formyl oxidation state, this group is transferred to another coenzyme unique to methanogens, methanopterin. The formyl group is then dehydrated and reduced by two 2-electron reductions to the methyl level. This reduction is analogous to C_1 metabolism in folate biochemistry. For at least two of these reductions, the electron carrier that directly supplies the electrons is another coenzyme unique to methanogens, factor F_{420} (Cheeseman et al., 1972). The methyl group is then transferred from methanopterin to coenzyme M (HS-CoM) (Poirot et al., 1987), and finally it is reductively cleaved to yield methane. This reaction is complex, requiring several proteins and cofactors (Gunsalus and Wolfe, 1978, 1980; Nagle and Wolfe, 1983) and is in some way linked to the carbon dioxide fixing step. Methane is apparently reductively cleaved from methyl-CoM by reaction with 7-mercaptoheptanoylthreonine phosphate (HS-HTP), and resulting in the disulfide HTP-CoM (see Blaut et al., 1990). The regeneration

Table 1. Genera of methanogenic and methanotrophic bacteria.

	Shape	Substrates[1]	Habitat
Methanogens			
Methanobacterium	rod	$H_2 + CO_2$, formate	Anaerobic digestors, animal feces and rumen contents, ciliate endosymbiont, compost soil of rice paddies, oil wells, and aquifers, and sediments of rivers, hot springs, and the ocean.
Methanobrevibacter	short rod	$H_2 + CO_2$, formate	Anaerobic digestors, human and animal feces, rumen, human vagina, termite gut.
Methanococcoides	coccoid	methyl amines and methanol	Marine sediments and production water from an oil well.
Methanococcus	coccus	$H_2 + CO_2$, formate	Anaerobic digestors, and thermal and non-thermal sediments of oceans estuaries, and salt marshes.
Methanocorpusculum	coccoid	$H_2 + CO_2$, formate	Anaerobic digestors, saline and freshwater sediments of lakes and rivers.
Methanohalobium	coccoid	methyl amines[2]	Hypersaline sediments of lakes and salterns.
Methanohalophilus	coccoid	methyl amines, methanol, methyl sulfides[3]	Hypersaline lake sediments and aquifer solids.
Methanolacinia	rod	$H_2 + CO_2$, formate	Marine sediments.

Table 1. Genera of methanogenic and methanotrophic bacteria.

	Shape	Substrates[1]	Habitat
Methanolobus	coccoid	methyl amines, methanol, methyl sulfides	Thermal and nonthermal marine sediments, oil wells.
Methanomicrobium	rod	$H_2 + CO_2$, formate	Bovine rumen.
Methanoplanus	coccoid	$H_2 + CO_2$, formate	Sediments, saline swamp of oil-drilling wastes, ciliate endosymbiont.
"Methanopyrus"	rod	$H_2 + CO_2$	Thermal marine sediments.
Methanosarcina	coccoid, psudo-sarcina grape-like aggrevates	acetate, methyl amines, methanol, methyl sulfides, $H_2 + CO_2$	Anaerobic digestors, wastewaters, thermal and nonthermal sediments of lakes, rivers, estuaries, and ocean, anaerobic soils, bovine rumen, and animal manure.
Methanosphaera	rod	H_2 + methanol[4]	Feces of humans and rabbit.
Methanospirillum	spiral rod	$H_2 + CO_2$, formate	Anaerobic digestors.
Methanothermus	rod	$H_2 + CO_2$	Thermal soils and sediments.
Methanothrix	sheathed rod	acetate	Anaerobic digestors.

Methanotrophs

Table 1. Genera of methanogenic and methanotrophic bacteria.

	Shape	Substrates[1]	Habitat
Methylobacter	rod	CH_4, CH_3OH	Soils, sediments, and waters.
Methylococcus	coccus	CH_4, CH_3OH	Soils, sediments, and waters.
Methylocystis	coccus	CH_4, CH_3OH	Soils, sediments, and waters.
Methylomonas	rod	CH_4, CH_3OH	Soils, sediments, and waters.
Methylosinus	rod or pear	CH_4, CH_3OH	Soils, sediments, and waters.

[1] Other substrates that may be used by methanogens are carbon monoxide and alcohols; the importance of these as methanogenic precursors in natural environments has not been established.

[2] Mono-, di- and trimethalamine.

[3] Methane thiol and dimethyl sulfide.

[4] Requires both H_2 and methanol and cannot grow on methanol alone as can other methanol-using methanogens.

[5] Other substrates that may be used by methanogens are carbon monoxide and alcohols; the importance of these as methanogenic precursors in natural environments has not been established.

[6] Mono-, di-, and trimethylamine.

[7] Methane thiol and dimethyl sulfide.

[8] Requires both H^2 and methanol and cannot grow on methanol alone as can other methanol-using methanogens.

of reduced HS-HTP and HS-CoM is likely the reaction coupled to energy generation for the cell.

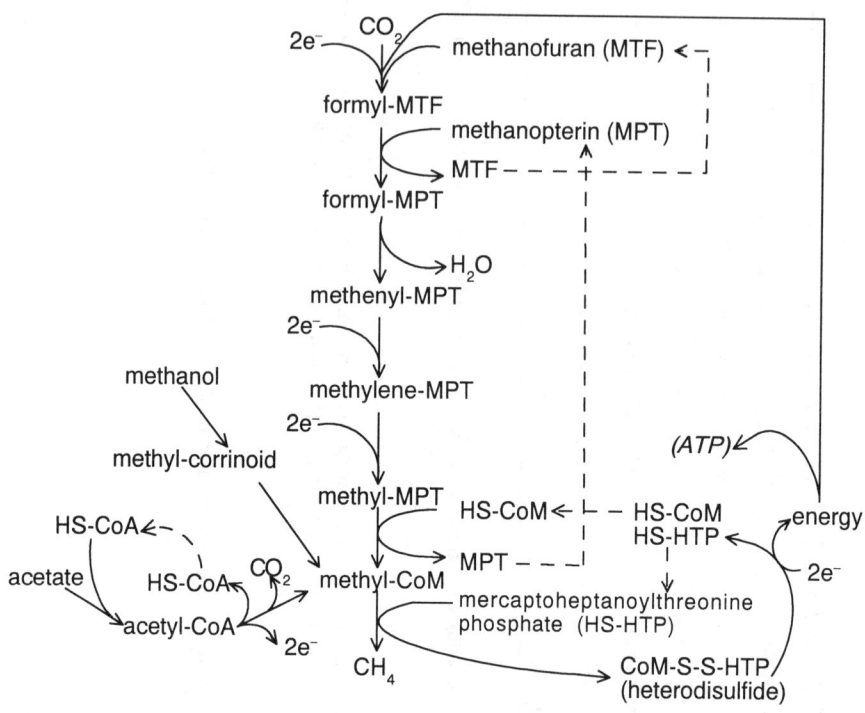

Fig. 3. Pathway of methane formation by methanogens. Carbon dioxide is reduced to methane by four 2-electron reductions; the electrons are carried on the electron carrier F_{420} for at least two of these reductions. The C_1 carriers in methanogenesis (methanofuran, methanopterin, and coenzyme M) are recycled, as shown by the dotted arrows. During methanogenesis from methyl groups of acetate, methanol, methyl amines, and methyl sulfides, C_1 groups are transferred to coenzyme M (forming methyl-CoM), as shown in the left half of the diagram.

When methyl compounds (acetate, methanol, trimethylamine, dimethyl sulfide) are the catabolic substrates, the methyl group of these compounds is transferred to one of the intermediates in the pathway for carbon dioxide reduction. For instance, the methyl group of methanol (and perhaps also dimethyl sulfide) is transferred to HS-CoM by a corrinoid methyltransferase; this methyltransferase is substrate specific. The methyl-CoM is then reductively demethylated to yield methane, as in carbon dioxide reduction. The source of the electrons for this reduction is not well known.

When H_2 is available, methanogens of the genus *Methanosarcina* or *Methanosphaera* use it as the electron donor for methyl-CoM reduction. In fact, H_2 is the only source of these reducing equivalents that *Methanosphaera* can use. *Methanosarcina* can, when no H_2 is available, disproportionate methanol or trimethylamine. That is, it can generate reducing equivalents from (for instance) methanol by oxidizing some methanol. For every molecule of methanol oxidized to carbon dioxide, 3 pairs of electrons are generated, and these electron pairs can each be used to reduce one methyl-coenzyme M to methane. Thus, the stoichiometry of methanol oxidation is that four molecules of methanol are degraded to three of methane and one of carbon dioxide (plus two of water). The mechanism of methanol oxidation is not well understood, but another novel cofactor (sarcinopterin) may be the C_1 carrier for the oxidation of methanol (van Beelen et al., 1984).

In ecosystems other than intestinal tracts and marine or estuarine sediments, acetate is the major precursor to methane. However, acetate can be catabolized by only two genera of methanogens, *Methanosarcina* and *Methanothrix* (= "*Methanosaeta*"), via the aceticlastic reaction. These organisms transfer the methyl group of acetate to HS-CoM (making methyl-CoM) and oxidize the carboxyl group to carbon dioxide. The oxidation of the carboxyl group to carbon dioxide provides electrons for the reductive release of methane from methyl-CoM. Lysates of aceticlastic methanogens form methane from acetate (Krzycki and Zeikus, 1984) but at rates much lower than whole cells and then only when H_2 is added. However, when the membrane fraction is included in cell-free lysates, the rate of aceticlastic methanogenesis approaches that of whole cells, even without added H_2 (Baresi, 1984). In fact, H_2 inhibits methane production from acetate by these particulate fractions (Baresi, 1984).

8. Influence of Environmental Factors on Methane Production

The presence of alternate electron acceptors (O_2, nitrate, Fe(III), sulfate, etc.) is a major factor affecting methane production. Methane is rarely found in oxic environments, and the main reason is probably that O_2-using bacteria out-compete methanogens for catabolic substrates, as described above. In addition, methanogens are very strict anaerobes, being poisoned by even very tiny concentrations of O_2. This extreme sensitivity to O_2 is readily apparent in pure cultures of methanogens, but it may not be important in natural environments where other bacteria remove either O_2 itself or toxic reaction products of O_2. Methanogens may thrive in microenvironments, such as dead-end pores in soil, where O_2 does not penetrate, and methane that is formed there may be oxidized by methanotrophic bacteria as the methane diffuses out into oxic zones. Like O_2, nitrate, ferric, sulfate, and other electron acceptors tend to block the formation of methane by supporting the activities of bacteria, which out-compete methanogens for their substrates.

Like those of most bacteria, the activities of methanogens increase approximately linearly with temperature except within a few degrees centigrade of the maximum temperature of that organism. The maximum growth temperature of most methanogens is about 37° to 45°C, although many grow up to 65°C, and some even grow near the boiling point of water. These latter organisms may be found at hydrothermal vents, where they grow by using the H_2 emanating from the vent. It has long been known that biogenic methane may be formed in permanently cold environments (Conrad et al., 1989; Conrad and Wetter, 1990). Only recently two new species of methanogens whose maximum growth temperatures are less than 25/26°C were isolated from Ace Lake in Antarctica (Franzmann et al., 1992, 1997).

The optimum pH for methanogens is generally near neutral (6 to 8), although some grow well at pH values as low as 4 (Patel et al., 1990; Maestrojuán and Boone, 1991) and as high as 10 (Worakit et al., 1985; Boone et al., 1986; Mathrani et al., 1988; Liu et al., 1990).

9. Factors Affecting Methane Transport to the Atmosphere

Probably the most important factor affecting the transport of methane to the atmosphere is the intervention of aerobic methane-oxidizing bacteria (Kiene, 1991). These bacteria can grow on methane as the sole carbon source in oxic environments, obtaining their energy by oxidizing methane to carbon dioxide. Table 1 shows the major genera of methane-oxidizing bacteria (see review: Lidstrom, 1992; Hansen et al., 1992). The importance of methane-oxidizing bacteria in diminishing methane inputs to the atmosphere may be seen in soils above sanitary landfills, where methane diffusing from underlying anaerobic zones enters aerobic zones. The high rates of methane oxidation in soils such as these may intercept most or all of the methane before it can reach the surface (Whalen and Reeburgh, 1988; Striegl and Ishii, 1989; Whalen et al., 1990).

Some mechanisms of transport avoid the activities of methanotrophs almost completely. When methane is formed slowly in the deep sediments of an oxic lake, the methane that diffuses upwards into the oxic upper sediments may be partly or completely oxidized before it reaches the surface. On the other hand, if methane is formed in the sediments more rapidly than it can leave via diffusion, the methane accumulates until its partial pressure is greater than the ambient pressure, and large bubbles can form and rise through the sediments. This methane passes through the oxic, methanotrophic sediments rapidly, and because methane is poorly soluble, most of the methane in the bubbles is released into the atmosphere. It is difficult to quantify methane loss to the atmosphere via bubbling, and it has been estimated that 80% of the methane formed in rice paddies is oxidized before it reaches the atmosphere (Craig, 1957). A more important mechanism by which methane can elude oxidation in rice paddies is by diffusion through plants (Dacey and Klug, 1979).

Another way methane may pass from the anoxic zones where it is formed into the atmosphere without exposure to methanotrophic environments is by eructation (belching) of ungulates. The methane accumulated in the anoxic rumen is discharged into the atmosphere without passage through an oxic environment that can support the activities of methanotrophs. Methane formed in groundwater or in landfills must pass through the soil before it enters the atmosphere, so some or most of the methane may be oxidized there (Striegl and Ishii, 1989; Whalen et al., 1990). Methane formed

in ocean sediments must often pass through an anoxic, sulfate-containing zone and then an oxic zone before it reaches the atmosphere. The abilities of O_2-using methanotrophs to oxidize methane is well documented, but cultures of anaerobic (sulfate-reducing) methane-oxidizers has not been confirmed. However, geochemical evidence strongly suggests that methane oxidation occurs in the anaerobic waters of the ocean (Reeburgh, 1976; Reeburgh and Heggie, 1977). Catabolic oxidation of alkane was recently thought to require a monooxygenase as the first step, making alkane oxidation impossible in anoxic ecosystems, but Aeckersberg et al. (1991) demonstrated that higher molecular-weight alkanes are
catabolized by anaerobic sulfate reducers. The possibility that low-molecular-weight alkanes such as methane may be similarly catabolized, or whether it is co-metabolized, remains to be proven.

10. Signatures of Biological Methane Formation

Two types of signatures are used to distinguish the source of methane: the ratio of methane to ethane plus propane; and isotopic ratios.

10.1 Ratio of Methane to Ethane Plus Propane

This ratio of methane to light hydrocarbons is very high for biologically-produced methane (ratios much greater than 100 [Cicerone and Oremland, 1988]) because the biochemical mechanisms for methanogenesis are very specific (Oremland et al., 1988). In contrast, thermogenic reactions that produce methane always form substantial amounts ethane and propane as well (ratios typically less than 50 [Oremland et al., 1988]). These ratios may be modified by differential rates of microbial alkane oxidation. Thus, some results may be ambiguous, especially when the methane is derived from multiple sources and the ratio is between the normal values for biogenic and abiogenic methane.

10.2 Stable Isotope Ratios

10.2.1 $^{13}C/^{12}C$ Ratio. Two stable isotopic ratios give useful information of the source of methane, $^{13}C/^{12}C$ and D/H. The isotopic ratio of $^{13}C/^{12}C$ (recently reviewed by Tyler [1991]) is the most commonly used evidence of methane source, normally calculated relative to the standard, carbonate in Pee Dee Belemnite (Craig, 1957), expressed as:

$$\delta^{13}C\ (\text{‰}) = [(^{13}C/^{12}C)_{sample}/(^{13}C/^{12}C)_{standard} -1] \times 1000 \tag{3}$$

Enzymatic reactions select for lighter isotopes, so biologically-produced methane tends to have less ^{13}C (i.e., a lower $\delta^{13}C$). Divergence of the $\delta^{13}C$ of biologically produced methane could be due to one of two factors. The first factor is discrimination by methanogenic enzymes between the two carbon isotopes, which is referred to as α_C (Whiticar et al., 1986):

$$\alpha_C = (\delta^{13}C_{CO_2} + 10^3)/(\delta^{13}C_{CH_4} + 10^3) \tag{4}$$

The second factor that can lead to divergence of $\delta^{13}C$ is formation of methane from a precursor pool whose $\delta^{13}C$ has been modified. Probably both of these factors are at work, because the biomass from which methane is produced is depleted in ^{13}C with respect to geological CO_2, having $\delta^{13}C$ values of -10‰ to -35‰. Thermogenic methane, which is derived from biomass, is slightly lighter than biomass ($\delta^{13}C$ values of -25‰ to -50‰), but biogenic methane is lighter still ($\delta^{13}C$ values of -45‰ to -80‰ [Tyler, 1991]).

This divergence of the $\delta^{13}C$ of the produced methane relative to the biomass from which it was produced is normally attributed to fractionation by enzymes of the methanogens themselves (α_C). However, there are at least two other mechanisms that may cause fractionation. The methane fermentation may selectively use certain carbons of the biomass (e.g., sediments preferentially convert the C-1 and C-6 carbons of glucose to methane [Krumböck and Conrad, 1991]), or the ultimate carbon substrates of methanogens may be preselected by other organisms, as illustrated by carbon flow in ruminants (see Hungate, 1966). In the rumen fermentation, biomass is fermented by bacteria to volatile fatty acids (mainly acetic, propionic, and butyric), bicarbonate/CO_2, and CH_4. The gases CO_2 and CH_4 (approximately 3:1) are released to the atmosphere by eructation. By swallowing saliva, the ruminant maintains a flow of liquids through the rumen, which buffers the rumen by its dissolved bicarbonate and transports the fatty acids to its next stomach (the abomasum) to facilitate absorption into the blood. The input of bicarbonate from the saliva to the rumen is a substantial source, and the isotopic ratio of that bicarbonate is affected by numerous enzymatic activities of the animal, most importantly by the carbonic anhydrase, which facilitates CO_2 removal in the lungs. Thus, the isotopic ratio of bicarbonate in the rumen may differ from that of the biomass that is eaten by the animal, and it is this bicarbonate that is the starting carbon for ruminal methane formation, but that may be further fractionated by the methanogens themselves.

Methane formation from CO_2 reduction accounts for about 30% of the methane from freshwater sediments and flooded soils, and nearly 100% of the methane from gastrointestinal tracts. The fraction of CH_4 in marine sediments that is due to CO_2 reduction is not well known; although most known halophilic methanogens catabolize methyl groups and do not reduce CO_2, the biogeochemical data of Whiticar et al. (1986) suggest that most methane comes from CO_2 reduction.

The precursor of about 70% or more of methane produced from flooded soils and freshwater sediments is acetate (Winfrey and Zeikus, 1979; Mackie and Bryant, 1981; Boone, 1982; Krumböck and Conrad, 1991). Whereas low $\delta^{13}C$ of methane from CO_2 reduction may be partially explained by isotopic discrimination by methanogens, such is not the case for methane formed from acetate. When methane is formed from a methyl group (such as the methyl group of acetate), methane formation requires an electron pair to reduce the methyl group. During methanogenesis from acetate, this electron pair is acquired by the oxidation of the carboxyl group to CO_2. Thus, the methyl group of acetate is converted intact to methane, and the carboxyl group is converted to carbon dioxide. Because these conversions are essentially complete, the $\delta^{13}C$ of the methane is the same as that of the methyl group. Thus, any measured differences between the $\delta^{13}C$ of the methane from acetate and that of the biomass are due to differences between the $\delta^{13}C$ of the methyl group of acetate and that of the biomass. In the most common fermentative pathway of hexoses, the Embden-Meyerhof-Parnas pathway, the C-1 and C-6 of glucose are converted to the methyl group of acetate, and hence to methane (Mah et al., 1978; Krumböck and Conrad, 1991).

During methanogenesis of other methyl compounds, such as methanol, trimethyl-amine, or dimethylsulfide, there is no carboxyl group to oxidize, so the electron pair needed to reduce the methyl group to methane must come from elsewhere. It may come from H_2 (Deppenmeier et al., 1988; Miller, 1991) or from oxidation of some of the methyl group to carbon dioxide (Deppenmeier et al., 1988). In the latter case, where there are two possible fates for the carbon of the methyl group, isotopic fractionation of the methane relative to that of the methyl group may occur. In addition, the $\delta^{13}C$ of the methyl group may be different than that of the biomass.

Thus, $\delta^{13}C$ of methane formed is determined in part by the isotopic make-up of the biomass from which it is produced and in part by isotopic discrimination of methanogenic enzymes. The $\delta^{13}C$ of methane released into the atmosphere may be affected by a third factor, isotopic discrimination by another group of bacteria, the methanotrophic bacteria, which oxidize methane. Aerobic methanotrophs selectively oxidize the lighter isotope, so their activity would raise the $\delta^{13}C$ of residual methane. It should be pointed out that this altered $\delta^{13}C$ would only be significant if a substantial fraction of produced methane were oxidized, therefore diminishing its impact on regional or global isotopic ratios. This may be the case in the ocean, where anaerobic methanotrophs oxidize methane. Anaerobic methane oxidation as a catabolic process has not been demonstrated in the laboratory, but there is convincing evidence in the shapes of methane gradients in the ocean. Also, the finding by Aeckersberg et al. (1991) of sulfate-reducing bacteria capable of oxidizing alkanes longer than methane suggests that slow-growing methane-oxidizing sulfate reducers may exist in the oceans.

10.2.2 D/H Ratio. The ratio of deuterium to hydrogen in methane may also be measured. Analogous to $\delta^{13}C$ for carbon, $\delta(D/H)$ is calculated relative to ocean water (Hoefs, 1987, in Tyler, 1991). When CH_4 is formed by CO_2 reduction, the hydrogens are derived from water (or protons). However, when methane is formed from methyl groups (acetate, methanol, etc.), the three hydrogens of the methyl group remain intact, and only the fourth hydrogen is derived from water. Thus, the $\delta(D/H)$ of methane from CO_2 reduction is determined by a methanogen's fractionation of the hydrogen of water, but most of the $\delta(D/H)$ of methane from methyl groups may be predetermined by the $\delta(D/H)$ of the methyl groups, based on the finding of Walther et al. (1981) that the methyl groups are reduced to methane with the hydrogens intact.

10.2.3 ^{14}C Content of Methane. Another isotopic measure of CH_4 which yields information about its source is the ratio of $^{14}C/^{12}C$. Because ^{14}C is radioactive, it is not present in very old carbon. Modern atmospheric carbon receives a constant input of ^{14}C which is derived from cosmic ray-derived neutrons, which are captured by ^{14}N which then ejects a proton to become ^{14}C. However, anomalies in atmospheric ^{14}C content caused by detonation of atomic weapons and escape of gases from nuclear power plants complicates ^{14}C dating of methane sources. Modern biomass, formed from atmospheric CO_2, has the same ^{14}C content as atmospheric CO_2, but old carbon loses its ^{14}C. The fraction of ^{14}C in CH_4 is expressed as a percentage of the fraction in modern carbon. Methane in geological formations can be dated by its ^{14}C content, but other measures (such as $\delta^{13}C$ or ratio of methane to ethane plus propane) must be used to differentiate old biogenic from old thermogenic methane (see also Whiticar, this volume, and Stevens, this volume).

11. Summary

Methane production occurs when organic matter is degraded in environments where light and inorganic electron acceptors such as O_2, Fe(III), Mn(IV), nitrate, sulfate, and sulfur are limiting. Under these conditions organic matter decomposition is catalyzed by consortia of bacteria including specialized bacteria that form methane. Methane and carbon dioxide are the terminal products of metabolism, and these compounds are stable (except at geological pressures). The contribution of this methane to the atmosphere may be moderated by the activities of methanotrophic bacteria, which can use O_2 to oxidize methane to carbon dioxide and water.

Acknowledgments. This work was supported by section 105 grant #14-08-001-G1636 from the U. S. Geological Survey.

12. References

Aeckersberg, F., F. Bak, F. Widdel. 1991. Anaerobic oxidation of saturated hydrocarbons to CO_2 by a new type of sulfate-reducing bacterium. *Arch. Microbiol., 156*:5-14.

Bak, F., K. Finster. 1993. Formation of dimethyl sulfide and methane thiol from methoxylated aromatic compounds and inorganic sulfide by newly isolated anaerobic bacteria. In: R.S. Oremland (ed.) *Biogeochemistry of Global Change: Radiatively Active Trace Gases*, p. 782-795. Chapman & Hall, New York.

Baresi, L. 1984. Methanogenic cleavage of acetate by lysates of *Methanosarcina barkeri*. *J. Bacteriol., 160*:365-370.

Barik, S., W. J. Brulla, M. P. Bryant. 1985. PA-1, a versatile anaerobe obtained in pure culture, catabolizes benzenoids and other compounds in syntrophy with hydrogenotrophs, and P-2 plus *Wolinella* sp. degrades benzenoids. *Appl. Environ. Microbiol., 50*:304-310.

Barker, H. A. 1956. *Bacterial fermentations*, p. 1-27. Wiley, New York.

Blaut, M., V. Müller, G. Gottschalk. 1990. Energetics of methanogens. In: *The Bacteria, Vol. 12* (J.R. Sokatch and L. Nicholas Ornston, eds.), Academic Press, Inc., San Diego, 505-537.

Boone, D.R. 1982. Terminal reactions in the anaerobic digestion of animal waste. *Appl. Environ. Microbiol., 41*:57-61.

Boone, D.R., S. Worakit, I.M. Mathrani, R.A. Mah. 1986. Alkaliphilic methanogens from high-pH lake sediments. *J. Syst. Appl. Microbiol., 7*:230-234.

Boone, D.R., R.L. Johnson, Y. Liu. 1989. Diffusion of the interspecies electron carriers H_2 and formate in methanogenic ecosystems, and its implication in the measurement of K_m for H_2 or formate uptake. *Appl. Environ. Microbiol., 55*:,1735-1,741.

Bryant, M.P. 1979. Microbial methane production: theoretical aspects. *J. Anim. Sci., 48*:193-201.

Bryant, M.P., E.A. Wolin, M.J. Wolin, R.S. Wolfe. 1967. *Methanobacillus omelianskii*, a symbiotic association of two species of bacteria. *Arch. Mikrobiol., 59*:20-31.

Buswell, A.M., W.D. Hatfield. 1939. *Anaerobic fermentations*. Illinois State Water Survey, Urbana, Ill.

Cheeseman, P., A. Toms-Wood, R.S. Wolfe. 1972. Isolation and properties of a fluorescent compound, Factor F_{420}, from *Methanobacterium* strain M.o.H. *J. Bacteriol., 112*:527-531.

Cicerone, R.J., R.S. Oremland. 1988. Biogeochemical aspects of atmospheric methane. *Global Biogeochem. Cycles, 2*:299-327.

Conrad, R., B. Wetter. 1990. Influence of temperature on the energetics of hydrogen metabolism in homoacetogenic, methanogenic, and other bacteria. *Arch. Microbiol., 155*:94-98.

Conrad, R., B. Schink, T.J. Phelps. 1986. Thermodynamics of H_2-producing and H_2-consuming metabolic reactions in diverse methanogenic environments under in situ conditions. *FEMS Microbiol. Ecol., 38*:353-360.

Conrad, R., F. Bak, H.F. Seitz, B. Thebrath, H.P. Mayer, H. Schultz. 1989. Hydrogen turnover by psychrotrophic homoacetogenic and mesophilic methanogenic bcteria in anoxic paddy soil and lake sediment. *FEMS Microbiol. Ecol., 62*:285-294.

Craig, H. 1957. Isotopic standards for carbon and oxygen and correction factors for mass-spectroscopic analysis of carbon dioxide. *Geochim. Cosmochim. Acta, 12*:133-149.

Dacey, J. W. H., M. J. Klug. 1979. Methane efflux from lake sediments through water lilies. *Science 203*:1253-1255.

Deppenmeier, U., M. Blaut, A. Jussofie, G. Gottschalk. 1988. A methyl-coM methylreductase system from methanogenic bacterium strain Göl not requiring ATP for activity. *FEBS Lett., 241*:60-64.

DiMarco, A.A., T. A. Bobik, R.S. Wolfe. 1990. Unusual coenzymes of methanogenesis. *Annu. Rev. Biochem., 59*:355-394.

Dwyer, D.F., E. Weeg-Aessens, D. R. Shelton, J. M. Tiedje. 1988. Bioenergetic conditions of butyrate metabolism by a syntrophic, anaerobic bacterium in coculture with hydrogen-oxidizing methanogenic and sulfidogenic bacteria. *Appl. Environ. Microbiol., 54*:1,354-1,359.

Ferry, J. G. (ed.) 1993. *Methanogenesis*: Ecology, Physiology, Biochemistry, and Genetics. Chapman & Hall, New York.

Franzmann, P.D., Liu, Y., Balkwill, D.L., Aldrich, H.C., Conway de Macario, E., and Boone, D.R. 1997. *Methanogenium frigidum* sp. nov., a psychrophilic, H_2-using methanogen from Ace Lake, Antarctica. *Int. J. Syst. Bacteriol. 47*: 1068-1072.

Franzmann, P.D., Springer, N., Ludwig, W., Conway de Macario, E., and Rohde, M. 1992. A methanogenic archaeon from Ace Lake, Antarctica: *Methanococcoides burtonii* sp. nov. *Syst. Appl. Microbiol. 15*: 573-581.

Gunsalus, R. P., R. S. Wolfe. 1978. ATP activation and properties of the methyl coenzyme M reductase system in *Methanobacterium thermoautotrophicum*. *J. Bacteriol., 135*:851-857.

Gunsalus, R. P., R. S. Wolfe. 1980. Methyl coenzyme M reductase from *Methanobacterium thermoautotrophicum*: resolution and properties of the components. *J. Biol. Chem., 255*:1,891-1,895.

Hanson, R.S., A.I. Netrusov, K. Tsuji. 1992. The obligate methanotrophic bacteria: *Methylococcus, Methylomonas,* and *Methylosinus.* In: *The Prokaryotes, A Handbook on the Biology of Bacteria: Ecophysiology, Isolation, Identification, Applications* (A. Ballows, H.G. Trüper, M. Dworkin, W. Harder, and K.-H. Schleifer, eds.), second edition. Springer-Verlag, New York, p. 2,350-2,364.

Hoefs, J. 1987. *Stable Isotope Geochemistry,* 3rd edition, p. 22-24. Springer-Verlag, New York.

Houwen, F. P., C. Dijkema, C. C. H. Schoenmakers, A. J. M. Stams, A. J. B. Zehnder. 1987. ^{13}C-NMR study of propionate degradation by a methanogenic coculture. *FEMS Microbiol. Lett., 41:*269-274.

Hungate, R.E. 1966. *The Rumen and Its Microbes,* p. 1-533. Academic Press, New York.

Jones, W. J., D.P. Nagel Jr., W.B. Whitman. 1987. Methanogens and the diversity of archaebacteria. *Microbiol. Rev., 51:*135-177.

Kandler, O., and H. König. 1985. Cell envelopes of archaebacteria. In: *The Bacteria: Vol. VIII, Archaebacteria* (C.R. Woese and R.S. Wolfe, eds.), Academic Press, Orlando, Fla., p. 413-457.

Kiene, R. P. 1991. Production and consumption of methane in aquatic sediments. In: *Microbial Production and Consumption of Greenhouse Gases: Methane, Nitrogen Oxides, and Halomethanes* (J.E. Rogers and W.B. Whitman, eds.), American Society for Microbiology, Washington, D.C., p. 111-146.

Krumböck, M., R. Conrad. 1991. Metabolism of position-labeled glucose in anoxic methanogenic paddy soil and lake sediment. *FEMS Microbiol. Ecol., 85:*247-256.

Krzycki, J. A., J. G. Zeikus. 1984. Acetate catabolism by *Methanosarcina barkeri:* hydrogen-dependent methane production from acetate by a soluble cell protein fraction. *FEMS Microbiol. Lett., 25:*27-32.

Langworthy, T. A. 1985. Lipids of archaebacteria. In: *The Bacteria: Vol. VIII, Archaebacteria* (C.R. Woese and R.S. Wolfe, eds.), Academic Press, Orlando, Fla., p. 413-457.

Leisinger, T., W. Brunner. 1986. Poorly degradable substances. In: *Biotechnology: Microbial Degradations, Vol. 8* (W. Schönborn, ed.), VCH Verlagsgesellschaft, Weinheim, Germany, p. 475-513.

Lidstrom, M.E. 1992. The aerobic methylotrophic bacteria. In: *The Prokaryotes, a Handbook on the Biology of Bacteria: Ecophysiology, Isolation, Identification, Applications* (A. Ballows, H.G. Trüper, M. Dworkin, W. Harder, and K.-H. Schleifer, eds.), second edition. Springer-Verlag, New York, p. 432-445.

Liu, Y., D. R. Boone, C. Choy. 1990. *Methanohalophilus oregonense* sp. nov., a methylotrophic methanogen from an alkaline, saline aquifer. *Int. J. Syst. Bacteriol., 40:*111-116.

Ljungdahl, L.G. 1986. The autotrophic pathway of acetate synthesis in acetogenic bacteria. *Ann. Rev. Microbiol., 40:*415-450.

Lovley, D. R., S. Goodwin. 1988. Hydrogen concentration as an indicator of the predominant terminal electron acceptor reactions in aquatic sediments. *Geochim. Cosmochim. Acta, 52:*2,993-3,003.

Mackie, R. I., M. P. Bryant. 1981. Metabolic activity of fatty acid-oxidizing bacteria and the contribution of acetate, propionate, butyrate, and CO_2 to methanogenesis in cattle waste at 40 and 60°C. *Appl. Environ. Microbiol., 41:*1,363-1,373.

Maestrojuán, G.M., D. R. Boone, L. Xun, R.A. Mah, L. Zhang. 1990. Transfer of *Methanogenium bourgense, Methanogenium marisnigri, Methanogenium olentangyi,* and *Methanogenium thermophilicum* to the genus *Methanoculleus,* gen. nov., emendation of *Methanoculleus marisnigri* and *Methanogenium,* and description of new strains of *Methanoculleus bourgense* and *Methanoculleus marisnigri. Int. J. Syst. Bacteriol., 40:*117-122.

Mah, R.A., M. R. Smith, L. Baresi. 1978. Studies on an acetate-fermenting strain of

*Methanosarcina. Appl. Environ. Microbiol., 35:*1,174-1,184.

Mathrani, I. M., D. R. Boone, R.A. Mah, G. E. Fox, P.P. Lau. 1988. *Methanohalobium zhilinae*, gen. nov. sp. nov., an alkaliphilic, halophilic, methylotrophic methanogen. *Int. J. Syst. Bacteriol., 38:*139-142.

McCarty, P. L. 1964. The methane fermentation. In: *Principles and Applications in Aquatic Microbiology* (H. Heukelekian and N.C. Dondero, eds.), John Wiley & Sons, New York, p. 314-343.

McInerney, M. J. 1986. Transient and persistent associations among prokaryotes. In: *Bacteria in Nature, Vol. 2* (E. R. Leadbetter and J. S. Poindexter, eds.), Plenum Publishing Corp., New York, p. 293-338.

McInerney, M. J., M.P. Bryant, N. Pfennig. 1979. Anaerobic bacterium that degrades fatty acids in syntrophic association with methanogens. *Arch. Microbiol., 122:*129-135.

McInerney, M. J., M.P. Bryant, R. B. Hespell, J. W. Costerton. 1981. *Syntrophomonas wolfei* gen. nov. sp. nov., an anaerobic, syntrophic, fatty acid-oxidizing bacterium. *Appl. Environ. Microbiol., 41:*1,029-1,039.

Miller, T. L. 1991. Biogenic sources of methane. In: *Microbial Production and Consumption of Greenhouse Gases: Methane, Nitrogen Oxides, and Halomethanes* (J. E. Rogers and W. B. Whitman, eds.), Amer. Soc. Microbiol., Washington, D.C., p. 175-187.

Nagle, D. P., Jr., R. S. Wolfe. 1983. Component A of the methyl coenzyme M methylreductase system of *Methanobacterium*: resolution into four components. *Proc. Nat. Acad. Sci. USA, 80:*2,151-2,155.

Ni, S., D. R. Boone. 1991. Isolation and characterization of a dimethylsulfide-degrading methanogen from an oil well, characterization of *Methanolobus siciliae* T4/M^T, and emendation of *M. siciliae. Int. J. Syst. Bacteriol., 41:*410-416.

Ni, S., D. R. Boone. 1993. Catabolism of dimetnylsulfide and methane thiol by methylotrophic methanogens. In: R.S. Oremland (ed.) *Biogeochemistry of Global Change: Radiatively Active Trace Gases*, p. 796-810, Chapman & Hall, New York.

Oremland, R.S., L. M. Marsh, S. Polcin. 1982. Methane production and simultaneous sulphate reduction in anoxic, salt marsh sediments. *Nature* (London), *296*:143-145.

Oremland, R.S., M. J. Whiticar, F. S. Strohmaier, R. P. Kiene. 1988. Bacterial ethane formation from reduced, ethylated sulfur compounds in anoxic sediments. *Geochim. Cosmochim. Acta, 51:*1,895-1,904.

Oremland, R.S., R. P. Kiene, I. Mathrani, M. J. Whiticar, D. R. Boone. 1989. Description of an estuarine methylotrophic methanogen which grows on dimethylsulfide. *Appl. Environ. Microbiol., 55:*994-1002.

Patel, G.B., G.D. Sprott, J. E. Fein. 1990. Isolation and characterization of *Methanobacterium espanolae* sp. nov., a mesophilic, moderately acidophilic methanogen. *Int. J. Syst. Bacteriol., 40:*12-18.

Poirot, C.M., S. W. M. Kengen, E. Valk, J. T. Keltjens, C. van der Drift, G.D. Vogels. 1987. Formation of methylcoenzyme M from formaldehyde by cell-free extracts of *Methanobacterium thermoautotrophicum*: evidence for involvement of a corrinoid-containing methyltransferase. *FEMS Microbiol. Lett., 40:*7-13.

Reeburgh, W. S. 1976. Methane consumption in Cariaco Trench waters and sediments. *Earth Planetary Sci. Lett., 28:*337-344.

Reeburgh, W. S., D.T. Heggie. 1977. Microbial methane consumption reactions and their effect on methane distributions in freshwater and marine environments. *Limnol. Oceanogr., 22:*1-9.

Striegl, R. G., A. L. Ishii. 1989. Diffusion and consumption of methane in an unsaturated zone in north-central Illinois, U.S.A. *J. Hydrology, 111:*133-143.

Taylor, C. D., R. S. Wolfe. 1974. Structure and methylation of coenzyme M ($HSCH_2CH_2SO_3$). *J. Biol. Chem., 249:*4,879-4,885.

Thiele, J. H., J. G. Zeikus. 1988. Control of interspecies electron flow during anaerobic digestion: significance of formate transfer versus hydrogen transfer during syntrophic methanogenesis in flocs. *Appl. Environ. Microbiol., 54:*20-29.

Tyler, S. C. 1991. The global methane budget. In: *Microbial Production and Consumption of Greenhouse Gases: Methane, Nitrogen Oxides, and Halomethanes* (J.E. Rogers and W.B. Whitman, eds.), Amer. Soc. Microbiol., Washington, D.C., p. 7-38.

Van Beelen, P., J. F. A. Labro, J. T. Keltjens, W. J. Geerts, G.D. Vogels, W. H. Laarhoven, W. Guijt, C.A.G. Haasnoot. 1984. Derivatives of methanopterin, a coenzyme involved in methanogenesis. *Eur. J. Biochem., 139:*359-365.

Vogels, G.D., J. T. Keltjens, van der Drift. 1988. Biochemistry of methane production. In: *Biology of Anaerobic Microorganisms* (A. J. B. Zehnder, ed.), John Wiley & Sons, New York, p. 707-770.

Walther, R., K. Fahlbusch, R. Sievert, G. Gottschalk. 1981. Formation of trideuteromethane from deuterated trimethylamine or methylamine by *Methanosarcina barkeri. J. Bacteriol., 148:*371-373.

Whalen, S. C., W. S. Reeburgh. 1988. A methane flux time series for tundra environments. *Global Geochem. Cycles, 2:*399-409.

Whalen, S. C., W. S. Reeburgh, K. A. Sandbeck. 1990. Rapid methane oxidation in a landfill cover soil. *Appl. Environ. Microbiol., 56:*3,405-3,411.

Whiticar, M. J., E. Faber, M. Schoell. 1986. Biogenic methane formation in marine and freshwater environments: CO_2 reduction vs. acetate fermentation--isotope evidence. *Geochim. Cosmochim. Acta, 50:*693-709.

Whitman, W.B. 1985. Methanogenic bacteria. In: *The Bacteria: Archaebacteria, Vol. 8* (C.R. Woese and R.S. Wolfe, eds.), Academic Press, Inc., New York, p. 3-84.

Whitman, W. B., T. L. Bowen, D. R. Boone. 1992. The methanogenic bacteria. In: *The Prokaryotes, a Handbook on the Biology of Bacteria: Ecophysiology, Isolation, Identification, Applications,* second edition (A. Ballows, H.G. Trüper, M. Dworkin, W. Harder, and K.-H. Schleifer, eds.), Springer-Verlag, New York, p. 719-767.

Winfrey, M. R., J. G. Zeikus. 1979. Anaerobic metabolism of immediate methane precursors in Lake Mendota. *Appl. Environ. Microbiol., 37:*244-253.

Woese, C. R. 1987. Bacterial evolution. *Microbiol. Rev., 51:*221-271.

Wolin, M. J. 1982. Hydrogen transfer in microbial communities. In: *Microbial Interactions and Communities, Vol. 1* (A.T. Bull and J.H. Slater, eds.), Academic Press, London, p. 323-356.

Worakit, S., D. R. Boone, R. A. Mah, M.-E. Abdel-Samie, M. M. El-Halwagi. 1985. *Methanobacterium alcaliphilum* sp. nov., an H_2-utilizing methanogen which grows at high pH values. *Int. J. Syst. Bacteriol., 36:*380-382.

5. Can Stable Isotopes and Global Budgets Be Used to Constrain Atmospheric Methane Budgets?

Michael J. Whiticar
School of Earth and Ocean Sciences, University of Victoria, P.O. Box 3500,
Victoria, B.C. V8W 3P6, Canada

1. Introduction

Global climate change associated with the increasing atmospheric methane burden is an important societal concern. Today we can monitor with good precision the yearly 1% rise in lower tropospheric methane mixing ratios (e.g., Blake and Rowland, 1988), and we have adequate, basic global coverage of atmospheric methane latitudinal variation. Mesoscopically, we are able to roughly estimate the various source strengths, e.g., from wetlands, agriculture, fossil fuels, but there is considerable uncertainty in the actual magnitudes of the various individual fluxes of methane across the geosphere-biosphere-atmosphere interface. This knowledge deficit includes our understanding of both release and uptake process-groups. Control of methane emissions to the atmosphere requires that we reliably characterize these source-sink relationships.

Stable isotope signatures of natural gases, and in particular those of methane, are potentially a key tool to track the movement of volatile carbon compounds in geosphere-biosphere-atmosphere systems. However, for this tool to be successful, three prerequisites must be satisfied:

1. *Classification* - Can we reliably characterize the stable carbon and hydrogen isotope ratios of the various methane sources? Are their isotope signatures relatively distinct, consistent, diagnostic, and measurable?
2. *C-, H-isotope effects* - Do we know or can we reliably predict the magnitude of the carbon and hydrogen isotope effects leading to isotope fractionation at every step of the way from formation to destruction?
3. *Uncertainty in input and output flux magnitudes, influences on isotope budgets* - Will or do changes in the different source strengths of methane cause a significant (distinct and measurable) shift in the stable carbon and hydrogen isotope signatures of the atmospheric methane input? What influence do changing sink functions have on atmospheric methane isotope ratios?

The objective of this paper is to directly address these three concerns.

Mohammad Aslam Khan Khalil   (Ed.)
Atmospheric Methane
© *Springer-Verlag Berlin Heidelberg 2000*

2. Classification of Methane Sources

2.1 Natural Gas Types

The geosphere is the primary source of methane released to the atmosphere. Therefore, the isotope signal of the integrated source input to atmospheric methane will be determined by the relative proportions of the various types of methane leaving the geosphere. Natural gases are derived from biogenic and non-biogenic sources through diverse processes, including bacterial formation, catagenesis (thermogenic natural gas generation), hydrothermal and geothermal activity and, to an unknown degree, from primordial or mantle emissions. The relative sizes of most global natural gas reservoirs, shown in Figure 1, can generally be constrained to within an order of magnitude (Whiticar, 1990, 1992). Conventional subsurface natural gas accumulations (e.g., thermogenic, bacterial, coal gases) together are estimated to comprise around 0.16 million Tg hydrocarbons (120 Gt C), far larger than 3.6 Gt C of atmosphere methane and 3 Gt C terrestrial gas hydrates (5.6×10^{12} m^3). Biogas accumulations from anthropogenic sources, including rice paddies (5 Tg C, 1×10^{10} m^3) and land fill and from wetlands (75 Tg C, 14×10^{10} m^3), are the next largest reservoirs. But all of the above natural gas reservoirs are minute compared with the amount of hydrocarbons bound up as gas hydrates on continental margins (Table 1). Conservative estimates for ocean clathrates range from 3 to 11 million Tg hydrocarbons (2000 to 8000 Gt C, 4 to 15×10^{15} m^3), and projections of up to 5 billion Tg hydrocarbons (4×10^5 Gt C, 7×10^{18} m^3) have been made (see Kvenvolden, 1988). The large uncertainty regarding the extent of gas hydrates precludes any firm estimation of the total budget for global natural gas. Furthermore, our lack of knowledge concerning the amounts of hydrocarbons in geothermal, crystalline, and mantle systems complicates this budget estimation.

2.1.1 Diagenetic Gases. Trace hydrocarbons (<1 ppb CH$_4$ wt. gas/wt. sediment) are often encountered in near-surface soils or sediments. Their source can be autochthonous (in situ), low temperature diagenetic reactions (Hunt et al., 1980; Whelan et al., 1982) or allochthonous, e.g., carried in by migration of gases that were more deeply buried in sediments. These diagenetic gases are not necessarily mediated by microbial processes as compared with bacterial gases. In many instances, they are not primary gas types, rather they represent a mixture of residual, allochthonous, and diagenetic gases. As such, it is usually not possible to assign a specific source or history to these trace gases. Due to their low concentrations they are not a significant source of methane to the atmosphere.

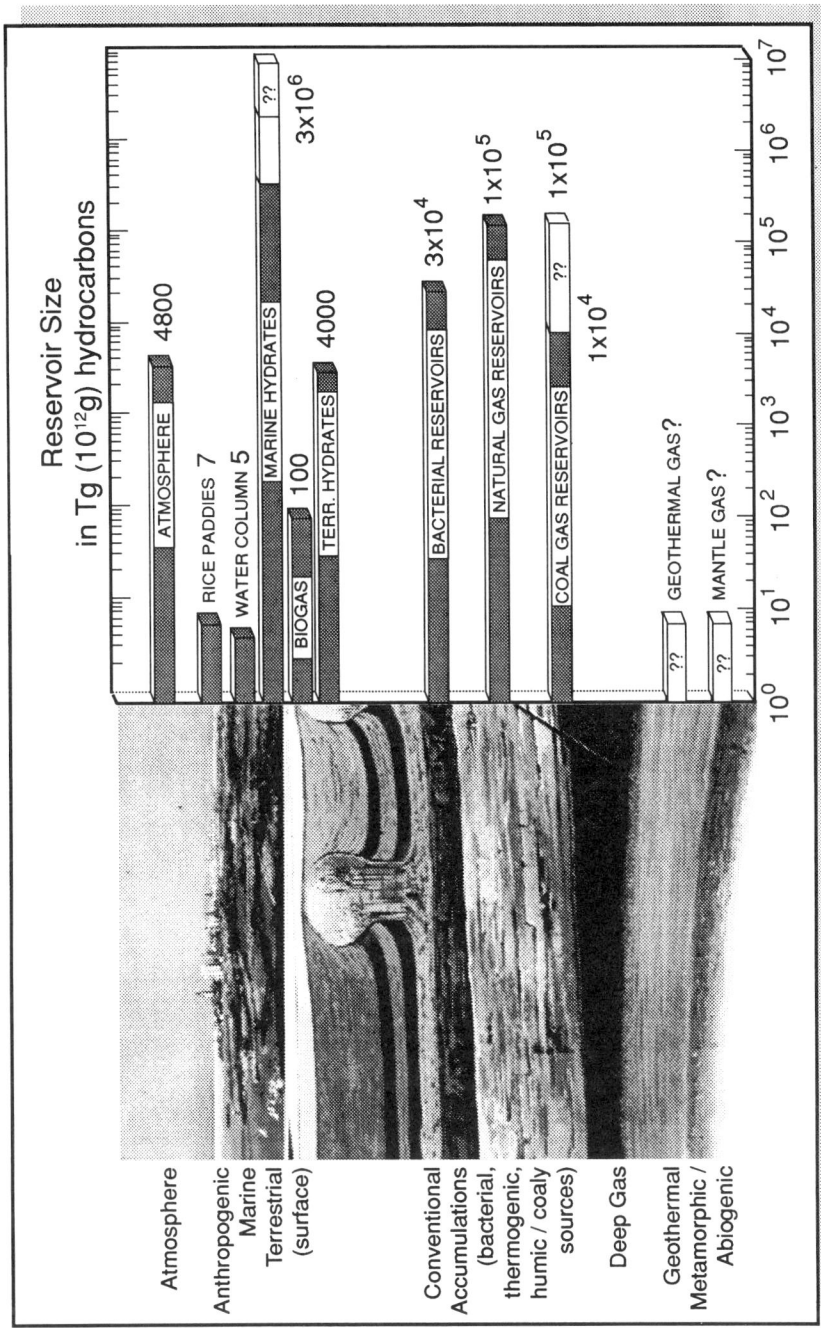

Fig. 1. A geochemist's view of natural gas types and their respective reservoir sizes. Hydrates are by far the largest hydrocarbon source type but their contribution to atmospheric methane is questionably thought to be minimal (<1% of total influx).

Table 1. Magnitudes of methane fluxes to the lower troposphere from various sources and their associated mean carbon and hydrogen isotope ratios

Source	flux (Tg CH_4 y^{-1})	% total emission	$\delta^{13}C$-CH_4 (‰, PDB)	δD-CH_4 (‰, SMOW)
rice paddies	110	20	-63	-390
cattle/enteric	80	15	-60	-330
natural gas	45	8	-44	-180
coal	35	7	-37	-110
biomass burning	55	10	-25	-90
natural wetlands	115	21	-58	-380
termites	40	7	-70	-390
landfill	40	7	-55	-380
oceans (water column)	10	2	-60	-220
freshwater	5	1	-58	-385
hydrates	5	1	-60	-240
mean input value			-54	-307

2.1.2 Bacterial Gases. Roughly 20% of the conventional natural gas reserves are of bacterial origin (Rice and Claypool, 1981; Rice, 1991). Methane is the major hydrocarbon constituent of this natural gas (usually <1% ΣC_{2+}, Figure 2) and have been formed, for the most part, by near-surface fermentation reactions by methanogenic bacteria as one of the final diagenetic remineralization stages. Methanogens are a broad consortium of obligate anaerobic microorganisms that can utilize a limited suite of precursor compounds such as 1) acetate and formate or 2) bicarbonate, i.e., CO_2 + H_2, to form methane. The carbonate reduction pathway, which predominates in marine environments (Whiticar et al., 1986), can be represented by the general reaction

$$CO_2 + 8H^+ + 8e^- \longrightarrow CH_4 + 2H_2O \qquad (1)$$

and the net reaction for the acetate fermentation pathway more common in freshwater settings is

$$*CH_3COOH \longrightarrow *CH_4 + CO_2 \qquad (2)$$

where the * indicates the intact transfer of the methyl position to CH_4. In addition to these so-called "competitive substrates," there are non-competitive substrates for methanogenesis including methanol, mono-, di- and tri-methylamines and certain organic sulphur compounds, i.e., dimethylsulphide, but their relative importance as the source for bacterial methane is uncertain. The microbiology and ecology of the various methanogenic pathways have been reviewed in several monographs (e.g., Zehnder, 1988; Boone, 1993).

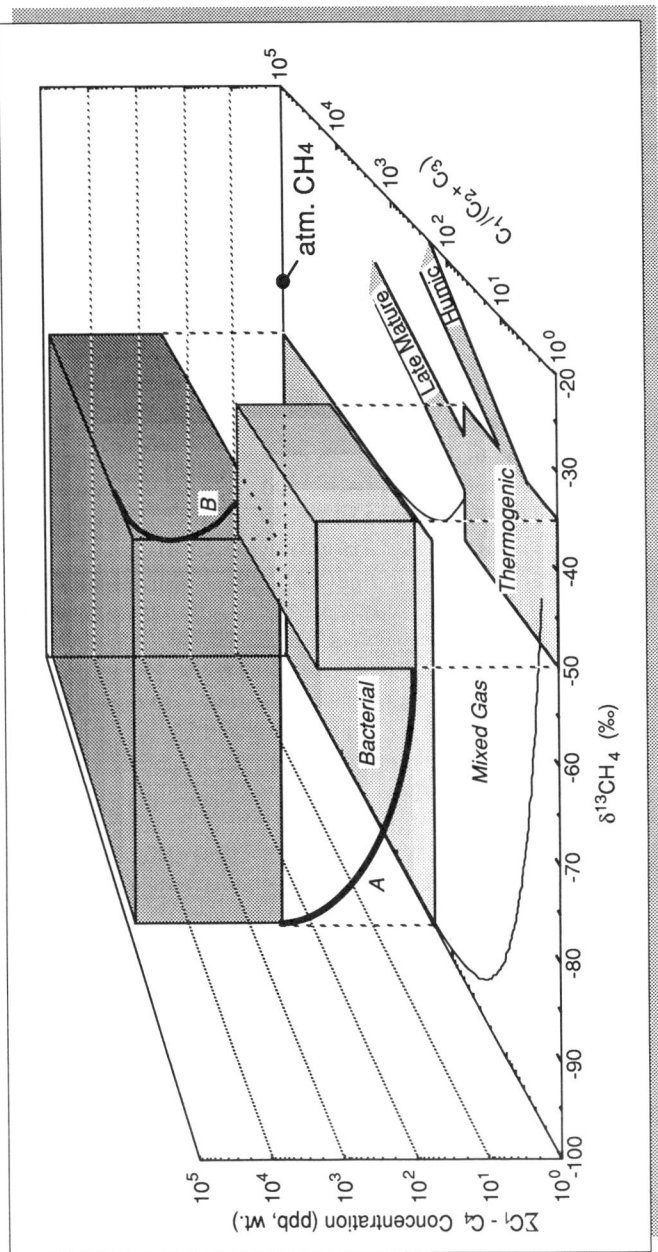

Fig. 2. Carbon isotope ratios of methane are alone often sufficient to define a natural gas type (i.e., bacterial, thermogenic, abiogenic), but sub-grouping is severely limited. The combination of concentration and molecular composition can often help, particularly to identify secondary alteration effects such as (1) gas mixtures or (2) microbial oxidation.

2.1.3 Thermogenic Natural Gas. During the catagenic transformation and reorganization of organic matter, various short-chained hydrocarbons, such as methyl groups, are being cleaved off higher molecular weight organic compounds and saturated to form the light hydrocarbons of conventional natural gas. The degree and type of product formed is dependent on several key factors including source kerogen type (sapropelic vs. humic) and richness (e.g., Total Organic Carbon = TOC, and molar H/C ratio), maturity, expulsion efficiency, and the presence of catalysts such as clays.

Sapropelic, Type I/II kerogen sources generate over suitable time periods significant quantities of thermogenic hydrocarbon gases at temperatures over $70\,°C$. The initial gases at low source maturity (Vitrinite Reflectance level: $R_m < 0.5\%$) will be relatively dry (<5% ΣC_{2+}), but the proportion of higher hydrocarbons will increase with maturation into and through the petroleum window (peak natural gas generation at $150 - 160\,°C$). Oil-associated natural gases can have over 80% ΣC_{2+} (e.g., Evans and Staplin, 1971). As maturation proceeds into late mature or condensate range ($R_m < 1.6\%$), subsequent kerogen transformation and cracking of hydrocarbons leads to a greater proportion of shorter-chained hydrocarbons and essentially a methane-rich gas at the base of the catagenic stage, roughly about $200\,°C$.

Natural gas generation profiles from humic Type III kerogens and coals are quite different from Type I/II kerogens. Significant hydrocarbon generation is traditionally thought to occur at higher maturity levels, i.e., $R_m > 0.7\%$, for humic kerogens than for sapropelic kerogens (Karweil, 1969; Tissot and Welte, 1978; Chung and Sackett, 1979). Throughout their maturation history, humic kerogens generate less C_{2+} hydrocarbons, and the total methane generative potential, including cracking of oil and condensates, is about 1 - 2 times lower than for a sapropelic source. Reliable characterization of humic kerogens is complicated by the high retention capacity of humic organic matter, which partitions gases, resulting in a time/maturity-dependent compositional fractionation of the gases released. These difficulties are compounded due to the high retention by humic organic matter of bacterial gases that were formed while the peat or fen deposit was at or near the surface.

2.1.4 Secondary Gases In addition to these "primary" gas sources, secondary alteration process can also influence the character of the natural gas emission. These alteration effects include microbial oxidation (aerobic and anaerobic), migration fractionation and mixing. Altered gases are an important type of natural gas and must be considered in the atmospheric methane budgets.

2.2 Natural Gas Classification Tools

2.2.1 Essential Parameters. Driven initially by the needs of the energy exploration sector, a suite of geochemical parameters has been developed to classify the various types of natural gas. Three categories of geochemical tools are commonly used to correlate natural gases to their sources:

1) gas concentration,
2) molecular composition, and

3) stable isotope ratios.

Analytical routines devised to find hydrocarbon fuels are and have been applied widely to samples from near-surface soils and seepages, cuttings from drill wells, and natural gas reservoirs, and thus a well-defined data set is available for environmental applications. Through the combination of parameters from the three categories, i.e., carbon isotope ratio of methane (δ^{13}C-CH$_4$), molecular ratio: methane to ethane + propane (C$_1$/(C$_2$+C$_3$), %vol.), and hydrocarbon concentration (ΣC$_1$-C$_4$) an interpretative schema can be applied as illustrated in Figure 2.

Concentration and molecular composition information can broadly characterize gas types, i.e., bacterial vs. thermogenic, but stable isotope ratios, such as carbon and hydrogen for hydrocarbons, are more specific and distinctive. In addition to identifying the natural gas type, the carbon and hydrogen isotope signature of a reservoired natural gas can often provide more detailed information about its source(s), for example, kerogen type (e.g., sapropelic Type I/II or coal/humic Type III), the level of maturity of the source, and sometimes the degree of generation. In addition to source typing and maturity estimates, the combination of molecular and isotope composition of a gas can serve to delineate altered (secondary) gases or gas mixtures from different sources. Molecular and concentration information alone is generally insufficient to do this.

For the purposes of this paper, emphasis is on the stable carbon and hydrogen isotope ratios to describe and discriminate the different methane sources, and formation/destruction pathways.

2.2.2 Stable Isotope Effects. Stable isotope data are determined as ratios, e.g. ^{13}C/^{12}C, rather than as absolute molecular abundances and are reported as the magnitude of excursion in per mil of the sample isotope ratio relative to a known standard isotope ratio. The usual d-notation generally used in the earth sciences is:

$$\delta R_X\ (‰) = B(F(R_a\ /\ R_b\text{-sample}, R_a/R_b\text{-standard}) - 1) \times 10^3 \qquad (3)$$

where R$_a$/R$_b$ are the isotope ratios, e.g. ^{13}C/^{12}C or D/H, referenced relative to the PDB or SMOW standards, respectively.

Numerous factors control the distribution of isotopes in natural gas components. The principal ones include the isotope ratios of the precursor material, e.g., the ^{13}C/^{12}C or D/H ratios of the source organic compounds, and the isotope effects associated with the processes of formation, retention, and destruction of natural gas. For the purpose of distinguishing the different natural gases, it is important that the isotope signatures and/or the isotope effects are distinctive and diagnostic for the different types of natural gas and for the mechanisms governing them.

Typical methane carbon isotope values are given in Figure 3 for the various bacterial and thermogenic natural gases (see also Stevens, 1993). However, this can only serve as an initial orientation. For example, methanogenesis by carbonate reduction can have δ^{13}C$_{CH4}$ values ranging from -20 ‰ to -110 ‰ due to (a) variations in the initial carbon isotope ratio of the substrate, (b) the degree of substrate utilization, or (c) the ambient temperature (Whiticar, 1992). In comparison with bacterial methane, thermogenic hydrocarbons have a much more constrained range of isotope compositions.

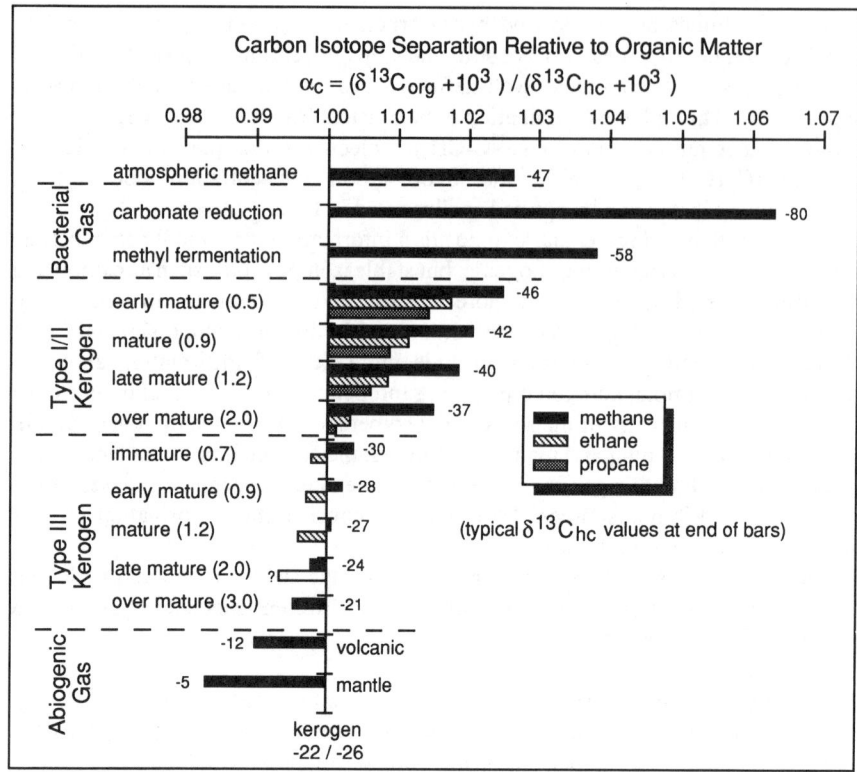

Fig. 3. Natural gas formation is typically the result of various bacterial or thermocatalytic kinetic reactions, which have isotope effects associated with the A -> B. Our knowledge of the magnitude of these isotope effects is paramount to our tracking the changes in isotope signals. The isotope offset relative to general organic matter (δ^{13}C-Corg = −22‰ and −26‰) for various products is presented in illustration of the range in offsets.

Initially equilibrium isotope effects (EIEs) had been proposed to explain the distribution of carbon isotopes in thermogenic hydrocarbons, most notably by Galimov and Petersil'ye (1967), Galimov and Ivlev (1973), and Galimov (1973), but Galimov (1985) has modified this equilibrium effect to a "thermodynamically ordered distribution," which is more consistent with the accepted view that kinetic isotope effects (KIEs) control the redistribution of isotopes (see for carbon: Sackett, 1968; Stahl, 1973; McCarty and Felbeck, 1986; Chung et al., 1988; and for hydrogen: Frank, 1972). For thermogenic hydrocarbon formation, kinetic isotope theory predicts that the light hydrocarbon formed by the saturation of an alkyl group cleaved from the kerogen molecule will be depleted in the heavier isotope relative to the remaining reactive kerogen. Similar KIE considerations apply to methanogenesis. Methanogens preferentially utilize isotopically lighter substrates; thus, the methane formed by methanogens is depleted in the heavier isotope relative to the precursor material (Whiticar, 1992).

Although the KIEs fractionate reactant and products pools, and thus can potentially generate a considerable range in stable isotope values for methane in natural gases, in many cases their carbon isotope signatures are sufficiently unambiguous to permit classification.

Isotopic fractionation or discrimination resulting from KIEs can be described by Rayleigh distillation relationships. The isotope ratio of the remaining reactant pool (e.g., a generating kerogen), which is being depleted in the lighter isotope, can be approximated by

$$R_r/R_o = f^{(1/\alpha - 1)} \tag{4}$$

and the progressive isotopic shift of the cumulative product pool (e.g., methane accumulation) by

$$R_\Sigma/R_o = (1 - f^{(1/\alpha 1)}) / (1 - f) \tag{5}$$

where R is the isotope ratio of the initial reactant (R_o), the residual reactant at a specified time(R_r), and the cumulative product (R_Σ), respectively (e.g., Claypool and Kaplan, 1974). The fraction of the reactant remaining is f, and α is the isotope fractionation factor for the conversion of the reactant to the product. Isotope shifts related to bacterial methane production (substrate depletion) and microbial methane oxidation, shown in Figures 4 and 6, are treated below. Although the KIEs partition isotopes between the reactant and products pools and can potentially generate a considerable range in stable isotope values for methane in natural gases, in many cases the carbon isotope signatures of different natural gas types are sufficiently unambiguous to permit classification.

Figure 3 illustrates the relative magnitudes of isotopic offset between various generalized types of natural gas ($\delta^{13}C_{hc}$) and organic matter ($\delta^{13}C_{org}$), according to the equation:

$$\alpha_{org-hc} = (\delta^{13}C_{org} + 10^3) / (\delta^{13}C_{hc} + 10^3) \tag{6}$$

The greatest isotope fractionation is observed for bacterial methane formed by the carbonate reduction ($\alpha_{org-hc} > 1.055$) and then methyl fermentation pathways ($\alpha_{org-hc} \sim 1.04$) (Whiticar et al., 1986). In general, thermogenic hydrocarbons formed from Type I or II kerogen ($\alpha_{org-hc} \sim 1.02$) have a larger KIE expressed than for Type III kerogens ($\alpha_{org-hc} \sim 1.003$). In both cases, the magnitude of the isotope separation is less than for methanogenesis (Figure 3). It also decreases with increasing maturity of the organic matter and with carbon number, i.e., α_{org-hc} of methane > ethane > propane > butane (Figure 3). Hydrocarbons more enriched in ^{13}C than their precursor organic matter are rarely measured in Type I/II kerogens but is common in humic or Type III sources. In Type I/II cases, this could signal the presence of secondary effects such as mixing of inorganic methane (e.g., volcanogenic, mantle), oxidation of hydrocarbons, or sometimes sampling and production artifacts.

2.2.3 Carbon Isotope Ratios Carbon isotope ratio of methane is the most common isotope measurement made to classify a natural gas (e.g., Colombo et al., 1965; Sackett, 1968; Silverman, 1971; Stahl, 1973; Schoell, 1980, 1988). Basic differences in carbon isotopes between bacterial and thermogenic natural gases are identified in Figures 2 and 3. As mentioned above, bacterial gases are commonly "lighter," i.e., more depleted in ^{13}C than thermogenic gases, and thermogenic gases become heavier with increasing maturity.

Owing to the multiple sources for methane and its susceptibility to secondary effects, the classification of thermogenic natural gases has recently placed more emphasis on the carbon isotope ratios of higher homologues, such as ethane, propane, and the butanes (e.g., James, 1983; Whiticar et al., 1984; Chung et al., 1988; Clayton, 1991). Furthermore, the differentiation of bacterial methane formation pathways can be improved by the carbon isotopes of co-existing species, including CO_2 or volatile organic acids (e.g., Whiticar et al., 1986). The latter is illustrated in Figure 4 (after Whiticar et al., 1986) where the combination of carbon isotope ratios co-existing CO_2 and CH_4 delineate the methanogenic formation pathways of carbonate reduction or methyl-fermentation. In addition, Figure 4 shows the $\delta^{13}C_{CO2}$-$\delta^{13}C_{CH4}$ trajectory of a microbially oxidized methane.

2.2.4 Hydrogen Isotope Ratios. Carbon isotopes, particularly those of methane, are the most common isotopic measurement made on natural gases, but hydrogen isotopes of methane are a diagnostic parameter in classifying its source and type. The work by Schoell (1980) provided one of the first rigorous treatments of carbon and hydrogen isotope variations of bacterial and thermogenic hydrocarbons. This work was extended later to include information on the different bacterial gas formation pathways, as shown in the CD-diagramme of Figure 5 (after Whiticar et al., 1986). The general zonation of C-, H-isotope signatures in the outlined areas of Figure 5 represent the regions occupied by the dominant natural gas types. The shaded regions of the diagramme outside the major gas types areas are less common or well defined gas signatures.

In contrast to carbon isotope ratios, the hydrogen isotope data of methane do not exhibit a clear dependency on maturity or oxidation effects; rather, they provide details on the depositional environment and pathway of formation. Hydrogen isotopes are particularly useful in distinguishing (1) methane from different methanogenic pathways, (2) bacterial from early mature thermogenic gas, (3) thermogenic from geothermal-hydrothermal gas, and (4) artificial or bit metamorphic sources. In these examples, methane carbon isotope ratios, if used alone, would deliver ambiguous results.

The geothermal-hydrothermal-crystalline zone in Figure 5 is defined, for example, by the gas data from open boreholes in the Canadian Shield crystalline rocks (Sherwood et al., 1988) and from the Gravberg-1 drill well (Schmidt, 1987; Laier, 1988; Jefferey and Kaplan, 1988). Examples of geothermal-hydrothermal data include Yellowstone and Lassen Parks (Welhan, 1988; Whiticar and Simoneit, 1992), New Zealand fields (Lyon and Hulston, 1984), Guaymas Basin (Welhan and Lupton, 1987, Simoneit et al., 1988) or Bransfield Strait (Whiticar and Suess, 1989).

Methane found at the East Pacific Rise, 21°N by Welhan (1988) and from the Zambales ophiolite (Abrajano et al., 1988) are maintained to be of abiogenic origin (Figure 5). Similarly, gas inclusions from south Greenland reported by Konnerup-Madsen et al. (1988) have ^{13}C-enriched and ^2H-depleted methane ($\delta^{13}C$-CH_4 = -1.0‰ to -5.1‰, δD-CH_4 ca. -100‰ to -150‰, Figure 5). These three sample locations are used to define the region of Figure 5 for abiogenic or mantle methane. The free gases in dolerites from the Gravberg-1 well (Schmidt, 1987; Laier, 1988; Jefferey and Kaplan, 1988) also have carbon isotope signatures approaching those of abiogenic gas, but their origin is unsure.

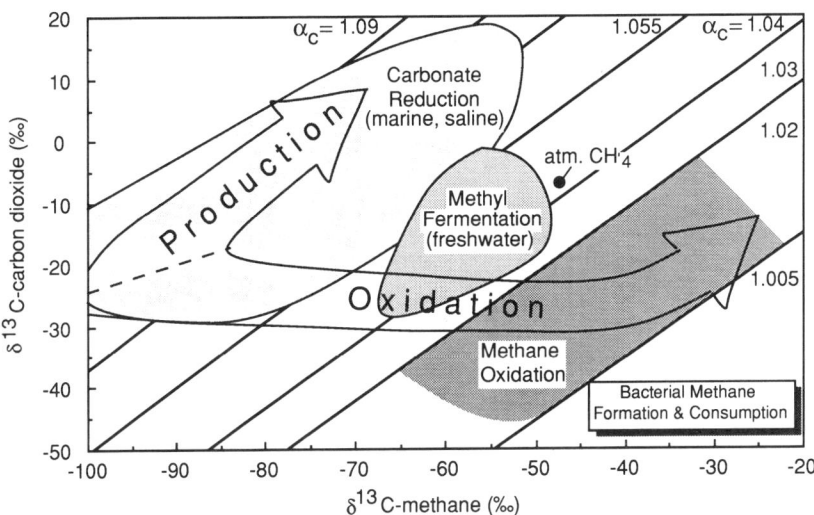

Fig. 4. Classification can be enhanced by isotope measurements on co-existing species, such as dissolved CO_2 or formation water. Methanogenesis follows strict relationships between the methane formed and these co-existing compounds. The example of $\delta^{13}C\text{-}CO_2$ vs. $\delta^{13}C\text{-}CH_4$ is shown; similar interpretative schemes are available between $\delta D\text{-}H_2O$ vs. $\delta D\text{-}CH_4$ and $d^{13}C$-substrate vs. $d^{13}C\text{-}CH_4$.

Bit metamorphism hydrocarbons noted in Figure 5 are formed artificially by drilling hard lithologies have distinctive methane isotope signatures enriched in ^{13}C ($\delta^{13}C\text{-}CH_4$ ca. -20 ‰) and strongly depleted in deuterium ($\delta D\text{-}CH_4$ ca. -750 ‰), (e.g., Gerling, 1985; Faber et al., 1987; Faber and Whiticar, 1989).

2.3 Resume

Stable carbon and hydrogen isotope signatures of methane are capable of reliably defining and distinguishing the various natural gas types and sources. In particular, C- and H-isotopes can clearly differentiate between thermogenic methane and methane from bacterial carbonate reduction and methyl-type fermentation pathways. The latter is important because these are the predominant sources of atmospheric methane. The isotopes are also capable of identifying secondarily altered methane, such as that subjected to microbial oxidation.

However, despite the promising usefulness of isotope signatures, two major difficulties complicate the situation. First, roughly 75% of atmospheric methane is derived from methyl-type fermentation, which limits the value knowing the isotope signatures of the other natural gases. Second, due to kinetic reactions, a range in values rather than unique carbon and hydrogen isotope ratios define bacterial methane. These caveats are treated in greater detail below.

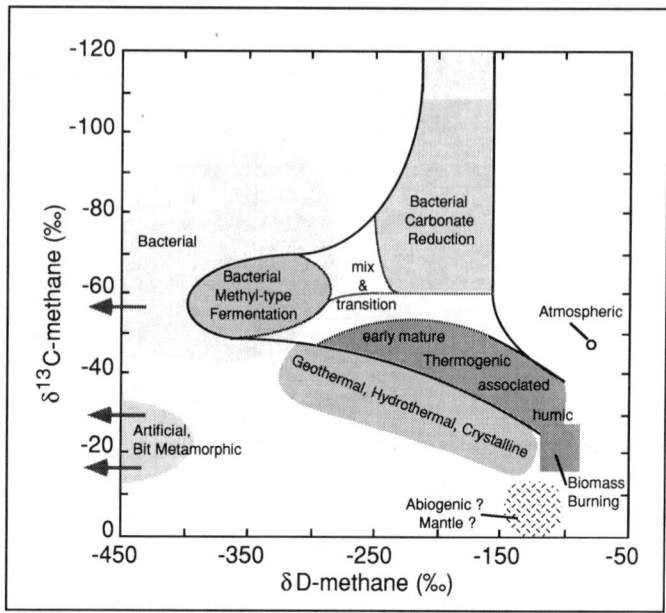

Fig. 5. Combination of carbon and hydrogen isotope ratios greatly improves our classification precision. Although less frequently measured, δD-CH₄ is well suited to distinguish between different methanogenic pathways and environments, which are the major source of methane to the atmosphere. Note: the mean atmospheric methane CD-value does not directly correspond with any of the known natural gas types, and the isotope offset associated with the removal of atmospheric methane needs to be considered (see Figure 7 for explanation).

3. C-, H-isotope Effects

3.1 C-D Isotope Variation of Atmospheric Methane Sources

In the course of the global carbon cycle, it is atmospheric carbon dioxide that sets the carbon isotope base for organic matter and ultimately the isotope ratios of atmospheric methane sources. However, kinetic isotope effects determine magnitude of the associated isotopic fractionation or offset of the methane from the precursor.

The CD-diagramme of Figure 6 shows the possible range of methane isotope pairs in response to (a) initial natural isotope variations in precursor substrates, (b) substrate depletion effects due to methanogenesis or thermocatalytic conversion, and (c) secondary, microbial methane oxidation. In addition, factors such as temperature can influence the magnitude of the isotope effects. While this confuses the issue of pegging unique isotope signatures to sources of atmospheric methane formed by similar processes, e.g., methyl-type fermentation, the isotope shifts provide key information on the degree to which the formation or oxidation process has proceeded.

Important in this context is that the CD-isotope ratio pair of the mean atmospheric methane does not fall within the CD-isotope signatures of any of the major natural gas types. On the face of it, the atmospheric $\delta^{13}C$-CH_4, δD-CH_4 pair of -47 ‰, -80 ‰ could not be created only by a direct mixture of these major natural gas types, and an isotope fractionation must be involved with the removal of atmospheric methane.

For comparison, mixing lines, which are not isotope effects, are drawn in Figure 6 between the current mean C- and H-isotope ratios of lower tropospheric methane and bacterial methane from both carbonate reduction and methyl-type fermentation pathways. Again, recognizing that the majority of atmospheric methane is generated by methyl-type fermentation, severe or even combined isotope shifts could not provide a methane source with the isotope signature close to that of atmospheric methane.

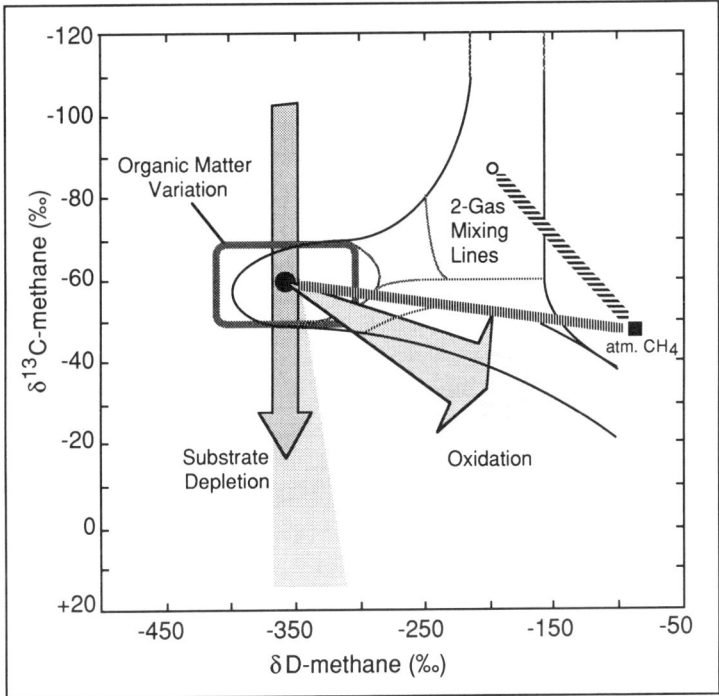

Fig. 6. How these various isotope effects combine to control the isotope signature of a particular gas is demonstrated for bacterial methane. The potential degree of the isotope shifts is shown for methanogenesis (pathway, substrate depletion), microbial oxidation, and isotope variations in the precursor organic matter (e.g., acetate, TMA, DMS). To confuse the issue, two mixing lines (are not isotope effects) are drawn to illustrate the mixing trajectories between atmospheric methane and two bacterial gas end-members. This mixing often takes place at soil-air and sea-air interfaces.

It is the isotope effect associated with the dominant atmospheric methane sinks, atmospheric OH-abstraction and soil uptake, which are the final fractionation steps responsible for the current atmospheric methane CD-isotope ratios.

3.2 Atmospheric Methane Sinks (Hydroxyl Abstraction, Microbial Oxidation)

The CD-isotope pair for atmospheric methane is not merely the averaged value of the gas entering the atmosphere; rather, it is the net atmospheric isotope value between the addition and removal terms. The removal of atmospheric methane is primarily by the hydroxyl abstraction reaction, although bacterial methane consumption in surface soils is thought to be an important atmospheric sink (see this volume). Isotope fractionation factors are known for both aerobic and anaerobic bacterial methane oxidation in soils and sediments ($\alpha C = 1.004$ to 1.02, $\alpha D = 1.2$, Whiticar and Faber, 1986; Alperin et al., 1988; Whiticar, 1992), but it is uncertain if these values are strictly applicable to this situation, or indeed to the role of soils as a regulator of atmospheric methane. are strictly applicable to this situation, or indeed to the role of soils as a regulator of atmospheric methane.

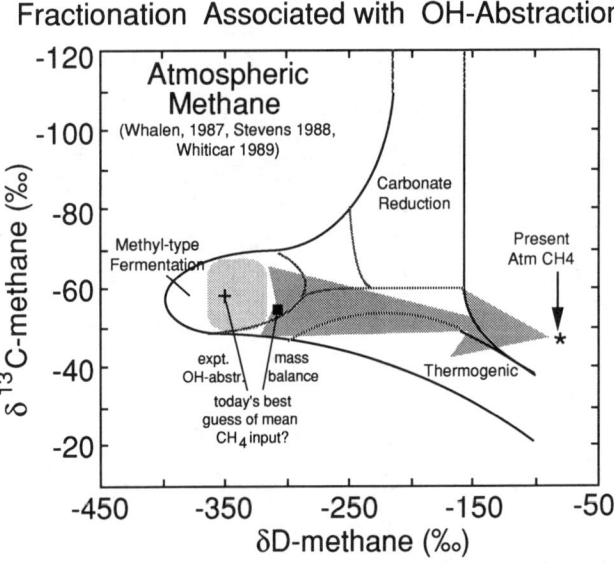

Fig. 7. Perhaps the weakest link in our ability to constrain atmospheric methane budgets by stable isotopes is the correctness of the isotope effects associated with OH radical removal of atmospheric methane. The isotope experimental base on OH-abstraction is limited, and the "best guesses" of the CD shift (experimental and calculated) are indicated on Figure 7. The shaded box is the range of values cited in the literature. We do not know if the fractionation is consistent or uniform temporally and geographically. What roles do temperature effects or OH-concentration variations play?

As mandated by isotope theory, the lighter isotope of methane species ($^{12}CH_4$) in the atmosphere react more rapidly with OH-radicals than the heavier species (e.g., $^{13}CH_4$ or $^{12}CH_3D$). As a result, the steady state CD-isotope signature of the residual atmospheric methane is enriched in ^{13}C and ^{2}H relative to the CD-isotope signature of the mean source input.

One of the greatest concerns in the application and suitability of stable isotopes to constrain estimates of atmospheric methane budgets is the uncertainty in the estimated magnitudes of carbon and hydrogen isotope effects associated with the hydroxyl abstraction reaction. Initial experimental studies by Rust and Stevens (1980) suggested that the carbon isotope fractionation factor for the OH-reaction (α_C) is approximately 1.003. Davidson et al. (1986, 1987) offered α_C values of 1.025 and 1.010. More recently, Cantrell et al. (1990) determined α_C to be 1.0054 ± 0.0009, which is comparable to the calculations of Lasaga and Gibbs (1991). The only value published for the hydrogen fractionation factor associated with the hydroxyl abstraction reaction (α_D) is 1.5 (Gordon and Mulac, 1975). Figure 7 shows the combined carbon and hydrogen isotope shift of the mean methane input to the mean atmospheric methane. Our limited understanding of these isotope effects, particularly in view of their degree of offset, curtails our use of isotope signatures in atmospheric methane mass balances. We may expect the isotope effects to be consistent but not necessarily constant. For example, do the isotope effects increase in magnitude with greater elevation due to lower temperatures?

Recognizing that several of the major atmospheric methane sources are of bacterial origin, i.e., by methyl-type fermentation, then their $\delta^{13}C$-CH_4 values lie around -55 ‰ to -65 ‰ and the δD-CH_4 around -300 ‰ to -400 ‰. The present uncertainty in the estimates of the C,H-isotope effects associated with hydroxyl abstraction is similar to the isotope range experienced for these sources.

3.3 Resume

Carbon and hydrogen isotope effects are generally predictable for the process controlling the bacterial or thermogenic formation of thermogenic natural gas. Despite the potentially large range in isotope values possible, the actual methane CD-pairs of typical gases are quite restricted in excursion. Metabolic rates or thermal activation energies are possible restrictors in these systems. Similarly, we have some preliminary understanding of isotope effects associated with methane oxidation in sediments, soils and water columns, but our confidence is considerably less in the estimation of the C-, H-isotope effects for the OH abstraction sink of atmospheric methane. However, if the estimated values of $\alpha_C = 1.005$ and $\alpha_D = 1.5$ for the OH-reaction are close to the actual values, then the globally averaged C-, H-isotope values of methane entering the atmosphere would be around $\delta^{13}C$-$CH_4 = -53$ ‰, δD-$CH_4 = -350$ ‰. As plotted in Figure 7, this mean source signal lies within and is dominated by the CD-zone defining methanogenesis by methyl-fermentation, i.e., bacterial methane from terrestrial sources.

4. Uncertainty in Flux Magnitudes

Provided that we are able to assign a definitive ranges of carbon and hydrogen isotope ratios to specific sources of atmospheric methane, and provided that we can make reasonable assumptions regarding the magnitudes of the operative isotope effects, we could then proceed to calculate mass balances for carbon and hydrogen isotopes of atmospheric methane based on the source input fluxes. Conversely, if we are able to ascertain changes or variations in the isotopes' ratios atmospheric methane, we may be able to ascribe them to changes in the flux strengths from different sources.

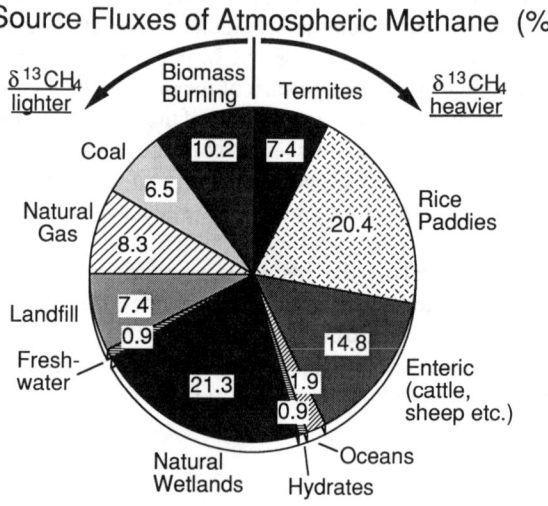

Fig. 8. Commonly referred to as the "Pie-of-Culprits," we can use the information on the input sizes and stable isotope signature of the different sources to calculate a mass balance. Perhaps more than "fine-tuning" of the pies wedges is still required (e.g., termite or rice controversies).

The success of this mass balance approach lies on dependable flux and isotope signal estimations. Table 1 lists generally accepted values of methane fluxes to the atmosphere from the major sources. Accompanying the flux magnitudes in Table 1 are crude estimates of mean carbon and hydrogen isotope ratios for these individual sources (after Schoell, 1980; Whiticar, 1990; Bundesanstalt für Geowissenschaften und Rohstoffe, (BGR) data base). As discussed earlier (Figure 6), a certain latitude in the assigned CD values must be tolerated at this stage. Figure 8 illustrates the relative proportions of the various source strengths. The sources are arranged in Figure 8 such that they are progressively heavier isotopically (^{13}C enriched) in a clockwise direction. Even though the flux estimates may be in error up to a factor of 2, it is readily apparent from Table 1 and Figure 8 that bacterial methane from rice paddies, wetlands and landfills constitute close to 50% of the emissions. Furthermore, the addition of livestock and insect releases bring the bacterial contribution to atmospheric methane to around 75%, i.e., as predicted by the CD diagramme (Figure 7). Important in this context is that both the carbon and hydrogen isotope ratios of these various sources differ very little between these sources. Sources of methane from coal and thermogenic natural gas comprise around 15%, probably similar to biomass burning, whereas marine and freshwater environments are currently not significant sources <5%.

Methane from termites illustrates the difficulty in obtaining representative flux values, and they continue to be a topic of considerable debate since Zimmerman et al. (1982) first suggested that termites could generate between 75 Tg and 310 Tg methane yearly. Taking the total annual global influx of methane to be 540 Tg (Table 1), then the termite contribution could comprise over 50%. More conservative estimates of termite methane emissions strengths of 40 to 50 Tg CH_4 were suggested by Rasmussen and Khalil (1983), Seiler et al. (1984), and Zimmerman et al. (1987), and more recently downwardly revised flux values of about 12 Tg/yr have been cited by Fraser et al. (1986) and Khalil et al. (1990). The conservative estimates are used in Figure 8.

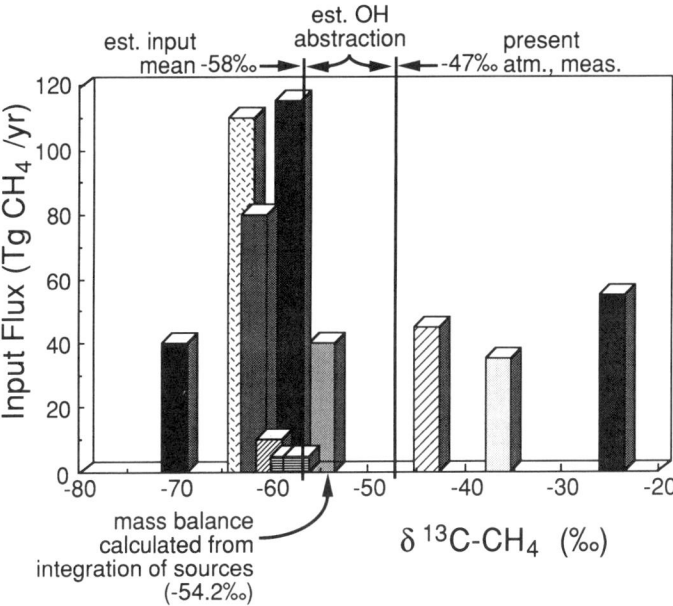

Fig. 9. A re-representation of the flux and isotope data of Figure 8. Based on an isotope effect causing an approximate 6 per mil offset in $\delta^{13}C$-CH_4 (α_C ca. 1.0054) from the steady state atmospheric methane value, a value of $-53‰$ is set for the mean $\delta^{13}C$-CH_4 entering the atmosphere (refer to Figure 7). The mass balance presented here gives an integrated $\delta^{13}C$-CH_4 input value of $-54.2‰$. The difference between the measured and calculated carbon isotope effect for the methane reaction is now much less than found previously, but the discrepancy for the hydrogen isotope effect remains significant.

Accepting the uncertainties in the flux and isotope estimates, we calculate by mass balance a mean (integrated) input value of methane entering the atmosphere to have a $\delta^{13}C$-CH_4 = -54.2 ‰, and δD-CH_4 = -307 ‰ (Table 1, Figure 9). The carbon isotope value is similar to that of $\delta^{13}C$-CH_4 = -53 ‰ predicted by the OH-reaction, but the hydrogen isotope ratio is significantly heavier than δD-CH_4 = -350 ‰ estimated for the abstraction. Assuming steady state, the revised fractionation factors based on the mass balance approach would be values α_C = 1.0076 and α_D = 1.33. The discrepancy in methane isotope values between the calculated mass balance and the expected isotope input, particularly in D/H, points to our weakness in identifying, estimating, or characterizing the various sources and sinks for atmospheric methane.

Because the majority of the sources have similar isotope ratios, uncertainties in the estimates of the flux magnitudes of various sources may not significantly affect the mass balance, nor change the overall story. Figure 10 illustrates the sensitivity of the mean $\delta^{13}C$-CH_4 value of methane entering the atmosphere on the correctness of the flux magnitude estimate for four different sources (rice, wetlands, natural gas and biomass burning). In Figure 10, the 540 Tg/yr mass balance of methane entering the atmosphere is maintained and the flux increase for any one species is proportionately reduced amongst the remaining sources.

In the cases of 100% error, i.e., where the single source is actually twice that normally estimated, then for an extreme case a shift of over 4 ‰ (only for the ^{13}C-rich biomass burning) could be realized. Typically changes in ^{13}C-CH_4 for atmospheric methane for a 100% single source revision would be 1 ‰ or less. More conservative revisions of individual flux strengths of 10% (see expanded scale, Figure 10) would generally result in $\delta^{13}C$-CH_4 shifts of less than 0.5‰.

Perhaps our greatest need of isotope constraints of atmospheric methane budgets is to recognize which source(s) is(are) contributing to the annual 1% increase in the atmospheric methane. Unfortunately, most of the important sources have similar or identical isotope ratios, so that "catching the culprit" may require further evidence, such as improved flux estimations, or $^{14}CH_4$ constraints. Figure 11 attempts to illustrate the point that if the entire annual 1% atmospheric methane increase were due to a single changing source (e.g., rice or biomass burning, etc.), then even after 5 years many of these sources would offset the current mean input $\delta^{13}C$-CH_4 value of -54.2 ‰ less than ± 0.5 ‰. Only biomass burning, coal-bed methane, or perhaps termites would generate $\delta^{13}C$-CH_4 changes > 0.5 ‰ after 5 years. If multiple sources are changing, which is likely, then these extreme excursions would be dampened considerably, and possibly below the analytical threshold for recognizing the isotope shift. The recent improvement in analytical precision by isotope ratio monitoring (GC-C-IRMS) may provide us with the improved sensitivity necessary.

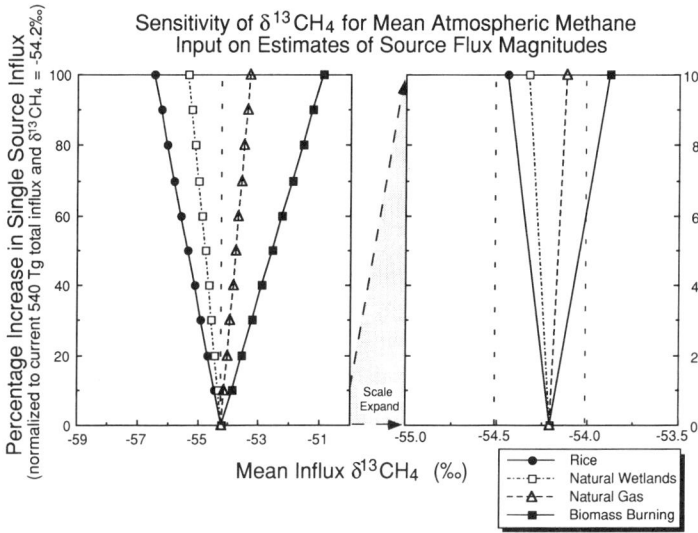

Fig. 10. The sensitivity of the integrated δ^{13}C-CH$_4$ input value to adjustments in our source strengths is demonstrated. The left-hand figure shows a relative increase in the magnitude of the source flux from 0 - 100% (up to twice now estimated) for 4 candidates and the corresponding shift in integrated δ^{13}C-CH$_4$ input value (total emission held to 540 Tg, loss distributed proportionally amongst remaining sources). The right-hand figure is an expanded 0-10% scale. In all cases, the isotope offset is well within our measurement capability, but obviously the system is probably not so simple

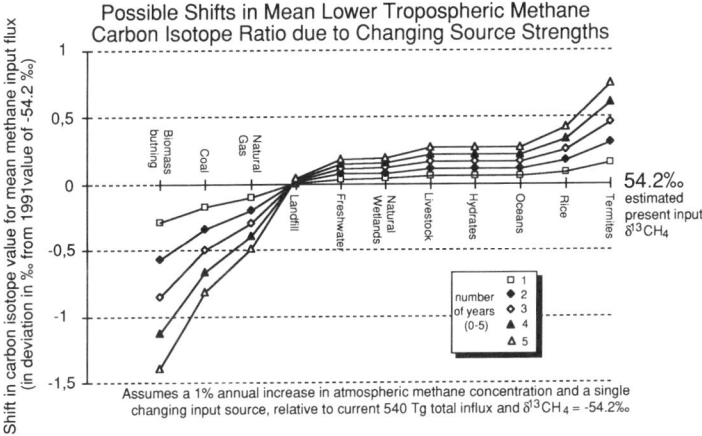

Fig. 11. The shift of the integrated δ^{13}C-CH$_4$ input value, based on the present 1% annual increase in atmospheric methane, that could be experienced over the next 5 years. As for Figure 10, the relative isotope shift calculation only accounts for single source strength adjustments, e.g., only biomass burning or rice paddies are responsible for the atmospheric methane rise. Again, measurable isotope offsets result, but multiple source changes would severely complicate the story.

5. Summary

The three prerequisites, which needed to be satisfied in order that we can use stable isotopes to constrain estimates of the atmospheric methane budgets, as set out at the start of the paper (i.e., 1. classification, 2. C-, H-isotope effects, and 3. uncertainty in input/output flux magnitudes) have been only partially met. We can discriminate the primary methanogenic and thermogenic (catagenic) formation and oxidation pathways. However, our knowledge of isotope fractionation during migration/transport is weak and our understanding of the isotope effects associated with the atmospheric OH-abstraction of methane insufficient. Based on mass balance calculations, the isotope effects for the OH-reaction are less than predicted by experimental data. Some of this discrepancy may be due to the impact of other methane sinks. Soils are now recognized as important sinks of atmospheric methane, and the possibility that the oceans may also be large scale sinks is currently being tested. The consequence of these new sinks for atmospheric methane, is that the magnitude of the total source fluxes may have to be adjusted upward to accommodate the annual 1% increase in atmospheric methane burden.

If the rise in atmospheric methane concentration is due to essentially a single source, and that source has an isotope signature dissimilar from the present mean $\delta^{13}C$-CH_4 (-54.2 ‰), then the isotope shift could constrain the budget estimates. Multiple sources definitely confuse the issue.

Stable isotopes are excellently suited to track the various process of formation, migration, and consumption, but in short time spans (i.e., next 5 years) they appear to lack sufficient resolution to constrain estimates of mean atmospheric methane budgets. Longer-term (30 - 50 years, or palaeo-atmospheric) changes in source strengths may be followed in the atmospheric methane isotope record, but again only if the changing components are significantly dissimilar from the isotope signature of mean methane input.

Finally, the intricate relationship between atmospheric methane and OH-radicals may permit us to estimate changes in atmospheric OH-radical abundance by monitoring the stable isotope ratios of methane, but more confidence in predicting and quantifying the operative isotope effects plays a key role in this development.

Acknowledgments. I would like to thank Keith Lassey for his incisive review, which improved the manuscript and alerted me to some recent investigations. This research effort is funded in Canada through the National Science and Engineering Research Council (NSERC) Strategic Research Grant No. 105389.

References

Abrajano, T.A., N.R. Sturchio, J.H. Bohlke, G.L. Lyon, R.J. Poreda, C.M. Stevens. 1988. Methane-hydrogen gas seeps, Zambales Ophiolite, Philippines: Deep or shallow origin? *Chem. Geol., 71*:211-222.

Alperin, M.J., W.S. Reeburgh, M.J. Whiticar. 1988. Carbon and hydrogen isotope fractionation resulting from anaerobic methane oxidation. *Global Biogeochem. Cycles,*

2:279-288.

Blake, D.R., F.S. Rowland. 1988. Continuing worldwide increase in tropospheric methane, 1978 to 1987. *Science, 239*:1,129-1,131.

Boone, D.R. 1993. Biological Formation and Consumption of Methane. In: *Atmospheric Methane: Sources, Sinks, and Role in Global Change*, edited by M.A.K. Khalil, Springer-Verlag, 102-127.

Cantrell, C.A., R.E. Shetter, A.H. McDaniel, J.G. Calvert, J.A. Davidson, D.C. Lowe, S.C. Tyler, R.J. Cicerone, J.P. Greenberg. 1990. Carbon kinetic isotope effect in the oxidation of methane by hydroxyl radicals. *J. Geophys. Res., 95*:22,455-22,462.

Chung, H.M., W.M. Sackett. 1979. Use of stable isotope compositions of pyrolytically derived methane as a maturity indices for carbonaceous materials. *Geochim. Cosmochim. Acta., 43*:1,979-1,988.

Chung, H.M., J.R. Gromly, R.M. Squires. 1988. Origin of gaseous hydrocarbons in subsurface environments: Theoretical considerations of carbon isotope distribution. *Chem. Geol., 71*:97-103.

Claypool, G.E., I.R. Kaplan. 1974. The origin and distribution of methane in marine sediments. In: *Natural Gases in Marine Sediments* (I.R. Kaplan, ed.), Plenum, New York, pp.99-139.

Clayton, C. 1991. Carbon isotope fractionation during natural gas generation from kerogen. *Mar. and Petrol. Geol., 8*:232-240.

Colombo, U., F. Gazzarini, G. Sironi, R. Gonfiantini, E. Tongiorni. 1965. Carbon isotope composition of individual hydrocarbons from Italian natural gases. *Nature, 205*:1,303-1,304.

Davidson, J.A., C.A. Cantrell, S.C. Tyler, J.G. Calvert, R.J. Cicerone, R.E. Shetter. 1986. The carbon kinetic isotope effect in the CH_4 + OH reaction. *EOS, 67*:245.

Davidson, J.A., C.A. Cantrell, S.C. Tyler, R.E. Shetter, R.J. Cicerone, J.G. Calvert. 1987. Carbon kinetic isotope effect in the reaction of CH_4 with HO. *J. Geophy. Res., 92*:2,195-2,199.

Evans, C.R., F.L. Staplin. 1971. Regional facies of organic metamorphism in geochemical exploration. In: *3rd Intl. Geochem. Explor. Symp.*, Proc. Can. Inst. Mining and Metallurgy, *Spec. Vol., 11*:517-520.

Faber, E., M.J. Whiticar. 1989. C- und -Isotope in leichtfluchtige Kohlenwasserstoffen der KTB. *KTB Reports*.

Faber, E., P. Gerling, I. Dumke. 1987. Gaseous hydrocarbons of unknown origin found while drilling. *Org. Geochem., 13*:875-879.

Frank, D.J. 1972. Deuterium variations in the Gulf of Mexico and selected organic materials. Ph.D. thesis, Texas A&M Univ.

Galimov, E.M. 1973. *Carbon Isotopes in Oil and Gas Geology Nauka, Moscow*, Engl. trans: *NASA TT-682*, Washington, D.C. 1975, 395 p.

Galimov, E.M. 1985. *The Biological Fractionation of Isotopes*, Academic Press, N.Y., 261 p.

Galimov, E.M., I.A. Petersil'ye. 1967. Isotopic composition of the carbon of methane isolated in the pores and cavities of some igneous minerals. *Doklady AN SSSR 176 (4)*.

Galimov, E.M., A.A. Ivlev. 1973. Thermodynamic isotope effects in organic compounds: 1. Carbon isotope effects in straight-chained alkanes. *Russian J. Phys. Chem. 47*:1564-1566.

Gerling, P. 1985. Isotopengeochemische Oberflächenprospektion Onshore. *BGR Internal Report No. 98576*, 36 pp.

Gordon, S., W.A. Mulac. 1975. Reaction of the OH ($X2\pi$) radical produced by the pulse radiolysis of water vapour. *Int. J. Chem. Kinetics*:289-299, Proc. Symp. on Chemical Kinetics Data for the Upper and Lower Atmosphere.

Hunt, J.M., A.Y. Huc, J.K. Whelan. 1980. Generation of light hydrocarbons in sedimentary rocks. *Nature, 288*:688-690.

James, A.T. 1983. Correlation of natural gas by use of carbon isotope distribution between hydrocarbon components. *Am. Assoc. Petrol. Geol., 67*:1,176-1,191.

Jefferey, A.W.A., I.R. Kaplan. 1988. Hydrocarbons and inorganic gases in the Gravberg-1 well, Siljan Ring, Sweden. *Chem. Geol., 71*:237-255.

Karweil, J. 1969. Aktuelle Probleme der Geochemie der Kohle. In: *Advances in Organic Geochemistry 1968* (P.A. Schenk and I. Havenaar, eds.), Pergamon Press, Oxford, pp. 59-84.

Khalil, M.A.K., R.A. Rasmussen, J.R.J. French, J.A. Holt. 1990. The influence of termites on atmospheric trace gases: CH_4, CO_2, $CHCl_3$, N_2O, CO, H_2, and light hydrocarbons. *J. Geophys. Res., 95*:3,619-3,634.

Konnerup-Madsen, J., R. Kreulen, U. Rose-Hansen. 1988. Stable isotope characteristics of hydrocarbon gases in the alkaline Ilimaussaq complex, south Greenland. *Bull. Minéral., 111*:567-576.

Kvenvolden, K.A. 1988. Methane hydrate - A major reservoir of carbon in the shallow geosphere. *Chem. Geol., 71*:41-51.

Laier, T. 1988. Hydrocarbon gases in the crystalline rocks of the Gravberg-1 well, Swedish deep gas project. *Mar. and Petrol. Geol., 5*:370-377.

Lasaga, T., G.V. Gibbs. 1991. Ab initio studies of the kinetic isotope effect of the CH_4 + OH• atmospheric reaction. *Geophys. Res. Lett., 18*:1,217-1,220.

Lyon, G.L., J.R. Hulston. 1984. Carbon and hydrogen isotopic compositions of New Zealand geothermal gases. *Geochim. et Cosmochim. Acta, 48*:1,161-1,171.

McCarty, H.B., G.T. Felbeck, Jr. 1986. High temperature simulation of petroleum formation, IV. Stable carbon isotope studies of gaseous hydrocarbons. *Org. Geochem., 9*:183-192.

Rasmussen, R.A., M.A.K. Khalil. 1983. Global production of methane by termites. *Nature, 301*:704-705.

Rice, D.D. 1993. Controls, habitat, and resource potential of ancient bacterial gas. In: *Biogenic Natural Gas* (in press).

Rice, D.D., G.E. Claypool. 1981. Generation, accumulation and resource potential of biogenic gas. *AAPG Bull., 67*:1,199-1,218.

Rust, F.E., C.M. Stevens. 1980. Carbon kinetic isotopic effect in the oxidation of methane by hydroxyl. *Int. J. Chem. Kinetics, 12*:371-377.

Sackett W.M. 1968. Carbon isotope composition of natural methane occurrences. *AAPG Bull., 52*:853-857.

Schmidt, M. 1987. Isotope-geochemical analysis of dunk tank gases, headspace gases, desorbed gases of cuttings and cores. *Vattenfall Deep Gas Project, Init. Rpt., Nov. 1987*, 14p.

Schoell, M. 1980. The hydrogen and carbon isotopic composition of methane from natural gases of various origins. *Geochim. et Cosmochim. Acta, 44*:649-661.

Schoell, M. (ed.) 1988. Origins of Methane in the Earth. *Chem. Geol., 71*, 265 p.

Seiler, W., R. Conrad, D. Scharffe. 1984. Field studies of methane emission from termite nests into the atmosphere and measurements of methane uptake by tropical soils. *J. Atmos. Chem., 1*:171-186.

Sherwood, B., P. Fritz, S.K. Frape, S.A. Macko, S.M. Weise, J.A. Welhan. 1988. Methane occurrences in the Canadian Shield. *Chem. Geol., 71*:223-236.

Silverman, S.R. 1971. Influence of petroleum origin and transformation on its distribution and redistribution in sedimentary rocks. *Proc., 8th World Petrol. Congr. 2*, 47-54.

Simoneit, B.R.T., O.E. Kawka, M. Brault. 1988. Origin of gases and condensates in the Guaymas Basin hydrothermal system (Gulf of California). *Chem. Geol., 71*:169-182.

Stahl, W. 1973. Carbon isotope ratios of German natural gases in comparison with isotopic data of gaseous hydrocarbons from other parts of the world. In: *Advances in Organic Geochemistry 1973* (B. Tissot and F. Bienner, eds.), Pergamon Press, Oxford, pp. 453-462.

Stevens, C.M. 1993. Isotopic abundances in the atmosphere and sources. In: *Atmospheric Methane: Sources, Sinks, and Role in Global Change*, edited by M.A.K. Khalil, Springer-Verlag, 62-88.

Tissot, B.P., D.H. Welte. 1978. *Petroleum Formation and Occurrence.* Springer Verlag, Berlin, 538 p.

Welhan, J.A. 1988. Origins of methane in hydrothermal systems. *Chem. Geol., 71*:183-198.

Welhan, J.A., J.E. Lupton. 1987. Light hydrocarbon gases in Guaymas Basin hydrothermal fluids: Thermogenic versus abiogenic origin. *Bull. Am. Assoc. Petrol. Geol., 71*:215-223.

Whelan, J.K., M.E. Tarafa, J.M. Hunt. 1982. Volatile C_1-C_8 organic compounds in macroalgae. *Nature, 299*:50-52.

Whiticar, M.J. 1990. A geochemical perspective of natural gas and atmospheric methane. In: *Advances in Organic Geochemistry* (B. Durand and F. Behar, eds.), Org. Geochem. 16, 531-547.

Whiticar, M.J. 1992. Isotope tracking of microbial methane formation and oxidation. In: *Cycling of Reduced Gases in the Hydrosphere* (D.D. Adams, P.M. Crill, and S.P. Seitzinger, eds.), Mitteilung (Communications) v. 23, Internationalen Vereinigung für Theoretische und Angewandte Limnlogie, E. Schweizerbart'sche Verlagsbuchhandlung (Nägele u. Obermiller), Stuttgart, Germany

Whiticar, M.J., E. Faber. 1986. Methane oxidation in sediment and water column environments - isotope evidence. *Org. Geochem., 10*:759-768.

Whiticar, M.J., E. Suess. 1989. Hydrothermal hydrocarbon gases in the sediments of the King George Basin, Bransfield Strait, Antarctica. In: *Geochemistry of Hydrothermal Systems, Applied Geochemistry, 5* (B.R.T. Simoneit, ed.), 135-147.

Whiticar, M.J., B.R.T. Simoneit. 1993. Carbon and hydrogen isotope systematics of hydrothermal hydrocarbons at Yellowstone Park, USA. (in press)

Whiticar, M.J., E. Faber, M. Schoell. 1984. Carbon and hydrogen isotopes of C_1-C_5 hydrocarbons in natural gases. *AAPG Research Conference on Natural Gases*, San Antonio TX.

Whiticar, M.J., E. Faber, M. Schoell. 1986. Biogenic methane formation in marine and freshwater environments: CO_2 reduction vs. acetate fermentation - isotope evidence. *Geochim. et Cosmochim. Acta, 50*:693-709.

Zehnder, A.J.B. (ed.) 1988. *Biology of Anaerobic Microorganisms.* Wiley, N.Y.

Zimmerman, P.R., J.P. Greenberg, S.O. Wandiga, P.J. Crutzen. 1982. Termites: A potentially large source of atmospheric methane, carbon dioxide and molecular hydrogen. *Science, 218*:563-565.

Zimmerman, P.R., C. Westberg, J. Darlington. 1987. Global methane production by termites. *Div. of Geochem., 193rd National Mtg., Am. Chem. Soc.* (abstract), p. 55.

6. Methane Sinks, Distribution, and Trends

M.A.K. Khalil,[1] M.J. Shearer,[1] and R.A. Rasmussen[2]

[1] Department of Physics, Portland State University, P.O. Box 751, Portland, Oregon 97207
[2] Department of Environmental Science and Engineering, Oregon Graduate Institute, Portland, Oregon 97219

1. Introduction

At present the amount of methane removed from the atmosphere each year is about 500 Tg/yr or more than 90% of that released into the atmosphere each year. Most of the methane is removed by reacting with tropospheric OH radicals; lesser amounts are removed by soils and stratospheric oxidation by OH, O(^{1}D), and minor reactions. This chapter is on the removal rate of CH_4 and its variability in space and time.

The oxidation of methane in the troposphere depends on the concentrations of OH and thus varies greatly with latitude and altitude. Sixty-four percent of it is destroyed in the tropics where OH is more abundant throughout the year compared to any other part of the world. The removal rates vary seasonally, reflecting the variability of OH, and also may be affected by seasonal variations in soil conditions and of stratospheric processes, particularly at middle and higher latitudes. On still longer time scales, the destruction rate and hence the atmospheric lifetime of methane may vary because of secular changes in OH, the soil sink, and possibly stratospheric conditions.

The global mass balance of methane may be written as:

$$\frac{\partial C}{\partial t} = S - L - \frac{1}{n}\nabla \cdot n(CV - K \cdot \nabla C) \tag{1}$$

where C is the mixing ratio, S is the source, L is the destruction or loss rate, ∇ is the spatial gradient operator, n is the number density of air molecules, and the last term represents transport by mean (V) and turbulent (K) processes. The destruction rate L_R (molecules/yr or ML_R/N_A in gm/yr) and the average lifetime τ_R (yrs) in a region R, which includes a surface area A at ground level, are:

Mohammad Aslam Khan Khalil (Ed.)
Atmospheric Methane
© *Springer-Verlag Berlin Heidelberg 2000*

$$L_R = \int_R \sum_J \eta_J(x,t)\, \rho\,(x,t)\, C\,(x,t)\, dx$$

$$+ \int_A v_d(x,t)\, \rho\,(x,t)\, C\,(x,t)\, dA \tag{2}$$

$$\tau_R = \frac{\int_R C\, \rho\, dx}{L_R} \tag{3}$$

$$\eta_J = K_J\, [J] \tag{4}$$

M = Molecular weight (gm/mole).
N_A = Avogadro's number (molecules/mole).
ρ = Density of air at location **x** and time t (molecules/cm^3).
v_d = Deposition velocity (cm/sec).
K_J = Rate constant for reaction of CH_4 with chemical species J (cm^2/molecule-sec).
[J] = concentration of species J (molecules/cm^3).

To estimate the current annual and seasonal global removal rates, we use the measured concentrations of methane, deposition velocities, reaction rate constants, and calculated distributions of OH and other reactants in the equations mentioned above (with data from Crutzen and Schmailzl, 1983; Spivakovsky et al., 1990; Brasseur et al., 1990; Dlugokencky, 1994; Lu and Khalil, 1991; Vaghjiani and Ravishankara, 1991; DeMore et al., 1992; Weisenstein et al., 1992).

In the next section we discuss the global distribution of methane used in our calculations and trends of methane in recent times.. Section 3 is on the vertical, latitudinal, and seasonal destruction of methane under current climatic and environmental conditions. In Section 4 we discuss possible long-term changes of the lifetime, and in Section 5 we consider some emerging issues in evaluating the sinks of methane.

2. Spatial and Seasonal Concentration Distribution of Methane and Current Trends

The nature of the current atmospheric data is dramatically different from the ice core measurements described by Chappellaz et al. (this volume). Ice core measurements represent concentrations only in the polar regions and integrated over periods of a few years to hundreds of years. But the measurements go back thousands of years. In contrast, current systematic measurements extend over only the past 20 years but have been taken all over the world. There are some 28 sites in the NOAA/CMDL (National Oceanic and Atmospheric Administration, Climate Monitoring and Diagnostics Laboratory) network and no fewer than four sites in other networks. Similarly, in time, measurements are usually taken every week, but sometimes the measurement frequency could be an hour or less (Khalil et al., 1993). As the space and time scales of the ice core data are so different from the current data, so are the applications of the two data sets in understanding the methane cycle. The current data are the foundation for understanding the effect of human activities on the global methane cycle, including present emission rates and their trends (see also Khalil and Shearer, this volume). Moreover, current data provide details of seasonal variations and even sub-monthly and diurnal variability, all of which are useful in validating the understanding of sources and sinks. Beyond the few-year time scale of the recent trends, the current data are not long enough to show how slower environmental changes affect atmospheric methane.

Recent measurements are used to construct latitudinal and vertical distributions of methane. The vertical profile for methane at middle northern latitudes is a composite of several published data sets (Fabian et al., 1981; Schmidt et al., 1984, 1987; Taylor et al., 1989). It is shown in Figure 1 extending from the surface to nearly 70 Km. In the troposphere concentrations are nearly constant with height, but in the stratosphere methane concentrations fall rapidly. Above the tropopause, concentration can be described by $C = C_T \exp(\lambda z)$ ($z_T < z < z_S$) where C_T is the concentration at the tropopause, z is the altitude, z_T is the tropopause height, and z_S is the stratopause height. Much higher up in the stratosphere, between 30-60 Km, methane concentrations have been measured by satellites and spacecraft (results reported by Ehhalt et al., 1972; Jones and Pyle, 1984; Gunson et al., 1990). Stratospheric concentrations fall off more rapidly in the higher latitudes than in the tropical latitudes (Bush et al., 1978).

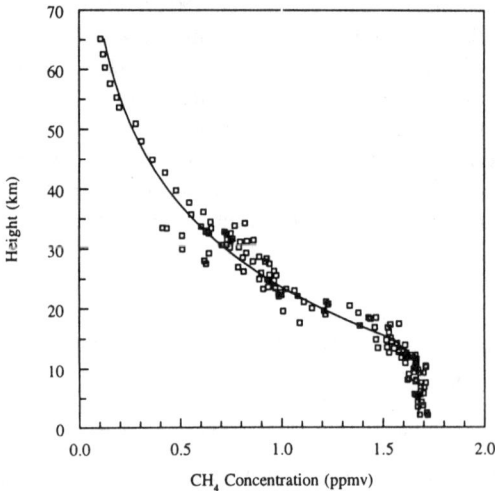

Fig. 1. Vertical distribution of CH$_4$ at 44°N latitude. Data from Fabian et al., 1981, Schmidt et al., 1984 and 1987, and Taylor et al., 1989. Data adjusted to base year 1990.

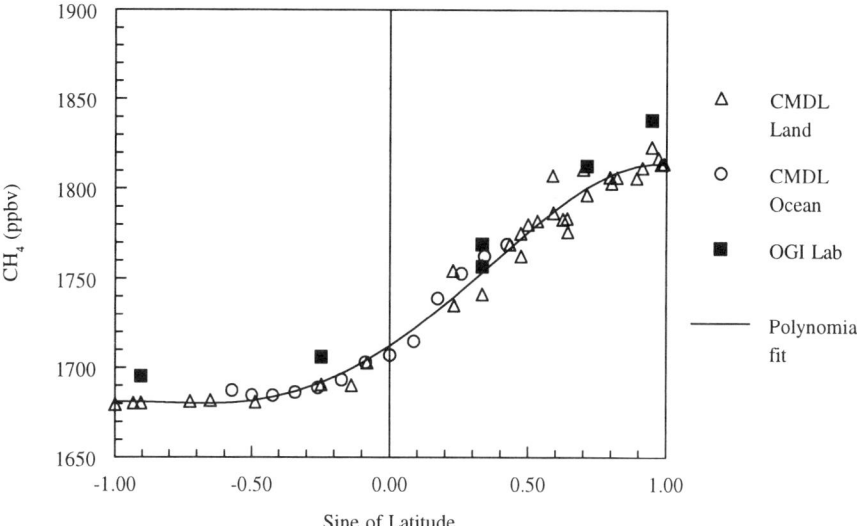

Fig. 2. Latitudinal distribution of CH_4. The 1996 NOAA/CMDL data (Dlugokencky et al., 1994) are shown next to our 1996 data from six sites. (Calibration difference: C(NOAA) = C(ours) + 12 ppbv.)

The latitudinal variations are taken from the NOAA/GMCC (now CMDL) data and our measurements (Steele et al., 1987; Dlugokencky, 1994; Khalil and Rasmussen, 1990, 1993, and previously unpublished data) and are shown in Figure 2. Concentrations are generally higher in the middle and higher latitudes of the northern hemisphere than elsewhere because of the similar global distribution of sources (Khalil and Shearer, this volume and references therein). The seasonal variations shown in Figures 3a and 3b are taken from our flask sampling network. Concentrations are lower in the summers compared to winters. The seasonal cycles of CH_4 concentrations at various latitudes are determined mostly by seasonal changes of OH, but substantial contributions from seasonal variations of sources are expected at some latitudes, particularly the middle northern latitudes (Khalil and Rasmussen, 1983, 1993; Fung et al., 1991).

When measurements of a trace gas are taken far from sources, it is accepted that the concentrations represent some usually unknown but large spatial scale. That the same idea applies to trends in time is not commonly recognized. When a trend is measured, it usually has an associated time scale over which the trend is expected to be valid. In longer time series, the trend changes, fluctuates, or undergoes cyclic variations over many time scales. Trends measured over short periods, therefore, do not reliably represent trends over longer periods extending either forward or backward in time from the period of the measurements. Of special interest are the longer-term changes of observed trends such as increases or decreases in the buildup of the gas in the atmosphere.

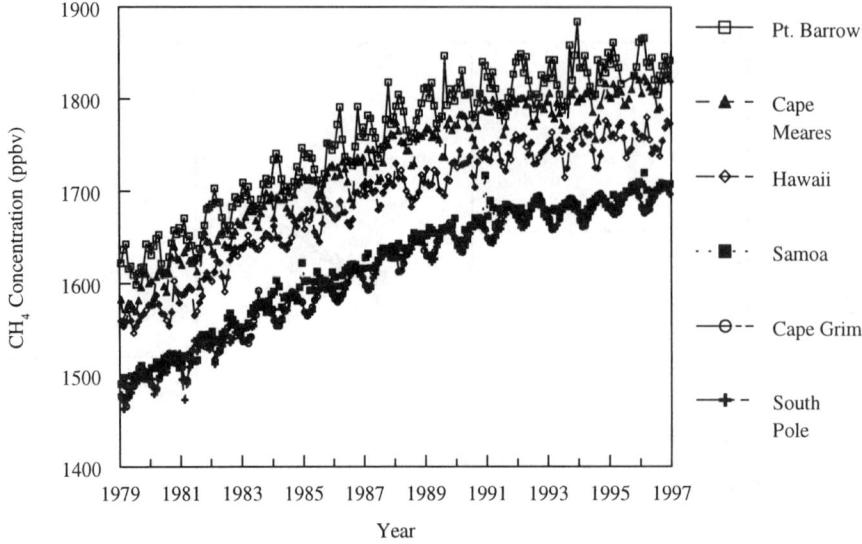

Fig. 3a. Time series of monthly measurements of CH₄ at six sites.

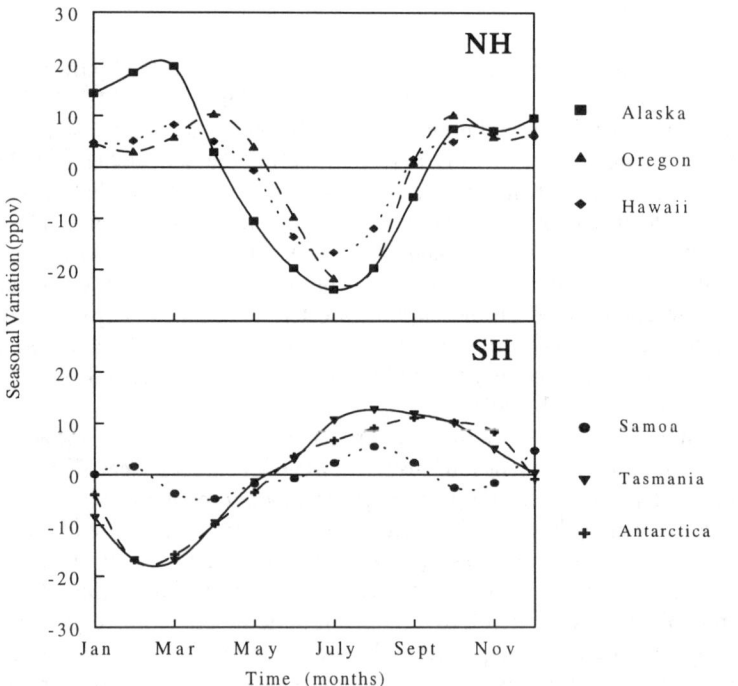

Fig. 3b. Average seasonal variations at the same six sites.

As mentioned earlier, detailed systematic measurements of methane have been taken only since 1979. Even over the relatively short period between then and now, the trends of methane have changed. The main feature of this change is the dramatic slowdown of trend between the early 1980s and recent times. This change in global trend has been apparent for some time (Khalil and Rasmussen, 1990) and has recently been documented for the large number of sites in the NOAA/CMDL network (Steele et al., 1992; Dlugokencky, 1994). The decrease of trend over a longer period than the NOAA/CMDL measurements but at fewer sites has been reported by Khalil and Rasmussen (1993).

In Figure 4 we show the decline of trend as the change in the rate of increase at various sites between 1980 and the present, represented as "moving slopes" of the measured concentrations over 3-year periods. Global average trends in the first 3 years (1980-83) were about 21 ppbv/yr and in the last 3 years (1993-96) are about 6 ppbv/yr with uncertainties of about ±1 ppbv/yr. Smaller changes have been reported by Steele et al. (1992) between 1983 and 1990.

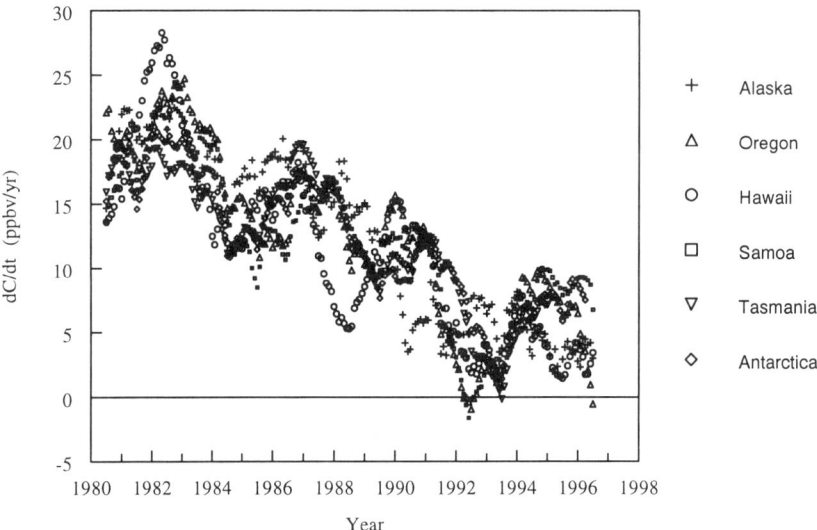

Fig. 4. The trends of methane over 3-year overlapping periods of time calculated by linear regression. (Adapted and extended from Khalil and Rasmussen, 1993.)

The causes of the slowdown of methane increase may come from increases of OH or a slowdown in the increases of anthropogenic emissions (Khalil and Shearer, this volume). There is evidence that the change of sources is perhaps the larger effect. Increases in historic anthropogenic sources such as rice agriculture and cattle have slowed, perhaps even stopped altogether during the last decade. Yet other, newer anthropogenic sources such as landfills, natural gas leakages, low-temperature coal burning, and the like may be increasing and eventually may cause methane trends to increase also. This represents a shift from agricultural sources to sources related to energy and waste disposal. Changes in sinks are discussed in section 4.

3. Seasonal and Spatial Destruction Rates and Lifetimes

The hydroxyl radical is the principle sink of atmospheric methane through a series of reactions that begins with $CH_4 + OH \rightarrow CH_3 + H_2O$ (see Wuebbles et al., this volume). The OH concentrations by latitude, altitude and season are taken from the tables published by Spivakovsky et al. (1990) with stratospheric data from Brasseur et al. (1990); concentrations in both studies are derived from model calculations. With the methane distribution described above, we estimate 440 Tg/yr of methane are destroyed in the troposphere by OH mostly during the daytime. Small amounts, around 5 Tg/yr of methane may be destroyed by nighttime OH (Lu and Khalil, 1992). At higher latitudes most of the annual destruction of methane occurs during summers compared to other seasons. An additional 10 Tg/yr is destroyed by OH in the stratosphere.

Other sinks believed to be important in the stratosphere are CH_4 reactions with $O(^1D)$ and Cl. We estimate about 5 Tg/yr of methane is destroyed by $O(^1D)$ and Cl destroys much less than 1 Tg/yr. (See Crutzen and Schmailzl, 1983, and Weisenstein et al., 1992, for base data.)

The soil sink was estimated from a global vegetation database (Matthews, 1983; 1984), using methane consumption factors compiled by Bartlett and Harriss (1993). Methane uptake by soils is mainly controlled by soil inundation and soil porosity (see Bartlett and Harriss, 1993, for a review). The total soil sink is estimated to be 25-30 Tg/yr.

The destruction rate by latitude in Figure 5 shows that most of the methane is destroyed in the tropics. A summary of the destruction rates and lifetimes is given in Table 1. Other estimates are given by Crutzen and Schmailzl (1983), Warneck (1988), Prather and Spivakovsky (1990), and Lelieveld et al. (1993).

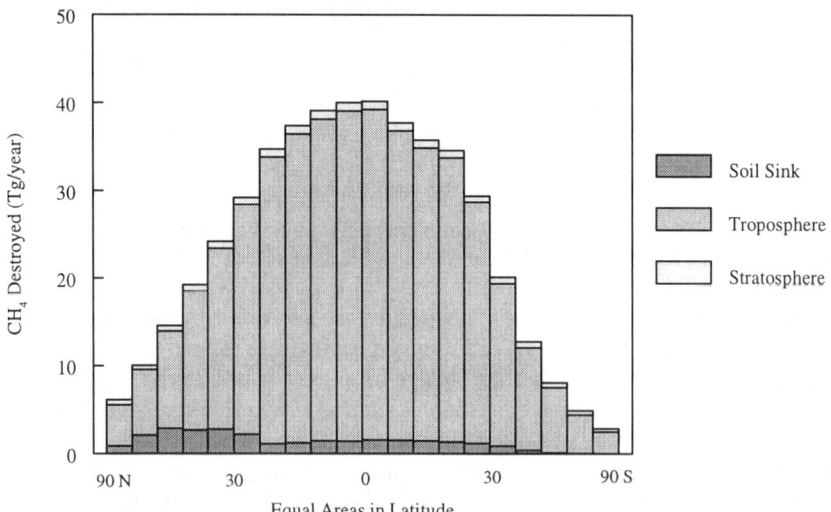

Fig. 5. Latitudinal distribution of CH_4 sinks by equal area latitude bands (increments of 0.1 in sine of latitude).

Table 1. Annual sinks of CH_4 and its mass balance for 1990.

	30-90° S	0-30° S	0-30° N	30-90° N	Global
C (Tg)	1166	1173	1206	1256	4865
dC/dt[§] (Tg/yr)	7.8	7.9	8.5	7.1	30
S (Tg/yr)	20	115	170	205	510
τ^* (yr)	24	7	7	17	10
L (Tg/yr)	48	178	180	74	480

Global values are rounded.

[§] Values for dC/dt are calculated from Khalil and Rasmussen (1993).

[*] Composite of destruction by oxidation and soil sink:

$$\tau = (\tau_{OH}^{-1} + \tau_{Soil}^{-1})^{-1}$$

4. Secular Trends of Lifetime

There has been much speculation on the long-term changes in atmospheric OH which is a principal oxidant of the global atmosphere. Such changes in the oxidation capacity affect not only the sinks of methane but of many other trace gases in the earth's atmosphere (for a current review see Thompson, 1992). If OH concentrations decrease over time, methane lifetime would increase and contribute to increasing trends. OH is produced mostly by the reaction of sunlight with O_3 ($O_3 + h\nu \rightarrow O_2 + O(^1D)$) followed by the reaction of the excited oxygen atoms with water vapor ($H_2O + O(^1D) \rightarrow 2\ OH$). It is destroyed mostly by reacting with CO and methane. There are other production and destruction processes as well, but they are individually of lesser importance.

It was thought that because CO and methane have been increasing, OH must be declining over the last century and particularly in recent decades (Levine et al., 1985; Khalil and Rasmussen, 1985; Thompson and Cicerone, 1986). Current research suggests however, that OH concentrations may be stabilized because at the same time that anthropogenic activities increase the removal rate of OH (by increasing CH_4 and CO), they also tend to increase production. This occurs from increasing tropospheric O_3 due to human activities and changes of water vapor in the transitions between climatic regimes (Pinto and Khalil, 1991; Lu and Khalil, 1991). In addition, as stratospheric O_3 is reduced it allows more uv to penetrate the troposphere thus increasing the production of $O(^1D)$ which also increases OH production (Madronich and Granier, 1992). Since it is possible that both the production and destruction of methane are increasing, the present concentrations of OH may be constant or even increasing slightly. Krol et al. (1998) deduced an OH trend of about 0.5% yr^{-1} between 1978 and 1993, based on measurements of methyl chloroform.

It is also possible that the soil sink may be slowing down due to human activities as suggested by Steudler et al. (1989) who showed that nitrogen deposition onto the soils reduces the uptake of methane. Ojima et al. (1993) estimate that land use and management changes have led to a 30% decrease in the methane soil sink in the northern temperate latitudes. The global significance of this mechanism is not yet understood.

5. Current Issues

The study of the OH sink is dependent mostly on laboratory studies of the reaction rate constants and model calculations of tropospheric OH. Since there is no global climatology of measured OH concentrations there are a number of theoretical factors that affect the calculation of the oxidation of methane, and many other gases. Two identified factors that affect OH but are not included in current models are the non-methane hydrocarbons and the effects of clouds.

The non-methane hydrocarbons have short lifetimes compared to methane and most have sources that are much greater in the northern hemisphere compared to southern latitudes. Over the vast expanses of the oceans the total non-methane hydrocarbon concentrations are practically nothing compared to concentrations over populated or forested land. Only the light non-methane hydrocarbons exist over most of the Earth's surface. These hydrocarbons are likely to reduce OH concentrations both by direct reactions and the reactions of OH with the organic by-products. Qualitatively the effect of nonmethane hydrocarbons is reduced on the global scale by their low concentrations that decline with altitude and latitude into the southern hemisphere and the lack of significant concentrations of the higher molecular weight hydrocarbons (see Khalil and Rasmussen, 1992). The magnitude of the effect of non-methane hydrocarbons on tropospheric OH is not well known but may well be important since a 10% reduction in calculated tropospheric OH would lead to a reduction of about 50 Tg/yr in the calculated destruction rate of methane. The model of Houweling et al. (1998) indicates that OH becomes very low in the boundary layer at continental sites with high biogenic emissions. However, this is balanced by concentrations over the ocean. A significant global change could not be quantified.

The effect of clouds on OH may be substantial also. Actinic flux is larger above the clouds and smaller below compared to clear sky conditions. Recent calculations show that low clouds at 1.5 Km altitude tend to increase the average vertical column concentration of OH and middle clouds at 4.5 Km altitude reduce the OH concentration (Lu, 1993). In both cases the concentration of OH below the cloud is about an order of magnitude smaller than under clear sky conditions. In the case of low clouds, there is an increase of OH above the cloud due to increase of actinic flux and water vapor, that spans a large altitude and overcomes the effect of reduced OH below the clouds. For higher clouds, the concentration of OH is reduced below the cloud, but is not compensated by the slightly higher calculated OH concentration above the cloud.

For the oxidation of trace gases, and particularly methane these results translate into a net reduction of methane removal from the atmosphere compared to clear sky conditions. At present we are not able to put definitive numbers behind the cloud effect since many theoretical issues remain unresolved so we can only report the first approximations. OH concentrations were calculated by a detailed photochemical model with and without the effect of clouds. Taking the average annual global cloud cover to be 35% (Hahn et al.,1987), annual methane destruction with only low clouds (< 2 Km) would increase to 460 Tg/yr, and with only middle clouds (3-6 Km) methane destruction would decrease to 340 Tg/yr. This is a sizeable range compared to calculations for clear sky conditions and warrants further study.

It is generally believed that methane oxidation is greatest in the boundary layer. This may not be so because the effects of both clouds and nonmethane hydrocarbons are likely to be greatest in the boundary layer and both tend to reduce the sink of methane.

6. Summary

Most of the annual removal of methane from the atmosphere is by reactions with tropospheric OH radicals (440 Tg/yr). Methane is also destroyed by deposition onto dry soils (30 Tg/yr) and 15 Tg/yr is destroyed in the stratosphere by reactions with OH, $O(^1D)$ and minor reactions. Non-methane hydrocarbons and clouds may substantially lower the total annual removal rate of methane by OH.

Acknowledgments. The data in Figure 2 are provided by the National Oceanic and Atmospheric Administration Climate Monitoring and Diagnostics Laboratory, Carbon Cycle Group. Support for some portions of this work was provided in part a grant from the Department of Energy (DOE DE-FG06-97ER62401). Additional support was provided by the Andarz Co.

References

Bartlett, K. B., R. C. Harriss. 1993. Review and assessment of methane emissions from wetlands. *Chemosphere, 26*:261-320.

Brasseur, G., M. H. Hitchman, S. Walters, M. Dymek, E. Falise, M. Pirre. 1990. An interactive chemical dynamical radiative two-dimensional model of the middle atmosphere. *J. Geophys. Res., 95* (D5):5,639-5,655.

Bush, Y. A., A. L. Schmeltekopf, F. C. Fehsenfeld, D. L. Albritton, J. R. McAfee, P. D. Goldan, E. E. Ferguson. 1978. Stratospheric measurements of methane at several latitudes. *Geophys. Res. Lett., 5*:,1027-1,029.

Crutzen, P. J., U. Schmailzl. 1983. Chemical budgets of the stratosphere. *Planet. Space Sci., 31* (9):1,009-1,032.

DeMore, W. B., S. P. Sander, C. J. Howard, A. R. Ravishankara, D. M. Golden, C.E. Kolb, R. F. Hampson, M.J. Kurylo, M.J. Molina. 1992. *Chemical Kinetics and Photochemical Data for Use in Stratospheric Modeling*. NASA Evaluation No. 10.

Dlugokencky, E.J., L.P. Steele, P.M. Lang, K.A. Masarie. 1994. The growth rate and distribution of atmospheric methane. *J. Geophys. Res., 99*, 17021-17043.

Ehhalt, D. H., L. E. Heidt, E. A. Martell. 1972. The concentrations of atmospheric methane between 44 and 62 kilometers altitude. *J. Geophys. Res., 77*:2,193-2,196.

Fabian, P., R. Borchers, G. Flentje, W.A. Matthews, W. Seiler, H. Giehl, K. Bunse, F. Müller, U. Schmidt, A. Volz, A. Khedim, F.J. Johnen. 1981. The vertical distribution of stable trace gases at mid-latitudes. *J. Geophys. Res., 86* (C6):5,179-5,184.

Fung, I., J. John, J. Lerner, E. Matthews, M. Prather, L. P. Steele, P. J. Fraser. 1991. Three-dimensional model synthesis of the global methane cycle. *J. Geophys. Res., 96* (D7):13,033-13,065.

Gunson, M. R., C.B. Farmer, R. H. Norton, R. Zander, C. P. Rinsland, J. H. Shaw, B.-C. Gao. 1990. Measurements of CH_4, N_2O, CO, H_2O, and O_3 in the middle atmosphere by the Atmospheric Trace Molecule Spectroscopy Experiment on Spacelab 3. *J. Geophys. Res., 95* (D9):13,867-13,882.

Hahn, C.J., S.G. Warren, J. London, R. L. Jenne, R. M. Chervin. 1987. *Climatological Data for Clouds over the Globe from Surface Observations.* Report NDP-026, Carbon Dioxide Information Center, Oak Ridge, TN.

Houweling, S., F. Dentener, Jos Lelieveld. 1998. The impact of nonmethane hydrocarbon compounds on tropospheric photochemistry. *J. Geophys. Res., 108* (D9), 10673-10696.

Jones, R. L., J. A. Pyle. 1984. Observations of CH_4 and N_2O by the NIMBUS 7 SAMS: a comparison with in situ data and two-dimensional numerical model calculations. *J. Geophys. Res., 89* (D4):5,263-5,279.

Khalil, M. A. K., R. A. Rasmussen. 1983. Sources, sinks, and seasonal cycles of atmospheric methane. *J. Geophys. Res., 88* (C9):5,131-5,144.

Khalil, M. A. K., R. A. Rasmussen. 1985. Causes of increasing atmospheric methane: depletion of hydroxyl radicals and the rise of emissions. *Atmos. Environ., 19*:397-407.

Khalil, M. A. K., R. A. Rasmussen. 1990. Atmospheric methane: recent global trends. *Environ. Sci. Technol., 24*:549-553.

Khalil, M. A. K., R .A. Rasmussen. 1992. Forest hydrocarbon emissions: relationships between fluxes and ambient concentrations. *J. Air & Waste Manage. Assoc., 42*:810-813.

Khalil, M. A. K., R. A. Rasmussen. 1993. Decreasing trend of methane: unpredictability of future concentrations. *Chemosphere, 26*:803-814.

Khalil, M. A. K., R. A. Rasmussen, F. P. Moraes. 1993. Atmospheric methane at Cape Meares: Analysis of a high resolution data base and its environmental implications. *J. Geophys. Res., 98*, 14753-14770.

Krol, M. P. J. van Leeuwen, Jos Lelieveld. 1998. Global OH trend inferred from methylchloroform measurements. *J. Geophys. Res., 103* (D9), 10697-10711.

Lelieveld, J., P. J. Crutzen, C. Brühl. 1993. Climate effects of atmospheric methane. *Chemosphere, 26*:739-768.

Levine, J. S., C. P. Rinsland, G. M. Tennille. 1985. The photochemistry of methane and carbon monoxide in the troposphere in 1950 and 1985. *Nature, 318*:254-257.

Lu, Y. 1993. Model calculations of radiative transfer and tropospheric chemistry. Ph.D. dissertation, Oregon Graduate Institute, Beaverton, OR.

Lu, Y., M. A. K. Khalil. 1991. Tropospheric OH: model calculations of spatial, temporal, and secular variations. *Chemosphere, 23*:397-444.

Lu, Y., M. A. K. Khalil. 1992. Model calculation of night-time atmospheric OH. *Tellus, 44B*:106-113.

Madronich, S., C. Granier. 1992. Impact of recent total ozone changes on tropospheric ozone photodissociation, hydroxyl radicals and methane trends. *Geophys. Res. Lett., 19*:465-467.

Matthews, E. 1983. Global vegetation and land use: new high-resolution data bases for climate studies. *J. Climate & Appl. Met., 22*:474-487.

Matthews, E. 1984. Vegetation, land-use and seasonal albedo data sets: documentation of archived data tape. NASA Technical Memorandum 86107, Goddard Space Flight Center, New York, U.S.A.

Ojima, D. S., D. W. Valentine, A. R. Mosier, W. J. Parton, D. S. Schimel. 1993. Effect of land use change on methane oxidation in temperate forest and grassland soils. *Chemosphere, 26* (1-4):675-685.

Pinto, J., M. A. K. Khalil. 1991. The stability of tropospheric OH during ice ages, inter-glacial epochs and modern times. *Tellus, 43B*:347-352.

Prather, M., C. M. Spivakovsky. 1990. Tropospheric OH and the lifetimes of hydrochlorofluorocarbons. *J. Geophys. Res., 95* (D11):18,723-18,729.

Rasmussen, R. A., M. A. K. Khalil. 1986. Atmospheric trace gases: trends and distributions over the last decade. *Science, 232*:1623-1624.

Schmidt, U., A. Khedim, D. Knapsa, G. Kulessa, F. J. Johnen. 1984. Stratospheric trace gas distributions observed in different seasons. *Adv. Space Res., 4* (4):131-134.

Schmidt, U., G. Kulessa, E. Klein, E.-P. Röth, P. Fabian, and R. Borchers. 1987. Intercomparison of balloon-borne cryogenic whole air samplers during the MAP/GLOBUS 1983 campaign. *Planet. Space Sci., 35*:647-656.

Spivakovsky, C. M., R. Yevich, J. A. Logan, S. C. Wofsy, M. B. McElroy, M. J. Prather. 1990. Tropospheric OH in a three-dimensional chemical tracer model: an assessment based on observations of CH_3CCl_3. *J. Geophys. Res., 95* (D11):18,441-18,471.

Steele, L. P., P. J. Fraser, R. A. Rasmussen, M. A. K. Khalil, T. J. Conway, A. J. Crawford, R. H. Gammon, K. A. Masarie, K. W. Thoning. 1987. The global distribution of methane in the troposphere. *J. Atmos. Chem., 5*:125-171.

Steudler, P. A., R. D. Bowden, J. M. Melilo, J. D. Aber. 1989. Influence of nitrogen fertilization on methane uptake in temperate forest soils. *Nature, 341*:314-316.

Taylor, F. W. A. Dudhia, C. D. Rodgers. 1989. Proposed reference models for nitrous oxide and methane in the middle atmosphere. In: *Handbook for MAP, Vol. 31.* (G.M. Keating, ed.), 67-79.

Thompson, A. M., R. J. Cicerone. 1986. Possible perturbations to atmospheric CO, CH_4, and OH. *J. Geophys. Res., 91* (D10):10,853-10,864.

Thompson, A. M. 1992. The oxidizing capacity of the Earth's atmosphere: probable past and future changes. *Science, 256*:1,157-1,165.

Vaghjiani, G. L., A. R. Ravishankara. 1991. New measurement of the rate coefficient for the reaction of OH with methane. *Nature, 350*:406-408.

Warneck, P. 1988. *Chemistry of the Natural Atmosphere.* Vol. 41, International Geophysics Series, Academic Press, Inc., San Diego, CA, USA.

Weisenstein, D. K., M. K. W. Ko, N.-D. Sze. 1992. The chlorine budget of the present-day atmosphere: a modeling study. *J. Geophys. Res., 97* (D2):2,547-2,559.

7. Sources of Methane: An Overview

M.A.K. Khalil and M.J. Shearer
Department of Physics, Portland State University, P.O. Box 751, Portland, Oregon 97207

1. Introduction

The sources of methane are the most complex and critical element in understanding the concentrations of atmospheric methane and their trends. For those who want to reduce methane in the atmosphere or prevent it from increasing, controlling the sources is perhaps the only practical approach. Accordingly, a significant portion of this book is devoted to estimating the global and regional emission rates. The purpose of this chapter is to introduce the subsequent chapters on individual sources and to lay the foundation for the common elements of determining global emission rates from the many and varied sources of methane.

There are three major sources (> 50 Tg/yr), all biogenic, namely rice agriculture, ruminants (particularly cattle), and the natural wetlands. There are many more minor sources that each emit between 10-50 Tg/yr but collectively are a significant fraction of the global budget. These sources include landfills, coal mines, biomass burning, urban areas, sewage disposal, natural gas leakages, lakes, oceans, termites, and tundra. Finally there are yet smaller sources including biogas pits, asphalt, several industrial sources, and possibly others that have not yet been identified (~ < 5 Tg/yr) (see Judd , this volume; Lacroix, 1993). The very small sources are thought to be a small fraction of the annual emissions even when taken together. There is still enough uncertainty in the estimates of emission rates from individual sources that some may go from minor to major or vice versa, but it is very unlikely that there are any unknown major sources. Research has concentrated more on the major sources so less is known about the minor sources, particularly the smaller of the minor sources.

2. Estimating Global Emissions: Issues and Procedures

The process of evaluating global emissions usually consists of two fundamental pieces of information. The first is data on fluxes or emission factors, often measured directly under field or laboratory conditions. The second is an extrapolating factor or extrapolant that, when associated (or multiplied) with the measured fluxes, results in the global emission rate.

Mohammad Aslam Khan Khalil (Ed.)
Atmospheric Methane
© *Springer-Verlag Berlin Heidelberg 2000*

$$S_{Global} = \phi \cdot G \qquad (1)$$

Here ϕ is the average emission rate (Tg/yr/Unit), G is the extrapolant (the number of Units in the world) and S_{Global} is the global emission rate. For instance, the emission factor may be the methane emitted per year by a global average cow, and the extrapolant may be the number of such cows in the world.

While this appears straightforward enough, in practice neither the flux from the standard unit nor the extrapolant is accurately known. At the same time, direct laboratory or field measurements can give rather precise (and probably accurate) estimates of emission factors from the source under the conditions of the experiment. This high precision and repeatability of the measurements in the field or laboratory can sometimes lead to a false sense of accuracy in the estimate of global emission rates. The conditions of any experiment, however, necessarily limit the representativeness of the measured emission factors because there usually are many variables that alter the flux under different environmental (or experimental) conditions. To approach the problem experimentally requires that we know the main factors that control emission rates from a particular source. For the major sources (rice agriculture, cattle, and wetlands) these factors are only now becoming known (see Neue and Roger, Johnson et al., and Matthews, this volume).

While the flux can be measured experimentally, the extrapolant is often obtained from geographical, economic, or political data; this introduces a new class of uncertainties and systematic biases (see Mitchell, 1982). The number of cattle, for instance, is obtained from country censuses, which are better for some regions and worse for others and more accurate in some years and less so in others. Finally, the correctness of associating a measured flux with an extrapolant is also not always apparent. For instance, when measurements of methane emissions are taken from rice fields in one area, how large an area of rice fields has similar emission rates? To obtain a global emission rate, fluxes measured in one region may be multiplied by the world-wide area of rice fields (extrapolant), as indeed had to be done when no other information was available. Such an estimate is not likely to be accurate, however. To go further, we must account for the factors that affect emission rates differently in different regions. For instance, rice agriculture may be defined as continuously flooded, intermittently flooded, or dryland (see Shearer and Khalil, this volume). Annual emission rates of methane from these regimes are very different. The global flux may then be calculated as:

$$S_{Global} = \sum_{i=1}^{N} \phi_i A_i \qquad (2)$$

where ϕ_i are the measured annual fluxes and A_i are the areas of rice harvested under the three regimes (N = 3). The annual flux ϕ_i may itself may be a product or function of other factors. For instance, it may consist of average (measured) emission rates per day times the number of days in the growing season. In this way measurements taken in regions with one length of the growing season may be extrapolated to regions with a longer or shorter growing season (and assuming all other factors are the same). This processes can be continued further. For each irrigation regime ϕ_i one may add the effects of characteristic soil types so that ϕ_{ij} is the measured emission rate (or determined by some other means) under irrigation regime i and soil type j. ϕ_{ij} may be further sub-divided to include the effect of fertilizers to obtain ϕ_{ijk} and extrapolant G_{ijk} (the area of rice harvested, under irrigation regime i, soil type j, and fertilizer type k) and so on. In general, then, the annual flux from any source may be represented as:

$$S_{Global} = \sum_{i_1=1}^{N_1} \sum_{i_2=1}^{N_2} \cdots \sum_{i_J=1}^{N_J} \phi_{i_1 i_2 \cdots i_J} G_{i_1 i_2 \cdots i_J} \tag{3}$$

Progression to apparently more and more detailed global estimates does not come without a price, however; the amount of information necessary increases geometrically, and if that information is highly uncertain, a more complicated calculation may be less accurate than a simpler one!

This procedure is essentially a search to find regions over which the methane flux is constant and can be represented adequately by measured or estimated fluxes. There are other equivalent methods for calculating global emission rates. For some variables there are better ways of expressing the flux. This formulation is general and allows us a common foundation for assessing emissions from many different types of sources (and for many gases), but there are some complications. One is that the effect of two or more factors may be synergistic, and we then cannot represent the factors additively as in Eqn. (3). In order for readers to reproduce calculated global emissions, the factors mentioned above must be specified. We have attempted to include such factors in the chapters on individual sources.

3. Constraints

There are several constraints that work at different levels of the global budget of methane. For the total emission rates from all sources, there is a constraint imposed by the global mass balance:

$$S = dC/dt + C/\tau \tag{4}$$

Here, S (Tg/yr) are the global emissions, τ is the average atmospheric lifetime (years) (see Khalil et al., this volume), and C is the global burden (Tg). dC/dt and C can be determined quite accurately from atmospheric measurements. The atmospheric lifetime is dominated by reaction with tropospheric OH, with lesser removal by soils and other chemical processes in the stratosphere so that:

$$1/\tau = K_R[OH] + 1/\tau_{other} \qquad (5)$$

where K_R is a suitably averaged reaction rate constant (see, for example, Khalil and Rasmussen, 1985). If we can calculate τ accurately, we can estimate the total emissions of methane from Eqn. (4). Fortunately, the effective global average concentration of OH may be estimated by empirical models from which τ can be obtained using Eqn. (5). A common method is to use the mass balance of methylchloroform (CH_3CCl_3) to determine the effective average OH (Lovelock, 1977; Singh, 1977; Khalil and Rasmussen, 1984; Prinn et al., 1987; Butler et al., 1991; Krol et al., 1998). Methylchloroform is an industrial de-greasing solvent that has been used for many years. It is emitted to the atmosphere soon after production (and sale), so that the well-documented industrial production records can be used as the emission rate (Midgley, 1989; Midgley and McCulloch, 1995). It is removed from the atmosphere primarily by reacting with OH, although there may also be other lesser sinks (Khalil and Rasmussen, 1984; Butler et al., 1991). Knowing the emission rate of CH_3CCl_3 from the industry records and the concentration and rate of change of CH_3CCl_3 from direct atmospheric measurements, we can determine average OH by:

$$<[OH]> \approx <1/K_R[S/C - dlnC/dt]> \qquad (6)$$

where $<\cdots>$ is an average over a suitable period. This average OH can be used in Eqns. (4)-(5) to determine the total annual emissions of CH_4. This method has many associated uncertainties, but it seems that it can limit the possible range of total methane emissions to within ± 15%. There are some who would argue that the method is more accurate than this, and others who believe that the OH from CH_3CCl_3 is but a very weak constraint. One of our early budgets relied on this constraint to produce the total emission of about 500 Tg/yr, which has persisted for over a decade (see Khalil and Rasmussen, 1983).

The method we have discussed involves two parameters for estimating total annual removal. Both parameters are subject to systematic errors. In addition to knowing the effective global OH concentration discussed above, we also need to know the reaction rate constant between OH and CH_4. This has undergone a recent revision that reduced its value by 20%, resulting in a lower estimate for the total methane sink (Vaghjiani and Ravishankara, 1991). There are other uncertainties, particularly systematic errors in estimates of average OH, that may offset the adjustment to the rate constant (see Rasmussen and Khalil, 1981; Khalil and Rasmussen, 1984; Butler et al., 1991).

For many reasons we need to know the methane budget in more detail than just the global average balance. Measurements of methane in polar ice core have shown that concentrations now are 2.5 to 3 times higher than 200 to 300 years ago (Khalil and Rasmussen, 1987; Chappellaz et al., 1990; Etheridge et al., 1992; Blunier et al., 1993; Chappellaz et al., this volume). If we assume that several hundred years ago human activities had not greatly disturbed the global methane cycle, we can regard the measured concentrations as representative of natural sources. The change of concentrations between then and now can be attributed to anthropogenic sources. Using a mass balance model, this process tells us what fractions of the present emissions are natural and what fractions are anthropogenic (Khalil and Rasmussen, 1990b). The results show that 40%-70% may come from anthropogenic sources, although this fraction continues to increase in recent times. This constraint allows us to obtain important information about the effect of human activities on the global methane cycle.

Calculations based on the ice core data that separate anthropogenic and natural emissions raise questions about how we define what is natural and what is not. For a long time methane emissions from rice fields and cattle were not regarded as anthropogenic. Clearly, however, the number of cattle in the world and the hectares of land under rice agriculture are determined by human needs. Archived data show that cattle and rice areas have increased dramatically, at first in some proportion to increasing human population but much less so now. Nonetheless, the number of domestic cattle has increased almost three-fold over the past century, and the hectares of rice harvested have increased about two-fold (Mitchell, 1980, 1982, 1983; U.N. FAO Yearbook, various years). These increases are linked directly to human activities and population growth. The increase of these sources has probably caused major increases in the concentration of atmospheric methane (Khalil and Rasmussen, 1985). On the other hand, wetlands have been drained over the past century, and bison in North America have been nearly exterminated. These activities may have led to a decrease of methane emissions that can also be attributed to human factors. These decreases do not, however, come close to compensating for the increases in the many other sources. For some sources there is less doubt that they are anthropogenic – sources such as leakage from natural gas lines or automobile exhaust.

At the next stage, we have constraints on emissions from individual sources (or source groups) or from specific regions. The stable isotopes of C and H provide constraints on specific sources or source types because each source has a characteristic isotopic composition. There are a number of factors that cause uncertainties and variabilities in the measured isotopic fractions in each source. Moreover, the available data provide many fewer constraints than there are sources. This situation (under-determined system) and the associated variability of isotopic composition from each source produce only weak constraints but hold considerable promise as measurement techniques are advanced and if isotopic combinations can be distinguished. More can be found on this subject in the chapters by Whiticar and Stevens, both in this volume, and articles by Mroz (1993), Levin et al. (1993), and Lassey et al. (1993). Constraints on regional emissions may be obtained by meteorological techniques where measurements at the boundaries of a region are used to evaluate the potential emissions inside the region (Fan et al., 1992; Simpson et al., 1995; Moncrieff et al., 1998). These constraints may become more important in the future as measurement techniques advance further.

In this context it is worth mentioning that there is also a constraint that limits the uncertainties of the total global emissions or emissions from a collection of sources. Often, as emissions from each source are evaluated based on the type of procedures mentioned in Section 2, a large range of emissions is calculated from each source.

This range may be justified based on the variability of measured emission rates or more likely based on the uncertainties in the knowledge of extrapolants. Frequently, all sources are added to arrive at the total emissions. Then the sum of the lower limits of all sources is used as the lower limit of the total emissions, and the sum of the upper limits from all the sources is taken as the upper limit of the total global emissions. This range often violates the first two constraints mentioned above, namely the OH-sink constraint and the natural and anthropogenic fractions constraint. In some cases the ranges may also violate isotopic constraints. Even a conservative statistical procedure shows that the range of the total global emissions is considerably smaller (by a factor of 2 or so) than the range obtained from the sum of minima and the sum of maxima. The procedure is described by Khalil (1992) and may lead to a reconciliation between the constraints on global emissions and the unreasonably large ranges of emissions that can be obtained by adding minima and maxima of individual sources.

4. Spatial and Temporal Aspects of Emissions

4.1 Spatial Variations

Methane emissions change appreciably from latitude to latitude and over oceans and land. The distribution of methane emissions on the Earth's surface is important to know both from a scientific perspective to understand the fundamental processes of emissions and removal and from a practical perspective for those who have an interest in mitigating global climate change. The oceans turn out to be a relatively small source (although there is some justifiable debate about this matter; Lambert and Schmidt, 1993; Bates et al., 1996; Judd , this volume); the majority of the emissions are confined to the biologically productive land, which excludes the polar regions, the deserts, mountains, and drylands.

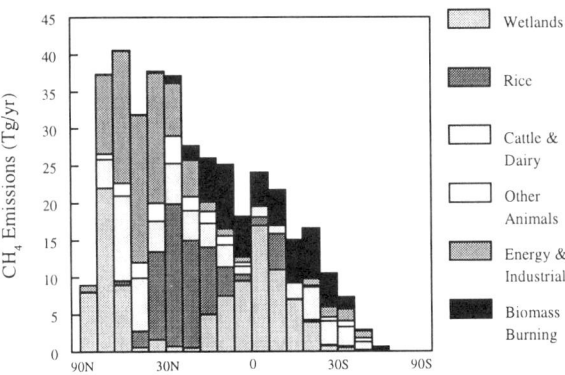

Fig. 1. The current latitudinal distribution of methane sources. About 57% of the sources are in the tropical and sub-tropical latitudes (30°N-30°S) and another 40% are in the middle and higher northern latitudes (> 30°N).

Estimates of latitudinal distributions are based on the same equations as in Section 2 but applied to each latitude or regional area. For the major sources such as rice agriculture, wetlands, and cattle such latitudinal information is available and adequate estimates can be made. For other sources the information is sketchy, unreliable, or unavailable, thus requiring many assumptions to obtain estimates. The spatial emissions rate as a function of latitude is shown in Figure 1 from our own work, with total emissions for some of the sources taken from this volume and additional information from Marland et al. (1985), Lerner et al. (1988), Cofer et al. (1991), Stocks (1991), Taylor and Zimmerman (1991), Fung et al. (1991), and Bartlett and Harriss (1993).

Emissions inventories on smaller spatial scales are currently being developed. It is worth noting that on small spatial scales there are no constraints available at present, and estimates may be substantially in error even though, when added over the whole world, they may be quite accurate. The summing procedure reduces uncertainties.

Regional estimates add another dimension leading to a matrix, S_{ij}, of emissions with one variable (i) representing the source type and the other the region (j). The region may be a country or a latitudinal and longitudinal box, depending on the intended use of the data. The sum $S_{*j} = \Sigma_i S_{ij}$ represent the total methane emissions from the all sources in a region, and $S_{i*} = \Sigma_j S_{ij}$ represents the global methane emissions from a given source (which is the focus of the subsequent chapters). The total methane emissions are $S = \Sigma_i \Sigma_j S_{ij}$. Uncertainties in S_{i*} or S_{*j} may be smaller than the average uncertainties in the S_{ij} that make up these sums because of cancellation of errors, provided the errors are not systematic. If there is systematic bias in the way emissions are calculated for a given source, then it would spread across all the regions and would persist in the sum representing the global emissions from that source. Constraints mentioned above can be brought to bear on the matrix S_{ij}, S_{i*}, S_{*j}, and S.

4.2 Temporal Variations

The changes of emissions in time can occur on several scales ranging from seasonal (annual), to inter-annual, to decadal, and to longer times. Observations have confirmed that there must be substantial seasonal variations of methane emissions to account for the atmospheric seasonal cycles, particularly at middle to higher northern latitudes (see Khalil and Rasmussen, 1983). Direct measurements of methane emissions from various sources such as rice fields, wetlands, or swamps have shown that emissions are confined to specific periods of the year (for example, Harriss et al., 1982; Seiler et al., 1984; Whalen and Reeburgh, 1992). At present there is no rigorous quantitative connection between known cycles of emissions from sources and the cycles of observed concentrations. A large fraction of the atmospheric seasonal cycle is driven by seasonal variations of OH (especially at higher latitudes), which complicates the evaluation of the effect of seasonal emissions on the atmospheric seasonal cycle.

There is perhaps a greater interest in the long-term changes of the sources because they tell us which sources have contributed to the doubling of methane over the past century and indeed whether it is the sources that have caused global increases of methane. The probable changes of emissions from the major sources over the past 100 years can be evaluated based on available geographical and agricultural data (Chappellaz et al., 1993; Kammen and Marino, 1993; Khalil et al., 1996). Figure 2 shows the increases in cattle populations and hectares of rice harvested. The increases are so large that they are bound to have a significant effect on the emissions and hence the atmospheric concentrations, regardless of the uncertain extrapolants that have to be applied to these data to calculate methane emissions. The growth of these and other sources suggests that the major increases of methane over the past 100 to 300 years have been caused by increasing emissions as opposed to being caused by decreasing OH (see Khalil and Rasmussen, 1985; Levine et al., 1985; Thompson and Cicerone, 1986; Lu and Khalil, 1991; Pinto and Khalil, 1991; Krol et al., 1998). There are still important sources such as biomass burning for which we have no direct means of estimating trends, though changes in biomass burning have been inferred from isotopic analysis of methane in the southern hemisphere (Lowe et al., 1994). It is interesting that the major sources such as rice fields and cattle have not increased much over the last decade – certainly not comparable to the rapid rates of increase in the 1950s. This slowdown of major sources may be affecting atmospheric trends, which also show a deceleration (Khalil and Rasmussen, 1990a, 1993, and elsewhere; Steele et al., 1992).

The connection between human population and methane emissions may be written as:

$$S_{ij} = e_{ij} \, P_j \tag{7}$$

where e_{ij} is the per capita emission of methane from source i (such as rice fields) in region j (China, for example) and P_j is the population of region j, making S_{ij} the emissions from region j due to source i. To have a constant relationship between population and emissions from source i, $\Sigma_j \, e_{ij} P_j / \Sigma_j \, P_j$ must be constant in time; for a constant relationship between emissions and population from a certain region j, $\Sigma_i \, e_{ij} P_j / \Sigma_i \, P_j$ must be constant; and to have a constant relationship between emissions from all anthropogenic sources and population, $\Sigma_{ij} \, e_{ij} P_{ij} / \Sigma_j \, P_j$ must be constant.

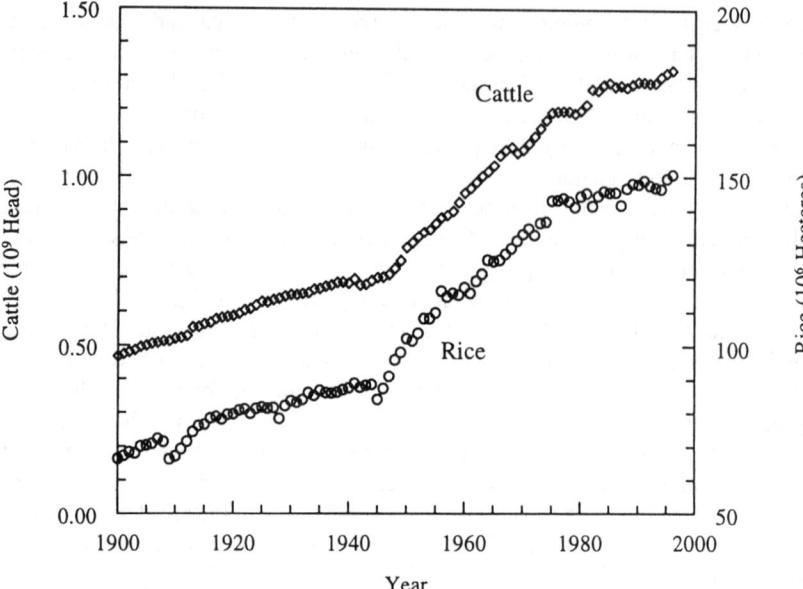

Fig. 2. Increases in cattle populations and area of rice harvested for the turn of the century tohe present. Cattle populations have increased by a factor of almost 3 and rice fields by 2 during this time. These agricultural sources represent a large role of human activities on the global methane cycle.

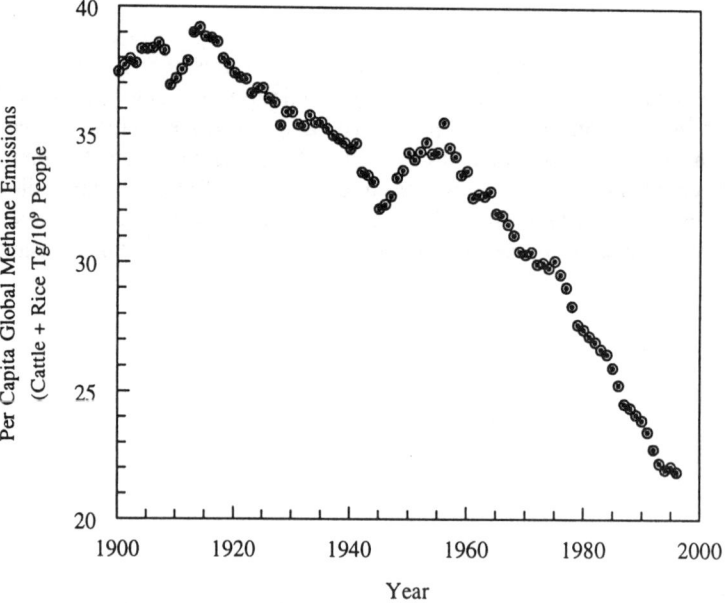

Fig. 3. The global per capita emission rate of methane from rice and cattle since the turn of the century. The per capita emission rates are not constant and show substantial declines in the recent decade.

In fact, the anthropogenic sources increase or decrease according to complex economic, social, and technological factors, which makes it difficult or perhaps impossible to predict future emissions. Generally, sources such as rice fields and cattle are related to human population since a larger population requires more food. This is not the only connection, however, since emissions are also related to "per capita" demand for the commodity. Also, there may be physical limits to land available for cattle grazing or rice agriculture, particularly on a country basis. According to Eqn. (7), the emissions are a product of the per capita consumption rate (from which per capita emissions are calculated) and the population, but both the per capita emission rate and population can change in time. It is likely that population increases are more predictable than the per capita demand or per capita consumption rates. For instance, if people are generally poor and undernourished, there will be a demand to grow more rice, causing increases of methane emissions even if the population is not increasing. After a certain point, however, as people become richer, the per capita demand for rice may decline as other foods are substituted and preferred, so methane emissions may not increase even if the population does. (See, for example, Ito et al., 1989.) Also, populations that have the greatest contribution to one source or another rise at different rates. The shifting nature of the connection between population and agricultural emissions makes population an unreliable predictor of the future even if it works well to explain the past. For instance, rice fields and cattle rose steadily in the past, keeping pace with population. As mentioned earlier, methane from rice and cattle may not be increasing rapidly compared to the previous decades, and yet the population is continuing to rise at close to the same rates as a decade ago, thus breaking the link between population and methane emissions. This effect is demonstrated in Figure 3 where we have plotted the expected methane emissions from cattle and rice (as shown in Figure 2) divided by the world population. Per capita emissions from these sources have fallen from about 39 Tg/billion people around the turn of the century to almost 20 Tg/billion people; this is a decrease of about 45% since 1900. Most of this change is very recent, about 19% over the last decade and 29% since 1960. Before 1950, population was a good indicator of agricultural CH_4 emissions, but now it is not.

Another way to evaluate trends in global emissions (possibly also over large regions such as semi-hemispheres) is to use Eqn. (4) and calculate the emissions necessary to balance the time series of atmospheric concentrations and trends (C and dC/dt from measurements). To do this requires a calculation not only of the mean OH but also its possible trends over decadal periods. One such calculation is shown in Figure 4. Here we assumed several different possible trends for OH during the period of measurements extending from decreases at -1.5 %/yr to increases at 2.0 %/yr. The results show that if OH is increasing at a rate faster than about 0.5%/yr, it requires a rapid increase of emissions, which appears untenable given the slowdown in the major sources mentioned earlier. We favor the scenario with no OH change over the period of measurements, which gives modest rises in sources during the past decade. These calculations also show that the average emissions are about 500 Tg/yr as expected from the constraints mentioned earlier. It is this 500 Tg/yr that has to be distributed among the various sources. These calculations also show that inter-annual variations are quite small at around ± 10 Tg/yr or ± 2% from the mean.

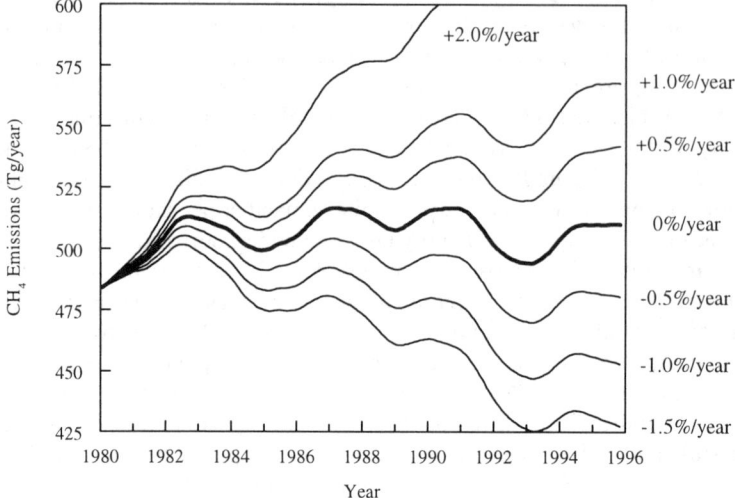

Fig. 4. The global emissions of methane during the period of systematic atmospheric measurements spanning more than a decade. The emissions are calculated so as to balance the observed concentrations and the removal rates. The effect of changing removal rates (particularly OH) is shown by the various lines. The bold line is for the case when the sinks are constant (0% change).

There is active research on the spatial and temporal emissions of methane, which will eventually lead to credible estimates. In most of the chapters on individual sources, spatial and temporal aspects have not been discussed.

5. Summary

According to current assessments, the major sources of methane are rice agriculture, domestic animals, mostly cattle, and the wetlands, each emitting more than 50 Tg/yr. Cattle and rice are directly linked to population and agricultural productivity, and even the global wetland emissions are affected by human activities. Another group of sources each emit lesser amounts (10-50 Tg/yr), but there are so many that together they constitute a large fraction of the total emissions. These sources include landfills, coal mines, biomass burning, urban areas, sewage disposal, natural gas leakages, lakes, oceans, termites, and tundra. Finally, there are yet smaller sources including biogas pits, asphalt, several industrial sources, and possibly others that have not yet been identified (~ < 5 Tg/yr). In this chapter we discussed the procedures for calculating global emissions from various sources and the inherent uncertainties. It shows some of the common foundations of the calculations in subsequent chapters on individual sources.

Acknowledgments. We have benefitted from discussions with R.A. Rasmussen, E. Matthews, J. Pinto, S. Thorneloe, K. Smith, V. Aneja, and D. Johnson. This work was supported in part by a grant from the U.S. Department of Energy (DE-FG06-97ER62401) and the resources of Andarz Co.

References

Aselmann, I., and P.J. Crutzen. 1989. Global distribution of natural freshwater wetlands and rice paddies, their net primary productivity, seasonality and possible methane emissions. *J. Atmos. Chem., 8*:307-358.

Bartlett, K., and R.C. Harriss. 1993. Review and assessment of methane emissions from wetlands. *Chemosphere, 26*:261-320.

Bates, T.S., K.C. Kelly, J.E. Johnson, and R.H. Gammon. 1996. A reevaluation of the open ocean source of methane to the atmosphere. *J. Geophys. Res.* 101(D3): 6953-6961.

Blunier, T., J.A. Chappellaz, J. Schwander, J.-M. Barnola, T. Desperts, B. Stauffer, and D. Raynaud. 1993. *Geophys. Res. Lett.* 20(20):2219-2222.

Butler, J.H., J.W. Elkins, T.M. Thompson, B.D. Hall, T.H. Swanson, and V. Koropolov. 1991. Oceanic consumption of CH_3CCl_3: implications for tropospheric OH. *J. Geophys. Res., 96*:22,347-22,355.

Chappellaz, J., J.M. Barnola, D. Raynaud, Y.S. Korotkevich, and C. Lorius. 1990. *Nature, 345*:127-131.

Chappellaz, J.A, I.Y. Fung, and A.M. Thompson. 1993. The atmospheric CH_4 increase since the Last Glacial Maximum, 1. Source estimates. *Tellus*, in press.

Cofer, W.R. III, J.S. Levine, E.L. Winstead, and B.J. Stocks. 1991. Trace gas and particulate emissions from biomass burning in temperate ecosystems. In: *Global Biomass Burning, Atmospheric, Climatic, and Biospheric Implications* (J.S. Levine, ed.):203-208.

Etheridge, D.M., G.I. Pearman, and P.J. Fraser. 1992. Changes in tropospheric methane between 1841 and 1978 from a high accumulation Antarctic ice core. *Tellus* 44B: 282-294.

Fan, S.M., S.C. Wofsy, P.S. Bakwin, D.J. Jacob, S.M. Anderson, P.L. Kebarian, J.B. McManus, C.E. Kolb, and D.R. Fitzjarrald. 1992. Micrometeorological measurements of CH_4 and CO_2 exchange between the atmosphere and subarctic tundra. *J. Geophys. Res.* 97(D15): 16,627-16,643.

Fung, I., J. John, J. Lerner, E. Matthews, M. Prather, L.P. Steele, and P.J. Fraser. 1991. Three-dimensional model synthesis of the global methane cycle. *J. Geophys. Res., 96* (D7):13,033-13,065.

Harriss, R.C., D.I. Sebacher, and F.P. Day, Jr. 1982. Methane flux in the Great Dismal Swamp. *Nature, 297*:673-674.

Ito, S., E.W.F. Peterson, and W.R. Grant. 1989. Rice in Asia: is it becoming an inferior good? *Amer. J. Agr. Econ., 71*:32-42.

Kammen, D.M., and B.D. Marino. 1993. On the origin and magnitude of pre-industrial anthropogenic CO_2 and CH_4 emissions. *Chemosphere, 26*:69-86.

Khalil, M.A.K. 1992. A statistical method for estimating uncertainties in the total global budget of trace gases. *J. Environ. Sci. Health, A27* (3):755-770.

Khalil, M.A.K., and R.A. Rasmussen. 1983. Sources, sinks, and seasonal cycles of atmospheric methane. *J. Geophys. Res., 88*:5,131-5,144.

Khalil, M.A.K., and R.A. Rasmussen. 1984. The atmospheric lifetime of methylchloroform (CH_3CCl_3). *Tellus, 36B*:317-312.

Khalil, M.A.K., and R.A. Rasmussen. 1985. Causes of increasing atmospheric methane: depletion of hydroxyl radicals and the rise of emissions. *Atmos. Environ., 19*:397-407.

Khalil, M.A.K., and R.A. Rasmussen. 1987. Atmospheric methane: trends over the last 10,000 years. *Atmos. Environ., 21*:2,445-2,452.

Khalil, M.A.K., and R.A. Rasmussen. 1990a. Atmospheric methane: recent global trends. *Environ. Sci. Tech., 24*:549-553.

Khalil, M.A.K., and R.A. Rasmussen. 1990b. Constraints on the global sources of methane and an analysis of recent budgets. *Tellus, 42B*:229-236.

Khalil, M.A.K., and R.A. Rasmussen. 1993. Decreasing trend of methane: unpredictability of future concentrations. *Chemosphere, 26* (1-4):803-814.

Khalil, M.A.K., M.J. Shearer, and R.A.Rasmussen. 1996. Atmospheric methane over the last century. *World Resource Review* 8: 481-492.

Krol, M., P.J. van Leeuwen, and J. Lelieveld. 1998. Global OH trend inferred from methylchloroform measurements. *J. Geophys. Res.* 103(D9): 10,697-10,711.

Lacroix, A.V. 1993. Unaccounted-for sources of fossil and isotopically-enriched methane and their contribution to the emissions inventory: a review and synthesis. *Chemosphere, 26* (1-4):507-557.

Lambert, G., and S. Schmidt. 1993. Reevaluation of the oceanic flux of methane: Uncertainties and long term variations. *Chemosphere* 26:579-589.

Lassey, K.R., D.C. Lowe, C.A.M. Brenninkmeijer, and A.J. Gomez. 1993. Atmospheric methane and its carbon isotopes in the southern hemisphere: their time series and an instructive model. *Chemosphere, 26* (1-4):95-109.

Lerner, J., E. Matthews, and I. Fung. 1988. Methane emission from animals: a global high-resolution data base. *Global Biogeochem. Cycles, 2*:139-156.

Levin, I., P. Bergamaschi, H. Dörr, and D. Trapp. 1993. Stable isotopic signature of methane from major sources in Germany. *Chemosphere, 26* (1-4):161-178.

Levine, J.S., C.P. Rinsland, and G.M. Tennille. 1985. The photochemistry of methane and carbon monoxide in the troposphere in 1950 and 1985. *Nature, 318*:254-257.

Lovelock, J.E. 1977. Methyl chloroform in the troposphere as an indicator of OH radical abundance. *Nature, 267*:32-33.

Lowe, D.C., C.A.M. Brenninkmeijer, G.W. Brailsford, K.R. Lassey, and A.J. Gomez. 1994. Concentration and ^{13}C records of atmospheric methane in New Zealand and Antarctica: Evidence for changes in methane sources. *J. Geophys. Res.* 99(D8): 16,913-16,925.

Lu, Y., and M.A.K. Khalil. 1991. Tropospheric OH: model calculations of spatial, temporal, and secular variations. *Chemosphere, 23* (3):397-444.

Marland, G., R.M. Rotty, and N.L. Treat. 1985. CO_2 from fossil fuel burning: global distribution of emissions. *Tellus, 37B*:243-258.

Matthews, E., I. Fung, and J. Lerner. 1991. Methane emission from rice cultivation: geographic and seasonal distribution of cultivated areas and emissions. *Global Biogeochemical Cycles, 5* (1):3-24.

Midgley, P.M. 1989. The production and release to the atmosphere of 1,1,1-trichloroethane (methyl chloroform). *Atmos. Environ., 23*:2,663-2,665.

Midgley, P.M., and A. McCulloch. 1995. The Production and global distribution of emissions to the atmosphere of 1,1,1-trichloroethane (Methyl chloroform). *Atmos. Environ.* 29(14): 1601-1608.

Mitchell, B.R. 1980. *European Historical Statistics*, 2nd Rev. Ed. Facts on File, New York, U.S.A.

Mitchell, B.R. 1982. *International Historical Statistics, Africa and Asia.* New York University Press.

Mitchell, B.R. 1983. *International Historical Statistics, The Americas and Australasia.* Gale Research Co., Detroit, Michigan.

Moncrieff, J.B., I.J. Beverland, D.H. Ó Néill, and F.D. Cropley. 1998. Controls on trace gas exchange observed by a conditional sampling method. *Atmos. Environ.* 32: 3265-3274.

Mroz, E.J. 1993. Deuteromethanes: potential fingerprints of the sources of atmospheric methane. *Chemosphere, 26* (1-4):45-53.

Pinto, J., and M.A.K. Khalil. 1991. The stability of tropospheric OH during ice ages, interglacial epochs and modern times. *Tellus, 43B*:347-352.

Prinn, R.G., D. Cunnold, R.A. Rasmussen, P. Simmonds, F. Alyea, A. Crawford, P. Fraser, and R. Rosen. 1987. Atmospheric trends in methylchloroform and the global average for the hydroxyl radical. *Science,* 238:945-950.

Rasmussen, R.A., and M.A.K. Khalil. 1981. Interlaboratory comparison of fluorocarbons 11, 12, methylchloroform, and nitrous oxide measurements. *Atmos. Environ., 15*:1,559-1,568.

Seiler, W., A. Holzapfel-Pschorn, R. Conrad, and D. Scharffe. 1984. Methane emission from rice paddies. *J. Atmos. Chem., 1*:241-268.

Simpson, I.J., G.W. Thurtell, G.E. Kidd, M. Lin, T.H. Demetriades-Shah, I.D. Flitcroft, E.T. Kanemasu, D. Nie, K.F. Bronson, and H.U. Neue. 1995. Tunable diode laser measurements of methane fluxes from an irrigated rice paddy field in the Philippines. *J. Geophys. Res.* 100(D4): 7283-7290.

Singh, H.B. 1977. Preliminary estimation of average tropospheric HO concentrations in the northern and southern hemispheres. *Geophys. Res. Let., 4*:453-456.

Steele, L.P., E.J. Dlugokencky, P.M. Lang, P.P. Tans, R.C. Martin, and K.A. Masarie. 1992. Slowing down of the global accumulation of atmospheric methane during the 1980s. *Nature, 358*:313-316.

Stocks, B.J. 1991. The extent and impact of forest fires in northern circumpolar countries. In: *Global Biomass Burning, Atmospheric, Climatic, and Biospheric Implications*, J.S. Levine, ed.:197-202.

Taylor, J.A., and P.R. Zimmerman. 1991. Modeling trace gas emissions from biomass burning. In: *Global Biomass Burning, Atmospheric, Climatic, and Biospheric Implications* (J.S. Levine, ed.), 345-350.

Thompson, A.M., and R.J. Cicerone. 1986. Possible perturbations to atmospheric CO, CH_4, and OH. *J. Geophys. Res., 91*:10,853-10,864.

United Nations. 1977, 1978, 1980, 1982, 1984, 1986, 1989, 1990, 1991, 1992, 1994, 1996. *FAO Production Yearbook, vols. 30, 31, 33, 35, 37, 39, 42, 43, 44, 45, 47, 49.* Food and Agriculture Organization of the United Nations, Rome.

Vaghjiani, G.L., and A.R. Ravishankara. 1991. New measurement of the rate coefficient for the reaction of OH with methane. *Nature, 350*:406-408.

Wahlen, M., N. Tanaka, R. Henry, B. Deck, J. Zeglen, J.S. Vogel, J. Southon, A. Shemesh, R. Fairbanks, and W. Broecker. 1989. Carbon-14 in methane sources and in atmospheric methane: the contribution from fossil carbon. *Science, 245*:286-245.

Whalen, S.C., and W.S. Reeburgh. 1992. Interannual variations in tundra methane emission: a 4-year time series at fixed sites. *Global Biogeochem. Cycles, 6*:139-159.

8. Ruminants and Other Animals

D. E. Johnson[1], K. A. Johnson[2], G. M. Ward, and M. E. Branine[3]
[1]Department of Animal Sciences, Colorado State University, Fort Collins, CO 80523, U.S.A.
[2]Department of Animal Sciences, Washington State University, Pullman, WA 99164, U.S.A.
[3]Roche, Inc., Parker, CO 80134, U.S.A.

1. Introduction

Animal methane emissions originate from two sources. Methane is produced through microbial digestion of feeds in the animals digestive tract, where energy loss ranges from zero to nearly 12% of dietary energy. The second source is from microbial degradation of excreted residues in manure. This paper will explore explanations for the observed variation and provide estimates of total emissions. Emphasis will be placed on ruminants which account for approximately 95% of animal methane emissions.

2. Gastrointestinal Fermentation

Herbivores have developed gastrointestinal tracts to facilitate the breakdown of plant products into usable nutrients (Hackstein and van Alen, 1996). The development of a pre-gastric (ruminoreticulum) or hindgut (cecum) fermentation chamber allows these animals to house a population of facultative or obligately anaerobic microorganisms that can ferment plant biomass to nutritive products. These digestive structures create conditions advantageous to microbes and symbiotically of value to the host. The temperature is a relatively constant 39°C, there is a constant influx of nutrients, a constant efflux of end products and the pH and osmotic pressure are relatively well controlled (Bauchop, 1977). Alterations in any of these factors can alter the microbial population and impact fermentation efficiency.

The foregut in ruminants is a multicompartmental out-pocketing of the nonglandular portion of the stomach, lined with papillae. Copious salivary influx helps to maintain a relatively constant pH (6.5-6.8) and strong ruminal contractions provide constant mixing. Rate of passage of solids and liquid is relatively slow, allowing the fermentation of structural carbohydrates like cellulose. The hindgut of mammalian herbivores is usually sacculated and is lined with columnar epithelium and goblet cells (Stevens, 1988). The pH is not as well buffered as in the foregut and the rate of passage much more rapid. Thus the efficiency of cellulosic fermentation in the hindgut is lower than in the foregut (McBee, 1977).

The microbial population of the fore-stomach and hindgut is remarkably similar

Mohammad Aslam Khan Khalil   (Ed.)
Atmospheric Methane
© *Springer-Verlag Berlin Heidelberg 2000*

(McBee, 1977). Some of the resident bacteria, protozoa, and fungi possess the ability to hydrolyze the beta-linked carbohydrates of plant cell walls into constituent sugars through extracellular cellulase, hemicellulase, pectinase and xylanase activities. Microorganisms use the released cell solubles and constituent sugars to produce short-chain volatile fatty acids (VFA) which are absorbed and used by the animal for energy. Microbial fermentation of plant products also yields heat, carbon dioxide, methane, hydrogen in some cases, and microbial cells. These microbial cells, in turn, provides amino acids and vitamins, required by the host animal.

Methane results from the symbiotic relationship between those bacteria and protozoa that produce metabolic hydrogen as an end product and methanogens who couple hydrogen with carbon dioxide or formate. The result of this reaction yields ATP for the methanogens (Czerkawski, 1978) and the removal of metabolic hydrogen from the ruminal environment. The disposal of metabolic hydrogen by hydrogen-utilizers benefits fermentation by allowing more complete oxidation of plant carbohydrate, the production of acetic acid and greater yield of ATP (Wolin and Miller, 1988). The carbon dioxide and methane gases produced from this fermentation are removed from the ruminoreticulum by eructation or absorption into the blood stream and then subsequently cleared by the lungs (Hoernecke et al., 1964). Methane produced in the hindgut is also absorbed into the blood and removed by respiration (Murray et al., 1976). Only 13% of hindgut methane is eliminated via the anus (Murray et al., 1978).

While domestic ruminants (cattle and sheep) have very similar fractional losses of methane from a wide range of common feeds, other ruminant species may be unique. Deer consuming a browse and concentrate diet produced methane at rates that are one-fifth or less than expected from farm animals (Robbins, 1973). It is unclear whether this is diet- or species-related; however, the relatively fast rate of passage of ingesta noted in deer could account for lower methane loss (Milchunas et al., 1978). Methane losses in llamas were seven percent of gross energy intake, similar to sheep and cattle fed similar diets (Carmean et al., 1991). Additionally, chamber measurements of methane production by Sika deer in China have found methane losses of 6.6% of the dietary gross energy (Li et al., 1996).

Species which are hindgut fermenters include: fish, birds, rats, rabbits, pigs, horses, man and elephants. In many cases the evidence of fermentation occurring in the hindgut is indirect, being based on identification of volatile fatty acids and isolation of bacterial and protozoal species known to be involved with fiber fermentation. Rats, beavers, voles, porcupine, guinea pigs and rabbits also harbor important microbial populations which digest dietary cellulose to volatile fatty acids and appear able to absorb and use the fatty acids for their own energy needs.

The omnivores (pig and man) and herbivores (horses, elephants and ruminants) also have a cecal fermentation and are dependent on the fermentation end products to varying degrees to meet energy requirements. Pigs fed production diets (grains) have little necessity for the hindgut fermentation end products; however, when these animals are fed high roughage diets, volatile fatty acids from the cellulolytic fermentation can supply up to 24% of their energy supply (Yen and Nienaber, 1991). All fiber digestion occurs in the cecum or colon of the pony or horse regardless of the roughage-to-grain ratio in the diet (Hintz et al., 1971). The efficiency of cellulose digestion ranges from 39 to 50%. In any case, the gut fermentation of none of these species appears to make a significant contribution to global methane.

From the standpoint of the host animal, fermentation of otherwise unavailable fibrous plant carbohydrates to VFA and microbial protein is a highly desirable process.   However, the concomitant production of methane resulting from fermentation results in a net energetic loss to the animal. Therefore, understanding the factors associated with methane production is important for the development of effective nutritional strategies and livestock management practices to improve food and fiber production and energetic efficiency of ruminants in addition to providing information concerning the potential global warming consequences resulting from methane emissions to the atmosphere.

3. Methods for Measuring Ruminant Methane

Many methods exist to measure methane production from ruminant animals (Johnson and Johnson, 1995). The predominant method used is respiration calorimetry. Data collected using this technique provide the foundation of the current models and prediction equations. Descriptions and designs of whole animal calorimetry systems can be found in many sources (Flatt et al., 1958; Kleiber, 1958a; McLean and Tobin, 1987; Cammell et al., 1980; Johnson, 1986; Miller and Koos, 1988). In addition to whole animal systems, other variations include ventilated hoods or headboxes (Young et al., 1975; McLean and Tobin, 1987; Kelly et al., 1994) and face masks (Kleiber, 1958b; Liang et al., 1989).   The principle behind all of these methods is measurement of the concentration differential between outside air and air expired by the animal. When the concentration differential is multiplied by the corrected volume of airflow though the system, gas production or consumption is determined. Chamber calorimetry systems are both accurate and precise although animal behavior may be impacted in many systems. Head boxes and face masks suffer from the same limitations and can restrict eating behavior as well.

Tracer approaches have been developed to determine methane production from ruminants. Both [^{3}H-] and [^{14}C-] methane can be infused or injected intraruminally (Murray et al., 1975; France et al., 1993). Once the specific activity of sampled ruminal gases has been determined, methane production can be calculated. Non-isotopic tracer techniques include the use of sulfur hexafluoride gas (SF_6) emitting boluses placed in the ruminoreticulum (Johnson et al., 1994b). With this technique, air is continually sampled from around the mouth and nose of the animal into an evacuated canister. Once the concentration of methane and SF_6 in the canister has been determined, total emissions can be calculated as the product of the known release rate of SF_6 from the bolus and the ratio of CH_4 to SF_6. The advantage of this technique is that it allows unrestricted movement of the animal unlike the chamber approach.

Sulfur hexafluoride has been used as a gas tracer for measuring methane emissions from groups of animals (Marik and Levin, 1996; Johnson et al., 1997). Emissions from dairy cows in a barn or room can be made during a release of SF_6. The gas tracer provides an estimate of the ventilation rate of the room. Therefore, once room (barn) air is sampled and the concentrations of tracer and methane determined, methane emission can be calculated as for individual animals. Emissions from grazing or outdoor penned ruminants can also be made using a similar approach (Johnson et al., 1997). Sulfur hexafluoride gas released at a known rate upwind of grazing cattle will travel in the same wind plume as the methane the cattle release. Measuring the downwind concentration of tracer and methane allow emissions to be calculated. An important aspect to the use of this technique is the consistency of the environmental conditions. Alterations in the wind speed and direction can cause the plumes of methane and tracer to shift away from the sampling canisters. Also, large sources of methane upwind can overwhelm the source of interest.

Mass balance and micrometeorological approaches can be used to measure livestock methane (Denmead et al., 1998). Mass balance procedures require specialized enclosures for cattle so the vertical planes of methane gas entering and leaving the enclosure can be measured. Airflow is measured with conventional meteorological methods. Micrometeorological approaches also can be used, however, these techniques have limited usefulness because of the required equipment and other constraints.

4. Methane Emissions from Ruminants as Affected by Diet

Contrary to what shall be our conclusion concerning most world ruminants under practical circumstances, methane losses can be influenced by many dietary factors, including processing, digestibility or starch or lipid content, particularly when the diet is fed at restricted levels (Table 1). In general, fractional methane (%CH_4, expressed as a fraction of animal's dietary gross energy intake) will decrease as daily dietary intake increases. At restricted intakes %CH_4 yield will increase as digestibility of the diet increases. However, in commercial practice, highly digestible diets are seldom fed at restricted or at maintenance levels. Additionally, laboratory chamber measurements are often made with animals consuming less than they would eat under normal circumstances. Therefore, precautions are necessary in extrapolation of laboratory chamber measurements for prediction of emissions from animal agriculture, as practiced under commercial conditions.

Physical and chemical processing of low quality forages can improve nutritive value. Studies examining the effects of ammoniation and sodium hydroxide treatment of various cereal straws have shown either unchanged (Moss et al., 1994) or increased %CH_4 losses in proportion to the improved digestibility (Robb et al., 1979; Sundstol, 1982; Birkelo et al., 1986). Physical alteration of low quality forages by chopping, grinding, or pelleting also affect %CH_4 loss (Blaxter and McGraham, 1956; Wieser and Wenk, 1970). Generally, %CH_4 loss will be higher for more coarsely chopped diets compared to finely ground and pelleted diets. This effect is primarily due to increased intake and decreased digestibility resulting from an increased particulate passage rate and shorter ruminal retention times, decreasing %CH_4 loss on finely ground and pelleted diets. In studies examining the effects of grain processing methods on energy balance and %CH_4 loss, Johnson (1966) found greater methane energy loss when beef steers fed an 80% grain diet were provided cracked corn compared to a similar diet of steam-flaked corn. Moe and Tyrrell

(1977) observed no differences in %CH$_4$ loss among mixed diets consisting of either whole corn, cracked corn, or ground corn meal when fed at the same level of intake to non-lactating cows. However, for lactating cows, %CH$_4$, loss when expressed as a percentage of digestible energy intake, was lower for the cracked corn and ground corn meal diets and contributed to the increased energy availability ratios observed for these diets. With pelleted diets or increased dietary intake, the reduction in energy loss attributable to methane can at least partially compensate for the increased fecal energy loss observed under these conditions.

Table 1. Examples of measured methane yield by cattle and sheep fed a range of diets and treatments

References	Species	Basal diet and treatment	Methane yield (% CH$_4$)[a]
Birkelo et al., 1986	Cattle	Wheat straw	5.9
		Ammoniated straw	6.5
Blaxter and McGraham, 1956	Sheep	Chopped grass	7.6
		Medium ground/cubed grass	5.9
		Fine ground/cubed grass	4.6
Hutcheson, 1994	Cattle	8% roughage - ad lib.	3,2
		8% roughage - restricted	6.4
Blaxter and Wainman, 1964	Cattle	100% hay	7.5
		80% hay:20% flaked	7.8
		60% hay:40% flaked corn	8.2
		40% hay:60% flaked corn	8.1
		20% hay:80% flaked corn	5.7
		5% hay:95% flaked corn	3.4
Whitelaw et al., 1984	Cattle	faunated (85% barley diet)	11.5
		ciliate-free (85% barley diet)	6.7
Van der Honing et al., 1981	Lactating dairy cows	0% tallow or oil	5.9
		5% animal tallow	5.3
		5% soybean oil	5.0
Johnson, 1974	Sheep	Control (90% concentrate)	6.2
		Methane inhibitor (90% conc.)	3.7

a %CH$_4$ = methane enthalpy % total diet enthalpy.

Some studies have investigated the effects of feed preservation method, forage species, or maturity on %CH$_4$ loss. Armstrong (1964) observed when sheep were fed timothy grass hay harvested at different stages of maturity, %CH$_4$ loss decreased from

7.8% to 7.1% as the digestibility of the forage decreased from 71% to 57% with advancing maturity. Similar results have been obtained with timothy grass that was harvested at three stages of maturity and fed fresh-frozen, ensiled, or as hay (Sundstol and Ekern, 1976) and for clover grass silage ensiled at an immature and a mature stage of development (Sundstol et al., 1979). Corbett et al. (1966) observed a greater %CH$_4$ loss for late season compared to early season herbage. In this study, the greater methane yield associated with the late season herbage was related to a lower content of soluble carbohydrate and a greater proportion of cellulose despite a negligible difference in digestible energy between early and late season herbages. Varga et al. (1985) examined differences in energy balance between alfalfa silage and orchardgrass silages and observed a decreased %CH$_4$ loss (5.8 vs. 6.3) for cattle consuming alfalfa silage. Neither wilting nor inoculation of ryegrass silages changed CH$_4$ losses (Fyan et al., 1996). Tyrrell and Varga (1985) indicated a slight, but significant decrease in methane energy loss from non-lactating dairy cows fed restricted levels of diets containing 40% shelled corn that had been dried as compared to ensiled (8.7 vs 8.85 %CH$_4$ loss, respectively).

In many studies the common denominator that apparently exists for methane production is the relative proportions of soluble and structural carbohydrates within the plant tissue. The ratio of soluble to structural carbohydrate will decrease as a plant matures and likewise is generally lower for grass species compared to legumes (Van Soest, 1982). Varga et al. (1985) suggests that the shifts that occur in the ruminal fermentation either inhibit or enhance ruminal methanogenesis are a result of the type of carbohydrate existing within the plant. The major portion of the digestible organic matter in orchardgrass is structural carbohydrate, whereas a major fraction of the digestible organic matter from alfalfa silage is soluble carbohydrate.

Under ruminal conditions that favor propionate-yielding fermentation pathways, propionate serves as a competitive hydrogen sink, decreasing methanogenesis. Conversely, fermentation pathways that favor acetate formation will generate hydrogen and stimulate methanogenesis. The latter type of fermentation is typically characteristic of bacterial degradation of structural carbohydrate such as cellulose. These observations agree with results of Moe and Tyrrell (1980) who indicated that when cows are consuming diets at high levels of intake, the digestion of cellulose yields 3 to 5 times more methane per unit hexose fermented as compared to soluble carbohydrate and starch. In summary, the results of these studies suggest that methane yield from forage-based diets can be variable with stage of maturity, species, and methods of processing. However, forages fed ad libitum, will likely yield methane losses in a fairly narrow range of 5.5 to 7.0% except when finely ground for pelleting.

4.1 Effects of High Grain Diets on Methane Yield

In countries with excess agricultural production, animal feeding and management optimization practices for some segments of the industry utilize high proportions of grain and other concentrate feeds to maximize energy intake required for achieving high levels of production. Typically, U.S. beef feedlot diets contain about 10% roughage and 90% grain, or other concentrate feeds. In their classic study, Blaxter and Wainman (1964) indicated that as the percentage of steam-flaked corn in the diet increased, %CH$_4$ loss increased until the corn was incorporated into the diet at a level of 60-80%, at which point %CH$_4$ loss abruptly declined. For cattle and sheep fed at the higher intake level, %CH$_4$ loss ranged from 7.5% and 6.5% for the 100% hay diets to 3.4% and 3.7% for the all-grain diets, respectively. In general, where diets in excess of 80% grain (usually corn-based diets) have been fed at levels of intake substantially above maintenance, %CH$_4$ loss are typically below 5% (Johnson, 1966; Byers, 1974; Wedegaertner and Johnson, 1983). Steers consuming typical U.S. feedlot corn-based diets lost 2 to 2.5% of diet energy as CH$_4$ (Hutcheson, 1994). These and other experiments (Abo-Omar, 1989; Carmean, 1991), when extrapolated to industry intake levels, suggest an average of 3% methane loss over the feeding period. The percentage of methane loss for high grain feedlot diets cannot be accurately predicted from general relationships such as the Blaxter and Clapperton equation (Blaxter and Clapperton, 1965; see section 5).

Barley may be a special case for evaluating %CH$_4$ loss. As indicated by the results of Hashizume et al. (1967) and Whitelaw et al. (1984), %CH$_4$ losses from high-grain, barley-based diets ranges from 6.5% to 12%. These values are higher than would be expected when feeding a comparable corn-based diet at similar levels of intake. A primary reason for this difference in %CH$_4$ loss might be the greater fermentability of the starch of barley. Theurer (1987) found ruminal starch digestibility coefficients of 93% and 73% for barley and corn, respectively.

Low methane yields from high-grain diets result from factors that create a more hostile environment for hydrogen-producing microbes or the methanogenic bacteria that utilize the hydrogen (Van Kessel and Russell, 1996). Van Soest (1982) has suggested some of these factors to be an increased rate of passage and digestion, depression in rumination, decreased ruminal pH, and reduction or elimination of certain populations of ruminal protozoa.

Rumen protozoa can have an important role in ruminal methane production. Krumholtz et al. (1983) observed that methanogens were directly adhered to protozoa and that protozoa supported methanogenic activity of the attached methanogens via an interspecies transfer of metabolic hydrogen. Whitelaw et al. (1984) observed reductions of approximately 50% in methane yield for defaunated steers fed a high-grain, barley-based diet compared to faunated steers. However, the form of forage fed to animals did not reduce methane losses (Itabashi et al., 1984).

4.2 Effect of Dietary Fat on Methane Yield

Supplemental fats are often fed to high-producing dairy cows for the primary purpose of increasing energy density of the diet. Swift et al. (1948) observed decreased %CH$_4$ loss from sheep that were fed isocaloric diets containing from 3% to 8% fat. Van der Honing et al. (1981) also found decreases of 10 and 15% in methane yield from dairy cows receiving supplemental fat in the form of either animal tallow or soybean oil, as compared to non-supplemented controls. Haaland (1978) observed reductions in methane production for beef steers fed protected tallow. The reduction in methane yield in this case, as in some others, was attributed to decreased fermentable substrate rather than a direct inhibition of methanogenesis.

The practice of providing polyunsaturated fatty acids in the diet for the express purpose of reducing methane has been studied.   Czerkawski et al. (1966) demonstrated that dietary additions of long chain, polyunsaturated fatty acids reduced ruminal methanogenesis by providing an alternative receptor for metabolic hydrogen, rather than the reduction of carbon dioxide to methane.   In many plant tissues, particularly grasses, unsaturated fatty acids account for a large proportion of the total fatty acids.   However, the percentage of the total metabolic hydrogen produced during the fermentation process that is actually used for the biohydrogenation of endogenous polyunsaturated fatty acids is only 1% to 2% (Czerkawski, 1972). This is a very small amount compared to the typical 48% used for reduction of carbon dioxide to methane, 33% used for VFA synthesis, and 12% used for bacterial cell synthesis (Czerkawski, 1988).

4.3 Effects of Ionophores and Other Inhibitors on Methane Yield

The energetic loss represented by ruminal methane production has stimulated a considerable research effort directed at minimizing methanogenesis. This endeavor has been based on the premise that by decreasing methane energy loss, more energy will be available for productive purposes. During the past several years many compounds have been developed that have shown a suppressive effect on ruminal methanogenesis.   In general, these compounds can be subdivided into two groups: polyhalogenated methane inhibitors and polyether ionophores. The polyhalogenated inhibitors include the hemiacetal of chloral and starch (Johnson, 1972), chlorinated fatty acids (Czerkawski and Breckenridge, 1975), and bromochloromethane (Sawyer et al., 1974).   All of these compounds reduced %CH_4 loss by 20% to 80% compared to untreated animals.   Responses in animal performance to these compounds have been variable although improvements in digestibility and metabolizability (Sawyer et al., 1974), feed efficiency (Trei et al., 1971), and average daily gain (Johnson et al., 1972) have been reported.   The use of polyether ionophores such as monensin and lasalocid have found widespread acceptance for improving growth and feed efficiency in cattle and sheep while the polyhalogenated methane inhibitors have not been commercially successful.   The antimethanogenic properties of ionophores are not the result of a direct toxic effect on the methanogenic bacteria (Van Nevel and Demeyer, 1977); rather, methane losses are decreased in two ways. First, 5% to 6% less feed is consumed and fermented per unit of production, thus reducing the corresponding methanogenesis (Goodrich et al., 1984). Second, several studies have demonstrated suppression of methane yield when feeding monensin (Joyner et al., 1979; Thornton and Owens, 1981; Benz and Johnson, 1982; Wedegaertner and Johnson, 1983) or lasalocid (Delfino et al., 1988) through alteration of ruminal fermentation patterns shifting hydrogen away from methane; however. suppression of methane loss per unit of feed consumed may not persist for long periods.

A problem common to using either the polyhalogenated methane inhibitors or ionophores for the suppression of methane production has been the phenomena of an apparent microbial adaptation and return to baseline methane levels after a relatively short period of exposure to these compounds.   Johnson (1974) fed sheep a pelleted diet with and without a halogenated methane inhibitor and observed (initial) methane depression in inhibitor-fed sheep to 36% of controls. By day 30, however, methane production by inhibitor-fed sheep had increased to 88% of control values. Considering the sum of methane and hydrogen production, the inhibitor decreased total gas energy losses to 49% of control initially with a significant rise to 88% of control values by day 30. Similarly, when beef steers were fed a 70% cracked corn basal diet with or without the ionophores (monensin or lasalocid), methane loss for the monensin and lasalocid fed steers initially decreased but returned to control values by day 12 (Rumpler et al. (1986). Research by Abo-Omar (1989) and Carmean and Johnson (1990) and others summarized by Johnson et al. (1994a) have indicated an initial 25% to 30% decrease in methane production in response to singly or alternatively fed ionophores with high-grain diets. However, as shown by Rumpler et al. (1986), methane production essentially returned to control levels after 16 days of ionophore supplementation. Development of microbial resistance to tetronasin and other ionophores has also been seen in vitro (Newbold et al., 1993). The results of these studies indicate that overall feeding period reductions in methane yield (%CH$_4$ loss) by feeding ionophores to cattle consuming high grain diets may be small. The decrease in total methane production per animal/day will primarily result from the 5% to 6% lower feed consumption commonly observed for cattle fed ionophores and only secondarily from a methane yield savings.

5. Empirical Predictions of Ruminant Methane Production

The Blaxter and Clapperton (1965) equation is the source of most of the methane emissions estimates used in inventorying the contribution of ruminant livestock to the global livestock budget. This equation relates %CH$_4$ loss to level of intake (LOI) and percent digestible energy (DE) of the diet fed at maintenance. This equation, however, is not easily applied as originally intended. The LOI depends on assessing the fasting heat production, which was loosely defined for animal age categories. Also, the DE of the diet was determined at maintenance level intake. Neither age nor DE at maintenance is commonly reported in the studies on methane production. In addition, the original publication contains a sign error preceding the LOI term. Blaxter and Clapperton's (1965) equation, with sign corrected, is:

$$\%CH_4 \text{ loss} = 1.30 + 0.112 \text{ DE} + \text{LOI} (2.37 - 0.05 \text{ DE}).$$

A database was created to examine the ability of the Blaxter and Clapperton equation to accurately predict methane losses from published reports. Available data relating %CH$_4$ loss to variables such as LOI as multiples of maintenance, percent concentrate in the diet and percent DE in the diet for beef cattle and sheep were summarized. Extremes in %CH$_4$ ranged from 2% to 12% in beef cattle diets and both extremes occurred with diets of 90% or greater concentrates. These large variations for somewhat comparable diets cause difficulties in empirical prediction of %CH$_4$ loss. Variations observed in our database (Figure 1) were compared to variations predicted by the equation of Blaxter and Clapperton (1965).

The equation of Blaxter and Clapperton predicts a considerably narrower range of methane loss than those observed for beef cattle (Figure 1) and sheep, suggesting their equation is either inaccurate or does not account for other important factors that influence methane losses. Certainly, their equation was based on a small subset of data that did not cover as wide a range of dietary conditions. Regression coefficients also are generally biased low in proportion to errors of measurement of their independent variable; i.e., intake and DE in this case.

The relationship of % CH_4 loss to LOI or both LOI and digestible energy for beef cattle is shown in Table 2. Difficulties in predicting %CH_4 loss exacerbate with increasing percent concentrate, as reflected by r^2 values of one-third or less for high vs. low concentrate diets (Table 2). Methane losses from high concentrate diets are highly variable and are frequently between 2 and 5 %CH_4. Under most circumstances, increasing intake by one multiple of maintenance will decrease methane yield by 1.8 and 1.6 percentage units of GEI in cattle and sheep, respectively.

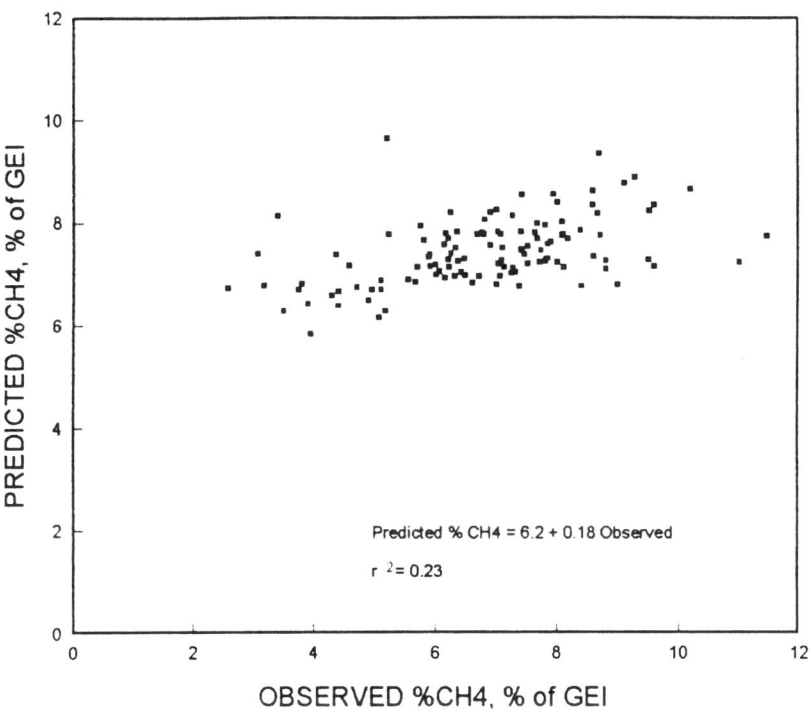

Fig. 1. Predicted vs. Observed methane production (using Blaxter and Clapperton, 1965, equation).

Table 2. Methane relationship to level of intake (LOI) and percent diet digestible energy (DE) for beef cattle[a]

Species/ % Forage in diet	%CH$_4$ vs. LOI	r^2	Equation No.
80-100	9.12 - 1.71 LOI	.57	1
20-80	9.80 - 1.84 LOI	.27	2
< 20	8.85 - 2.14 LOI	.17	3
all diets	9.49 - 1.84 LOI	.31	4
80-100	5.50 - 2.25 LOI + 0.06 DE	.72	5
20-80	11.33 - 1.81 LOI - 0.02 DE	.27	6
< 20	9.90 -1.54 LOI - 0.02 DE	.18	7
all diets	10.32 - 1.79 LOI - 0.01 DE	.31	8

[a] n = 118 experimental means

The relationship of %CH$_4$ loss to DE is much more variable than the relationship of %CH$_4$ loss to LOI. The high variability in the relationship of %CH$_4$ to digestible energy reflects the conflict between the increasing amounts of fermentable substrate versus the changing patterns of fermentation, (i.e., decreasing acetate-to-propionate ratios occurring as the percent digestible energy increases).

6. Estimates of Methane from U.S. Livestock

Methane output from U.S. livestock were estimated at 6.27 Tg/yr from the 1994 census (Table 3). Approximately 70% of these emissions were from beef cattle and 26% from dairy cattle. To develop these estimates, the approximately 100 million cattle in the U.S. were divided into 20 classes to describe numbers, body weights, death losses, days in class, feed consumption and methane output (Johnson and Johnson, 1995). The 6.0 Tg production of CH$_4$ from cattle compares well to independent estimates of 5.63 (Gibbs and Baldwin, 1994) and 6.19 Tg (Westberg et al., 1995). The estimates of Crutzen et al. (1986) adjusted to the current census, are similar although higher for beef and lower for dairy cattle. Direct measurements of methane emissions per cow were made from a 100 cow dairy barn with measurements occurring over 4 years in Eastern Ontario. The 600 kg dairy cows, yielding 29 kg of milk/d in this study produced 527 L CH$_4$/hd/d (Kinsman et al., 1997). Recent Washington State University SF$_6$ tracer measurements (Johnson et al., 1997) and European analyses (Kirchgessner et al., 1995; Vermorel, 1995; Marik and Levin, 1996) likewise substantiate these higher dairy cow production rates.

Table 3. Livestock methane emissions in the United States in 1996

	Average no. head x 10⁶	Methane % of GE	Methane Tg/yr.
Beef cattle			
Cows	35.2	6.2	2.38
Feedlot	10.2	3.5	0.35
Other	42.2	6.5	1.50
Dairy cattle			
Cows	9.2	5.8	1.14
Replacements	8.3	6.5	0.44
Sheep and goats	10.0	6.0	0.08
Swine	61.0	0.6	0.09
Horses	6.1	2.5	0.11
TOTAL			6.09

Beef and dairy extrapolated from Johnson et al., 1993; other species extrapolated from Crutzen et al., 1986.

To predict methane emissions from livestock, estimates are needed of feed intake adequate to support assumed functions of maintenance, growth, pregnancy, lactation and work. Estimated dry matter intake then can be multiplied by selected factors for gross energy and methane as a percent of gross energy. Default values to use at present would be those presented above for U.S. cattle (except for finishing or feedlot phase which only are valid for very high grain diets, not fed in most other areas of the world).

7. Estimates of Global Methane Production from Livestock

Earlier global animal methane emission estimates were 100 and 160 Tg/yr (Ehhalt, 1974; Ehhalt and Schmidt, 1978) with later estimates made in the early 1980s being similar, 90 Tg (Sheppard et al., 1982), 120 Tg (Khalil and Rasmussen, 1983), and 115 Tg (Blake, 1984). The extensive investigation and summary of 1983 livestock methane emissions compiled by Crutzen et al. (1986) resulted in lower estimates, even though animal numbers were increasing by about 1.2%/yr during this period. Crutzen et al. (1986) combined literature data and expert opinion to estimate the typical fraction of dietary energy lost as methane (5.5% to 9.0%) and the diet type consumed by animal classes with United Nations Food and Agricultural Organization estimates of animal numbers to arrive at global sums. They concluded that domestic animals emit 74 million metric tons annually with an additional four tons contributed by wild animals. Of the domestic animals, cattle, because of numbers, high feed intakes, combined with their extensive gastrointestinal tract fermentation, are the major contributors, producing 74% of animal emissions. Another bovine, the water buffalo, was estimated to contribute another eight percent. Smaller ruminant species, sheep and goats, produce approximately thirteen percent, while camels produce about one percent of animal emissions. Horses and mules appear to be the principal nonruminant methane emitters (two percent), followed by pigs (one percent). These authors also estimated that wild ruminants globally produce about 4 Tg of additional animal methane.

A more recent examination by largely independent methods (Gibbs and Johnson, 1993) estimated that the 1990 global livestock animal methane emissions were 79 Tg, supporting the Crutzen et al. (1986) estimates. While there are many differences in estimated emissions within groups of animals made by the two investigations, total methane emissions are very similar. Increased emissions largely result from the 5% increase in global numbers of cattle and buffalo between 1983 and 1990.

Livestock methane emissions appear to have increased from 17 to 79 Tg, i.e. nearly five-fold over the last century (Table 4), although the rate of increase has been slower during recent years. World cattle numbers have remained at approximately 1.3 billion since 1990 (FAO, 1996) although some increases have occurred in numbers of buffalo and goats. Livestock, particularly cattle, have contributed to the doubling of total, global atmospheric methane which has occurred during this time period; however, this is not the sole factor attributing to this increase.

The principal uncertainty factors and our subjective estimate of the degree of uncertainty are number of animals in each country (± 10%), their distribution into age, weight, and production class, the amount of feedstuffs consumed daily (± 25%), and the methane yield per unit of diet for each class (± 20%). Many errors may cancel, such as the beef-dairy tradeoff for the Crutzen et al. (1986) estimates mentioned previously. Internationally, two areas particularly warrant attention. One is the estimate of 6.3 Tg emissions by water buffalo with essentially no supporting methane measurements and the methane yield from cattle in developing countries (Crutzen et al., 1986). Currently this value is estimated to be nine percent but this estimate may be approximately 30% high judging from citations in our data base, particularly for emissions from animals fed low quality diets. An in-country examination of CH_4 emissions by livestock in Ukraine (Martinez et al., 1996) suggested that the Gibbs and Johnson (1993) estimate was 30% high because of inaccurate assumptions in the number of young cattle and breeding bulls.

Table 4. Changes in animal population and methane emissions during the last century

Item	1890[a]	1983[a]	1990[b]	1996
Population (x 10^9)				
Cattle and buffalo	0.34	1.35	1.42	1.46
Sheep and goats	0.84	1.61	1.75	1.71
Methane (Tg/yr)				
Cattle and buffalo	12	61	66	68
Sheep and goats	4	9	10	10
Other domestic	1	4	3	3
Total (Tg/yr)	17	74	79	81

[a] Extrapolated from Crutzen et al. (1986).
[b] Condensed from Gibbs and Johnson (1993).

8. Methane Emissions from Livestock Manure

Animal excreta can be further fermented to produce methane. This fermentation may be extensive, such as in anaerobic fermenters designed produce methane for energy, or it may be inconsequential, as when feces are quickly dried and/or burned. The potential contribution of animal excreta to the global methane pool has only recently been examined, and this examination is far from complete.

Intensive livestock production, such as practiced by the cattle feeding industry and in all phases of dairy, swine, and poultry production such as occur in the U.S., results in the accumulation of manure. This mass of manure, whether stockpiled or stored as a slurry in a lagoon, creates a suitable anaerobic environment for methane production.

Considerable work has been done to determine the potential methane production from organic matter (OM) in different manures used as substrate in anaerobic digesters to produce methane as a power source. Hashimoto et al. (1981) showed a potential of 0.33 and 0.17 m^3CH_4/kg OM, for manure from cattle fed forage/grain mixtures or pasture forages, respectively. Likewise, Morris (1976) determined methane production from dairy cattle manure to be 0.24 m^3 CH_4/kg OM.

The majority of the world's livestock are free-ranging. Manure deposited in these locations does not undergo the same degree of fermentation as is exhibited for manure produced in the confined livestock production systems. Research by Lodman et al. (1993) and Williams (1993) indicate that manure deposited in this manner is likely to produce less than one percent of the potential methane that would result from anaerobic lagoon disposal. Maximum loss rate was three percent of potential initially which decreased rapidly over the first and second day of drying and/or aeration.

Estimation of global methane production from manure is difficult due to uncertainty about manure handling practices. Manure from non-confined animals, representing the majority of the world's ruminants, most likely results in minor methane emissions. Likewise, manure handling practices designed to collect methane for fuel energy likely contribute only small amounts through leakage to the global methane pool, as indicated by research in China (Khalil et al., 1990). The principal problem is anaerobic manure storage or anaerobic lagoon disposal of animal manure (including nonruminant species, i.e., poultry and swine). Anaerobic lagoons can ferment almost all of the non-lignin organic matters thus releasing close to the maximum potential methane (Safley, 1989). Important modifying factors to the extent of anaerobic fermentation are temperature and moisture content of the manure (Safley and Westerman, 1987). Previous estimates of 28 Tg/yr (Safley et al., 1992) of methane emissions from livestock manure have been decreased reflecting the lower contribution resulting from pastured cattle (EPA, 1994).

Given the above assumptions along with others concerning numbers and routes of manure disposal, U.S. livestock produce about 2.5 million tons of methane (Table 5). Unlike gastrointestinally produced methane, the biggest share now comes from swine and poultry production (54%). Globally, swine and poultry account for 47% of the estimated 13.8 million tons of methane emissions from livestock manures.

Table 5. Estimates of methane emissions from livestock manure[a]; Tg/yr

	Tg/yr	
Species	United States	Global
Dairy	0.71	2.89
Beef	0.19	3.16
Swine	1.11	5.29
Sheep and goat	--	0.71
Poultry	0.23	1.28
Other	0.24	0.51
Total	2.48	13.84

[a] Data from the EPA, 1994

9. Amelioration Potential

The prior literature review and knowledge of management systems suggest several ways to reduce methane emissions (see IPCC, 1995 for more details) with many that could be initiated or more widely implemented with existing technologies. Implementation is complicated, however, by the many considerations of economic and social factors that would be involved in determining their adoption.

When cereal grain feeding to livestock is economically advantageous, several improvements result from the addition of starch-based feeds to animal diets. High starch diets shift gastrointestinal microbial populations to a fermentation utilizing hydrogen for the production of propionate, decreasing methane yield. Additionally, rate of animal gain efficiency and productivity is maximized, thereby decreasing total feed fermented and consumed, which all minimize methane per unit of product produced. This effect can be further enhanced by feeding more slowly fermented starches resulting in more starch bypassing the rumen to the small intestine where digestion of the starch proceeds without methane production.

Any feeding system that increases animal productivity results in lower methane production per unit of product (i.e., milk, meat, work, or wool). Methane emission per per unit of product will decrease and, in economies saturated with livestock products, will lower total livestock methane. Several feeding and management strategies have been effective in this regard. Providing high quality forages, harvesting, preserving, and feeding or grazing forage with high soluble carbohydrate contents, shifts fermentation toward propionate, away from acetate with its obligatory transfer of excess hydrogen to methane.

Eliminating nutrient deficiencies by the addition of a protein, mineral and vitamin supplements to an unbalanced forage diet can dramatically improve productivity (product per unit of time and per animal) by decreasing the fraction of feed and corresponding methane wasted in the nonproductive maintenance function. This is of particular importance to developing countries.

Additionally, many performance enhancers have been developed such as implants, growth promotants, additives, and vaccines which result in improved rate of gain, composition of gain (low fat), health, and reproduction. Many improvements stem from the application of genetics including implementation of crossbreeding programs, matching genotype to environment, selecting for improved body composition, and efficiency of feed to product conversion, all effective available tools.

Eliminating anaerobic pit or lagoon storage of manure or enhancing their conversion to collectors of biogas energy would be doubly effective by reducing emissions and replacing the burning of other energy sources.

There are several technological possibilities for future reduction of methane losses from livestock. Persistent ionophores or antibiotics that inhibit methane over the entire feeding period or production cycle would reduce ruminant methane by 20% to 25%. Protozoal inhibitors would decrease methane in many circumstances.

Enhancing ruminal acetogens is another intriguing possibility. This group of microbes produces acetic acid from excess hydrogen and carbon dioxide rather than producing methane as is the case with methanogens. They exist in the rumen as a minor species, predominate in the gut of some termites, and may be important in the lower gut of several animal species. Developing ways to make them competitive in the rumen or transferring the acetogenesis genes to already successful ruminal organisms could be very helpful to animal efficiency and the environment.

10. Conclusions

Livestock are significant contributors to global methane and producing approximately 81 Tg/yr from microbial activity in their gastrointestinal tracts. Approximately, another 14 Tg/yr is estimated to enter the atmosphere from livestock manure decomposition. Most of this methane originates from domestic ruminants of which cattle and domestic buffalo are the most important. Methane generated in the rumen, as a percentage of feed energy consumed, varies from 2% to 12%. On average, highly digestible diets fed ad libitum and containing large amounts (80-90%) of grain are at the lower end of the range (3-4%), while the average for all other forage and mixed diets is similar (6-6.5%) when fed to cattle and sheep ad libitum. Restricted feeding of highly digestible diets can result in the very high 10-12% methane yield, but this feeding strategy rarely occurs in commercial production..

Livestock-productivity-enhancing additives, supplements, or management practices can markedly decrease the methane per unit of livestock product produced. They are the major mitigation opportunity. Growth promotants, nutrient supplements, ionophores, vaccines, and improved reproductive rates are all included in this category. The world wide trend is towards improved feeding and management practices which will probably reduce methane output per unit of product as well as total methane.

At present, extensive literature on methane emissions as affected by diets exists for the developed countries, but little or no data are available for cattle or buffalo or other ruminants fed diets typical for developing countries. The value of 6% of feed energy is probably the best default value for those countries.

More precise estimates for the large populations of ruminants in developing countries requires more analysis of existing data and use of expert opinion to estimate feed types, intakes, and animal productivity. Actual data on CH_4 emission by these animals are needed.

Acknowledgment. This work was supported in part by NASA Interdisciplinary Research Program in Earth Sciences.

References

Abo-Omar, J. M. 1989. Methane losses by steers fed ionophores singly or alternatively, Ph.D. Dissertation, Colorado State University, Fort Collins.

Armstrong, D. G. 1964. Evaluation of artificially dried grass as a source of energy for sheep, II. The energy value of cocksfoot, timothy, and two strains of ryegrass at varying stages of maturity. *J. Agric. Sci. (Camb.), 62*:399.

Bauchop, T. Foregut fermentation. 1977. In: *Microbial Ecology of the Gut* (R.T.J. Clarke and T. Bauchop, eds.), Academic Press.

Benz, D. A., D. E. Johnson, 1982. The effect of monensin on energy partitioning by forage fed steers. *Proc. West Sec. Amer. Soc. Anim. Sci. 33:*60.

Birkelo, C. P., D. E. Johnson, and G. M. Ward. 1986. Net energy value of ammoniated wheat straw. *J. Anim. Sci. 63:*2044.

Blake, D. R. 1984. Increasing concentrations of atmospheric methane. Ph.D. Dissertation, University of California at Irvine. Pp. 213.

Blaxter, K. L., and N. McGraham. 1956. The effect of the grinding and cubing process on the utilization of the energy of dried grass. *J. Agric. Sci. (Camb.), 47*:207.

Blaxter, K. L., and F. W. Wainman. 1964. The utilization of the energy of different rations by sheep and cattle for maintenance and fitting. *J. Agric. Sci. (Camb.), 63*:113.

Blaxter, K. L., and J. L. Clapperton. 1965. Prediction of the amount of methane produced by ruminants. *Brit. J. Nutr., 19:*511.

Byers, F. 1974. The importance of associative effects of feeds on corn silage and corn grain net energy values. Ph.D. Dissertation, Colorado State University, Fort Collins.

Cammell, S. B., D. E. Beever, K. V. Skelton, and M. C. Spooner. 1980. The construction of open circuit chambers for measuring gaseous exchange and heat production on sheep and young cattle. *Lab. Pract. 30:*115.

Carmean, B. R. 1991. Persistence of monensin effects on nutrient flux in steers. M.S. Thesis, Colorado State University, Ft. Collins.

Carmean, B. R., and D. E. Johnson. 1990. Persistence of monensin-induced changes in methane emissions and ruminal protozoa numbers in cattle. *J. Anim. Sci., 65:(Supp. 1):*517.

Carmean, B. R., K. A. Johnson, and D. E. Johnson. 1991. Maintenance energy requirements of the llama. *Am. J. Vet. Res., 53:*1696.

Corbett, J. L., J. P. Langlands, I. McDonald and J. D. Pullar. 1966. Comparison by direct animal calorimetry of the net energy values of an early and a late season of growth of herbage. *Anim. Prod. 8:*13.

Crutzen, P. J., I. Aselmann, and W. Seiler. 1986. Methane production by domestic animals, wild ruminants, and other herbivorous fauna and humans. *TELLUS, 38B:*271.

Czerkawski, J. W. 1972. Fate of metabolic hydrogen in the rumen. *Proc. Nutr. Soc., 31:*141.

Czerkawski, J. W. 1978. Transfer of metabolic hydrogen in the rumen. In: J. H. Moore and A. F. Rook, Eds. The Hannah Research Institute 1928-1978. The Hannah Research Institute, Ayr, Scotland.

Czerkawski, J. W. 1988. *An Introduction to Rumen Studies*, Pergamon Press, Oxford

Czerkawski, J. W. and G. Breckenridge. 1975a. New inhibitors of methane production by rumen micro-organisms. Development and testing of inhibitors in vitro. *Brit. J. Nutr.*, *34*:429.

Czerkawski, J. W., and G. Breckenridge. 1975b. New inhibitors of methane production by rumen-micro-organisms. Development and testing of inhibitors in vitro. *Brit. J. Nutr.*, *34*:447.

Czerkawksi, J. W., K. L. Blaxter, and F. W. Wainman. 1966. The metabolism of oleic, linoleic, and linolenic acids by sheep with reference to their effects on methane production. *Brit. J. Nutr., 20*:349.

Delfino, J., G. W. Mathison, and M. W. Smith. 1988. Effect of lasalocid on feedlot performance and energy partitioning in cattle. *J. Anim. Sci.*, *66*:136.

Denmead, O. T., L. A. Harper, J. R. Freney, D. W. Griffeth, R. Leuning, and F. R. Sharpe. 1998. A mass balance method for non-intrusive measurements of surface-air trace gas exchange. *Atmos. Environ.,32* (21):3679-3688.

Ehhalt, D. H. 1974. The atmospheric cycle of methane. *TELLUS, 26*:58-69.

Ehhalt, D. H., and U. Schmidt. 1978. Sources and sinks of atmospheric methane. *PAGEOPH, 116*:452-464.

EPA. 1994. International Anthropogenic Methane Emissions: Estimates for 1990, EPA-230-R-93-010. U.S. EPA, Washington, D.C.

Fyan, T., D. C. Patterson, F. J. Gordon, and M. G. Porter. 1996. The effects of wilting of grass prior to ensiling on the response to bacterial inoculation. 1. Silage fermentation and nutrient utilization over three harvests. *An. Sci. (Pencaitland) 62*:405.

FAO. 1996. Food and Agricultural Production Yearbook, Vol. 50.

Flatt, W. P., P. J. Van Soest, J. F. Sykes, and L. A. Moore. 1958. A description of the energy metabolism laboratory at the U.S. Department of Agriculture, Agricultural Research Center in Beltsville, MD. In: G. Torbeck and H. Aersoe (Eds.) 1. A Symposium on energy metabolism: principles, methods and general aspects. Statens Husdyrbrugsudvalg, Copenhagen, Denmark. 53.

France, J., D. E. Beever, and R. C. Siddons. 1993. Compartmental schemes for estimating methanogenesis in ruminants from isotope dilution data. *J. Theor. Biol. 164*:207.

Gibbs, M. J., and D. E. Johnson. 1993. Livestock Emissions. In: *International Anthropogenic Emissions of Methane*, EPA 430-R-93-003. U.S. EPA, Washington, D.C.

Gibbs, M., and R. L. Baldwin. 1994. Inventory of U.S. greenhouse gas emissions and syncs. EPA 230-R-94-014.

Goodrich, R. D., J. E. Garrett, D. R. Gast, M. A. Kirick, D. A. Larson, and J. C.. Meiske. 1984. Influence of monensin on the performance of cattle. *J. Anim. Sci.*, *58*:1,484.

Haaland, G. L. 1978. Protected fat in bovine rations. Ph.D. Dissertation, Colorado State University, Fort Collins.

Hackstein, J. H. P., and T. A. van Alen. 1996. Fecal methanogens and vertebrate evolution. *Evolution. 50(2)*:559-572.

Hashimoto, A. G., V. H. Varel, and Y. R. Chen. 1981. Ultimate methane yield from beef cattle manure: Effect of temperature, ration instruments, antibiotics and manure age. *Agricultural Waste*, *3*:241, 1981.

Hashizume, T., H. Morimoto, T. Hayer, M. Itch, and S. Tanabe. 1967. Utilization of the energy of fattening rations containing ground or steam-rolled barley by Japanese Black Breed Cattle. *4th Energy Symp., EAAP, 12*:261.

Hintz, H. F., D. E. Hogue, E. F. Walker, J. E. Lowe, and H. F. Schryver. 1971. Apparent digestibility in various segments of the digestive tract of ponies fed diets with varying roughage-grain ratios. *J. Anim. Sci., 32*:245-248.

Hoernecke, H., W. F. Williams, D. R. Waldo, and W. P. Flyatt. 1964. Composition and absorption of rumen gases and their importance for the accuracy of respiration trials with tracheostomized ruminants. Symposium on Energy Metabolism, EAAP.

130   Johnson et al.

Hutcheson, J. P. 1994. Anabolic implant effects on body composition, visceral organ mass and energetics of beef steers. Dissertation, Colorado State University.

IPCC. 1995. Mitigation options in Agriculture, Coal, V. (ed.) Cambridge University Press.

Itabashi, H., T. Kobayashi, and M. Matsumoto. 1984. The effects of rumen protozoa on energy metabolism and some constituents in rumen fluid and blood plasma of goats. *Jap. J. Zootechnic Sci. 55*:248.

Johnson, D. E. 1966. Utilization of flaked corn by steers. Ph.D. Dissertation, Colorado State University, Fort Collins.

Johnson, D. E. 1972. Effects of a hemiacetal of chloral and starch on methane production and energy balance of sheep fed a pelleted diet. *J. Anim. Sci., 35:*1,064.

Johnson, D. E. 1974. Adaptational responses in nitrogen and energy balance of lambs fed a methane inhibitor. *J. Anim. Sci., 38:*154.

Johnson, D. E., A. S. Wood, J. B. Stone, and E. T. Moran, Jr. 1972. Some effects of methane inhibition in ruminants. *Can. J. Anim. Sci., 52:*703.

Johnson, K. A., and D. E. Johnson. 1995. Methane emissions from cattle. *J. Anim. Sci., 73*:2483.

Johnson, D. E. 1986. Fundamentals of whole animal calorimetry: use in monitoring body tissue deposition. *J. Anim. Sci. 63(Suppl. 2)*:111.

Johnson, D. E., T. M. Hill, G. M. Ward, K. A. Johnson, M. E. Branine, B. R. Carmean, and D. W. Lodman. 1993. Chapter. 11. Ruminants and Other Animals, In: *Atmospheric Methane: Sources, Sinks, and Role in Global Change*, ed. M.A.K. Khalil, Springer-Verlag, Berlin.

Johnson, D. E., J. S. Abo-Omar, C.F. Saa, and B. R. Carmean. 1994a. Persistence of methane suppression by propionate enhancers in cattle diets. In: *Energy Metabol. Of Farm Animals.*, Ed. J.F. Aguilera, EAAP Publ. 76, p 339.

Johnson, K. A., M. T. Huyler, H. H. Westberg, B. K. Lamb, and P. Zimmerman. 1994b. Measurement of methane emissions from ruminant livestock using a SF$_6$ tracer technique. *Environ. Sci. & Technol. 28*:359.

Johnson, K. A., H. H. Westberg, B. K. Lamb, and R. L. Kincaid. 1997. The use of sulfur hexafluoride for measuring methane production by cattle. In: *14th Symp. On Energy Metabolism of Farm Animals. Newcastle, No. Ireland.*

Joyner, A. E., L. J. Brown, T. J. Fogg, and R. T. Rossi. 1979. Effect of monensin on growth, feed efficiency, and energy metabolism of lambs. *J. Anim. Sci., 48:*1,065.

Kelly, J. M., B. Kerrigan, L. P. Milligan, and B. W. McBride. 1994. Development of a mobile open-circuit indirect calorimetry system. *Can. J. Anim. Sci. 74*:65.

Khalil, M. A. K., and R. A. Rasmussen. 1983. Sources, sinks, and seasonal cycles of atmospheric methane. *J. Geophys. Res., 88:*5,131-5,144.

Khalil, M. A. K., R. A. Rasmussen, M.-X. Wang, and L. Ren. 1990. Emissions of trace gases from Chinese rice fields and biogas generators. *Chemosphere, 20*:207.

Kinsman, R. G., F. D. Sauer, H. A. Jackson, and M. S. Wolynetz. 1997. Methane and carbon dioxide emissions from lactating dairy cows in full lactation monitored over a six-month period. *J. Dairy Sci. 78*:2760.

Kirchgessner, M., W. Windich, and H. L. Mueller. 1995. Nutritional factors for the quantification of methane production In. *Ruminant Physiology: Digestion Metabolism, Growth and Reproduction,* Engelhardt, W. V., et al. (ed) p 333.

Kleiber, M. 1958a. Some special features of the California appartus for respiration trials with large animals. In: In: G. Torbeck and H. Aersoe (Ed.) 1. A Symposium on energy metabolism: principles, methods and general aspects. Statens Husdyrbrugsudvalg, Copenhagen, Denmark. 53.

Kleiber, M. 1958b. A respiration apparatus for tracer trials with radiocarbon on cows, sows and sheep. In: G. Thorbeck and H. Aersoe (Ed.) 1. A Symposium on energy metabolism: principles, methods and general aspects. Statens Husdyrbrugsudvalg, Copenhagen, Denmark. 53.

Krumholtz, L. R., C. W. Forsberg, and D. M. Veira. 1983. Association of methanogenic bacteria with rumen protozoa. *Can. J. Microbiol, 29:*676.

Li, Z. X. Gao, H. Li, and X. Zhang. 1996. Studies on the metabolic rule of methane energy of Sika deer. *Acta Theriol. Sinica, 16:*100.

Liang, J. B., F. Terada, and I. Hamaguchi. 1989. Efficacy of using the face mask technique for the estimation of daily heat production of cattle. In: Y. Van Der Honing and W. H. Close (Ed.). *Energy Metabolism of Farm Animals.* Pudoc, Waginingen, The Netherlands.

Lodman, D. W., M. E. Branine, B. R. Carmean, P. Zimmerman, G. M. Ward, and D. E. Johnson. 1993. Estimates of methane emissions from manure of U.S. cattle. *Chemosphere, 26* (1-4):189-200.

Marik, T., and I. Levin. 1996. A new tracer experiment to estimate methane emissions from a dairy cow shed using sulfur hexafluoride (SF_6). *Global Biogeochemical Cycles. 10:*413.

Martinez, A., G. Bogdanof, D. Johnson, and J. Rust. 1996. Reducing methane emissions from ruminant livestock: Ukraine pre-feasibility study. Winrock International, Morrelton, AR

McBee, R. H. 1977. Hindgut fermentation. In: *Microbial Ecology of the Gut* (R.T.J. Clark and T. Bauchop, eds.), Academic Press.

McLean, J. A. and G. Tobin. 1987. *Animal and Human Calorimetry.* Cambridge University Press, New York.

Milchunas, D. G., M. I. Dyer, O. C. Wallmo and D.E. Johnson. 1978. In vivo/in vitro relationships of Colorado. Mule deer forages. *Spec. Report #43, Colo. Div. of Wildlife.*

Miller, W. H., and R. M. Koos. 1988. Construction and operation of an open-circuit indirect calorimetry system for small ruminants. *J. Anim. Sci. 66:1042.*

Moe, P. W., and H. F. Tyrrell. 1977. Effects of feed intake and physical form on energy value of corn in timothy hay diets for lactating cows. *J. Dairy Sci., 60:*752.

Moe, P. W., and H. F. Tyrrell. 1980. Methane production in dairy cows. *8th Energy Symp., EAAP, 26:*12.

Morris, G. R. 1976. Anaerobic fermentation of animal wastes: A kinetic and empirical design fermentation. M.S. Thesis, Cornell University.

Moss, A. R., D. I. Givens, and P. C. Garnsworth. 1994. The effect of alkali treatment of cereals straws on digestibility and methane production of sheep. *Animal Feed Sci. Tech. 49:245.*

Murray, R. M., A. M. Bryant, and R.A. Leng. 1975. Measurement of methane production in sheep. In: *Tracer Studies on Non-Protein Nitrogen for Ruminants II.* IAEA, Vienna, Austria.

Murray, R.M., A.M. Bryant, and R.A. Leng. 1976. Rates of production of methane in the rumen and large intestine of sheep. *Brit. J. Nutr. 36:1.*

Murray, R. M., A. M. Bryant, and R. A. Leng. 1978. Methane production in the rumen and lower gut of sheep given lucerne chaff: Effect of level of intake. *Brit .J. Nutr. 39:337-345.*

Newbold, C.J., R.J. Wallace, and N.D. Walker. 1993. The effect of tetronasin and monensin on fermemtation, microbial numbers and the development of ionophore resistant bacteria in the rumen. *J. Appl. Bacteriol. 75:129.*

Robb, J., P. J. Evans, and C. Fisher. 1979. A study of the nutritional energetics of sodium hydroxide-heated straw pellets in rations fed to growing lambs. *8th Energy Symp., EAAP, 26:*13.

Robbins, C. T. 1973. The biological basis for the determination of carrying capacity. Ph.D. Thesis, Cornell Univ., Ithaca, NY.

Rumpler, W. V., D. E. Johnson, and D. B. Bates. 1986. The effect of a dietary cation concentration on methanogenesis by steers fed diets with and without ionophores. *J. Anim. Sci. 62:*1,737.

Safley, L. M. 1989. Methane productions from animal wastes management systems. Methane *Emissions from Ruminants*, ICF/USEPA Workshop, Palm Springs, CA.

Safley, L. M, and P. W. Westerman. 1987. Biogas production from anaerobic lagoons. *Biological Wastes 23:*181.

Safley, L. M., Jr, M. E. Casada, J. W. Woodbury, and K. F. Roos. 1992. *Global Methane Emissions from Livestock and Poultry Manure.* U.S. EPA/400/1-91/048.

Sawyer, M. S., W. H. Hoover, and C. J. Sniffen. 1974. Effects of a ruminal methane inhibitor on growth and energy metabolism in the ovine. *J. Anim. Sci. 38*:908.

Sheppard, J. C., H. Westberg, J. F. Hopper, K. Genesan, and P. Zimmerman. 1982. Inventory of global methane sources and their production rates. *J. Geophys. Res.* 87:1,982.

Stevens, C. E. 1988. Comparative physiology of the vertebrate digestive system. Cambridge University Press, Cambridge, UK.

Sundstol, F. 1982. Energy utilization in sheep fed untreated straw, ammonia treated straw or sodium hydroxide heated straw. *9th Energy Symp., EAAP, 29*:120.

Sundstol, F., and A. Ekern. 1976. The nutritive value of frozen, dried, and ensiled grass cut at three different stages of growth. *7th Energy Symposium, EAAP, 19*:241.

Sundstol, F., A. Ekern, P. Lingvall, E. Lindgren, and J. Bertilsson. 1979. Energy utilization in sheep fed grass silage and hay. *8th Energy Symp., EAAP, 26*:17.

Swift, R. W., J. W. Bratzler, W. H. James, A. D. Tillman, and D. C. Meek. 1948. The effect of dietary fat on utilization of the energy and protein of rations by sheep. *J. Anim. Sci., 7*:475.

Theurer, C. B. 1987. Grain processing effects on starch utilization by ruminants. *J. Anim. Sci. 63:1,649.*

Thornton, J. H., and F. N. Owens. 1981. Monensin supplementation and in vivo methane production by steers. *J. Anim. Sci.,* 52:628.

Tyrrell, H. F., and G. A. Varga. 1985. Energy value for lactation of rations containing ground whole ear maize or maize meal both conserved dry or ensiled at high moisture. *10th Energy Symp., EAAP, 32:*306.

Trei, J. E., R. C. Parrish, Y. K. Singh, and G. C. Scott. 1971. Effect of methane inhibitors on rumen metabolism and feedlot performance of sheep. *J. Dairy Sci., 54*:536.

Van Kessel, J. A. S., and J. B. Russell. 1996. The effect of pH on ruminal methanogenesis. *FEMS Micro. Ecol. 20*:205.

Van Soest, P. J. 1982. *Nutritional Ecology of the Ruminant,* O. and B. Books, Corvallis, OR.

Van Nevel, C. J., and D. I. Demeyer. 1977. Effect of monensin on rumen metabolism in vitro. *Appl. Environ. Microbiol.,* 34:251.

Van der Honing, Y., B. J. Wieman, A. Steg, and B. van Donselaar. 1981. The effect of fat supplementation of concentrates on digestion and utilization of energy by productive dairy cows. *Neth. J. Agric. Sci., 29*:79.

Varga, G. A., H. F. Tyrrell, D. R. Waldo, G. B. Huntington, and B. P. Glenn. 1985. Effect of alfalfa orrchardgrass silages on energy and nitrogen utilization for growth by Holstein steers. *10th Energy Symp., EAAP,* 32:86.

Vermorel, M. 1995. Yearly methane emissions of digestible origin by cattle in France, variations with level and type of production, *INRA Prod. Anim., 8*:265.

Wedegaertner, T. C., and D. E. Johnson. 1983. Monensin effects on digestibility, methanogenesis and heat increment of a cracked corn-silage diet fed to steers. *J. Anim. Sci.* 57:168.

Westberg, H. K., A. Johnson, and B. Lamb. 1995. Ruminant methane emissions from livestock. Report to WESGEC, Davis CA.

Whitelaw, F. G., J. M. Eadie, L. A. Bruce, and W. J. Shand. 1984. Methane formation in faunated and ciliate-free cattle and its relationship with rumen volatile fatty acid production. *Brit. J. Nutr., 52*:261.

Wieser, M. F., and C. Wenk. 1970. Effect of plane of nutrition and physical form of ration on energy utilization and rumen fermentation in sheep. *5th Energy Symp., EAAP, 17*:53.

Williams, D. J. 1993. Methane emissions from the manure of free-range cows. *Chemosphere, 26* (1-4):179-188.

Wolin, M. J., and T. L. Miller. 1988. Microbes interactions in the rumen microbial ecosystems. In: P. N Hobson, ed., *The Rumen Microbial Ecosystem,* Elsevier Applied Science, New

York.

Yen, J. T., and J. Nienaber. 1991. Absorption of volatile fatty acids from the gastrointestinal tract of swine. *J. Anim. Sci.*, *69*:2,001.

Young, B. A., B. Kerrigan, and R. J. Christopherson. 1975. A versatile respiratory pattern analyzer for studies of energy metabolism of livestock. *Can. J. Anim. Sci. 55*:17.

9. Rice Agriculture: Factors Controlling Emissions

H.-U. Neue[1] and P.A. Roger[2]
[1] Department of Soil Sciences, UfZ-Environmental Research Center Leipzig-Halle GmbH, Theodor-Lieser-Str. 4, D-06120 Halle, Germany
[2] Laboratoire ORSTOM de Microbiologie des Anaerobies (LOMA), Université de Provence, CESB/ESIL, Case 925, 163 Avenue de Luminy, F-13288, Marseille Cedex 9, France

1. Introduction

Recent atmospheric measurements indicate that concentrations of greenhouse gases are increasing. Atmospheric methane concentration has increased at about 1% annually to 1.7 ppmV during the last decades (Khalil and Rasmussen, 1987). The resulting effect on global temperature is highly significant because the warming efficiency of methane is up to 30 times that of carbon dioxide (Dickinson and Cicerone, 1986). Data from polar ice cores indicate that tropospheric methane concentrations have increased by a factor of 2-3 over the past 200-300 years (Khalil and Rasmussen, 1989). The increase of methane concentrations in the troposphere correlate closely with global population growth and increased rice production (Figure 1), suggesting a strong link to anthropogenic activities. The total annual global emission of methane is estimated to be 420-620 Tg/yr (Khalil and Rasmussen, 1990), 70-80% of which is of biogenic origin (Bouwman, 1990). Methane emissions from wetland rice agriculture have been estimated up to 170 Tg/yr, which account for approximately 26% of the global anthropogenic methane budget. Flooded ricefields are probably the largest agricultural source of methane, followed by ruminant enteric digestion, biomass burning, and animal wastes (summarized by Bouwman, 1990).

Projected global population levels indicate that the demand for rice will increase by 65% over the next 30 years, from 460 million t/yr today to 760 million t/yr in the year 2020 (IRRI, 1989). The growing demand is most likely to be met by the existing cultivated wetland rice area through intensifying rice production in all rice ecologies, mainly in irrigated and rainfed rice. Coupled with existing rice production technologies, global methane emissions from wetland rice agriculture are likely to increase. Mitigation of methane emissions is needed to stabilize or even lower atmospheric methane concentrations.

This paper discusses principles and prospects of rice cultivation in view of methane formation, methane fluxes, and mitigation options.

Mohammad Aslam Khan Khalil   (Ed.)
Atmospheric Methane
© *Springer-Verlag Berlin Heidelberg 2000*

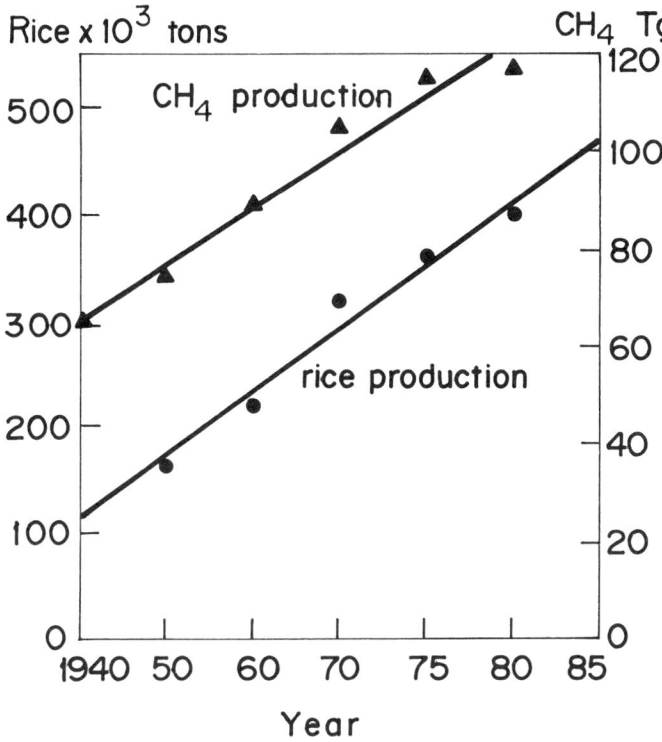

Fig. 1. Rice production and methane emission rates (IRRI, 1991; Bolle et al., 1986).

2. Rice Environments

Rice is cultivated under a wider variety of climatic, soil, and hydrological conditions than any other crop. It is grown from the equator to as far as 50°N and 40°S, and from sea level to altitudes of more than 2500 m. The temperature may be as low as 4°C during the seedling stage and as high as 40°C at flowering. Rice is irrigated in arid areas and is grown in rainfed areas with only 500 mm rain/yr. Rice is cultivated as an upland crop and in soils that are submerged more than 1 m. Rice is the only major crop grown on flooded soils.

Rice cultural systems have developed to suit the physical, biological, and socioeconomic conditions of different regions. Because the water regime during the growing season is the most discriminating physical factor, ricelands can be grouped into two main systems: wetlands and uplands. Terms used to differentiate rice cultures are, for example, lowland rice, irrigated rice, rainfed rice, deepwater rice, swamp rice, upland rice, hill rice, dryland rice, and pluvial rice. Many other terms have been evolved in different regions reflecting specific characteristics of and constraints to rice cultivation in these areas. These terms reflect the wide range of agroecologies in which rice is grown and are very reasonable in the context they evolved. But the general use of these terms, although understandable, is often semantically and technically incorrect (Moormann and Van Breemen, 1978). A comprehensive classification of rice ecologies has been outlined by Neue (1989). He defined three major rice ecologies with a total of seven subecologies by hierarchically applying floodwater source and floodwater depth as diagnostic criteria (Table 1). Further differentiation is done by modifiers related to climate, landform, floodwater regime, soil, and cropping system.

Table 1. Classification of rice ecologies

Flood-water source	Irrigation		Pluvial, phreatic, surface flow or tidal				
Flood-water depth (cm)	1-5	5-25	0-25	25-50	50-100	>100	<0
Rice ecologies	-- Irrigated rice --		-- Rainfed rice --				Upland rice
Sub-ecologies	Shallow	Medium	Shallow	Medium	Deep	Very deep	Upland rice
Land ecosystem			-- Wetland --				Upland

Rice ecologies are major discriminators for the potential of methane production in ricefields because of their distinct floodwater regimes. The potential of upland rice for methane production is not significant since upland rice is never flooded for a significant period of time. Aerobic soils, including upland rice soils, seem to be important sites for deposition and microbial oxidation of atmospheric CH_4 (Seiler and Conrad, 1987; Cicerone and Oremland, 1988). Irrigated rice has the highest potential to produce CH_4 because flooding and, consequently, anoxic conditions are assured and controlled. The potential for methane production in rainfed rice should vary widely in time and space since floodwater regimes are primarily controlled by rainfall within the watershed. Periods of severe droughts or floods during the growing season are characteristic for rainfed rice. Subecologies are determined by floodwater depth, which likely affects methane fluxes. Emission rates and harvested area of each rice ecology determine the global methane emission. Rice areas harvested in different regions of the world are given in Table 2.

Table 2. Distribution of harvested ricelands (million ha) by rice ecologies (FAO, 1988).

Region	Irrigated	Rain-fed	Deep water	Upland	Total area	Yield (t/ha)	Rough rice production (10⁶ ton)
East Asia[a]	34.0	2.8	-	-	36.8	5.4	200.0
Southeast Asia[b]	13.9	13.7	3.75	4.65	36.0	2.9	102.5
South Asia[c]	19.4	20.0	7.3	6.7	53.4	2.0	105.5
Near East[d]	1.25	-	-	-	1.25	3.3	4.1
South/ Central Am. Caribbean and USA	2.5	0.5	0.4	5.65	9.05	2.9	26.5
Africa	0.9	1.95	-	2.70	5.5	1.8	9.9
USSR	0.66	-	-	-	0.66	4.1	2.7
Europe	0.42	-	-	-	0.42	5.4	2.3
Oceania	0.12	-	-	-	0.12	6.6	0.79
Australia	0.11		-	-	0.11	7.1	0.76
World	73.26	38.95	11.45	19.70	143.4	3.2	455.05

[a]China Taiwan Korea DPR Korea RP Japan;
[b]Burma Cambodia Indonesia Laos Malaysia Philippines Thailand Vietnam;
[c]Bangladesh Bhutan India Nepal Pakistan Sri Lanka;
[d]Afghanistan Iran Iraq.

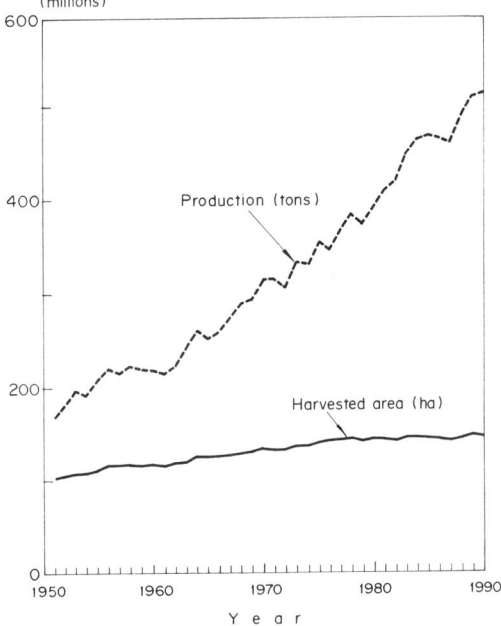

Fig. 2. Rough rice production and harvested area (IRRI, 1991)

Especially since the 1960s, rice production dramatically increased because of high-yielding rice cultivars, large investments in irrigation schemes, and improved soil, water, and crop management. The developed irrigation schemes and the shorter growth duration of modern cultivars increased the harvested area by allowing 2 to 3 crops per year. However, expansion of residential and industrial areas as well as diversification of crops resulted in only a slight increase in the total harvested area of rice (Figure 2). Though many factors determine the relative contribution of each rice ecology to rice supplies in the future, irrigated areas will continue to dominate rice production. At present, about 50% of the harvested area is in irrigated rice but it contributes about 70% of total production.

3. Microbiology of Methane Emission

3.1 Methanogens

Biogenic methane production is exclusively accomplished by methanogenic bacteria that can metabolize only in strict absence of oxygen and at redox potentials of less than -200 mV. Oxygen causes an irreversible disassociation of the F_{420}-hydrogenase enzyme complex probably due to the lack of protective superoxide dismutase (Schönheit et al., 1981). Methanogens are found in strictly anaerobic environments of freshwater, brackish and marine sediments, hot springs, mid-ocean ridges, decomposing algal mats, heart wood of living trees, intestinal tracts of man and animals (especially the rumen of herbivores), and sewage digesters. In ricefields, methanogenesis occurs in the reduced soil of wetland rice and, possibly, in anoxic water of deepwater rice.

Recent reviews on methanogenic bacteria deal with their biogeochemistry (Oremland and Capone, 1988; Boone, this volume), taxonomy, and ecology (Garcia, 1990).

Twenty genera of methane-producing bacteria have been described (Table 3) but only a few, including *Methanobacterium* and *Methanosarcina*, have been isolated from rice soils (Rajagopal et al., 1988). *Methanospirillum* and *Methanocorpusculum*, which were isolated from freshwater sediments, as well as methanogens found as endosymbionts in sapropelic amoeba should also be present in wetland ricefields (Garcia, ORSTOM, personal communication).

The distribution of methanogens in natural environments depends on their adaptation to temperature, pH, and salinity ranges. Most methanogens are mesophilic with temperature optima of 30-40°C. Thermophilic (40-70°C) species account for 20% of the strains (Garcia, 1990) and some extreme thermophilic (up to 97°C) are also known. Most methanogens are neutrophilic with a relatively narrow pH range of 6-8. A few alkaliphilic isolates with optimum growth at pH 8-9 have been reported in the genera *Methanosarcina, Methanobacterium* (Blotevogel et al., 1985; Worakit et al., 1986), and *Methanohalophilus* (Mathrani et al., 1988). No acidophilic strains have been reported. A strain isolated from peat tolerated a pH of 3, but its optimum was 6-7 (Williams and Crawford, 1984, 1985).

Table 3. Simplified classification of methanogenic bacteria (adapted from Garcia 1990) and their habitat (Garcia, ORSTOM, personal communication).

Methanobacteriales
Methanobacteriaceae

Methanobacterium	Various freshwater habitats. Half of the species are thermophilic; few are alkaliphilic.
Methanobrevibacter	Specialized habitats such as trees (Zeikus and Henning, 1975), rumen (Smith and Hungate, 1958), sewage sludge, intestinal tracts of animals (Miller and Wolin, 1985).
Methanosphaera	Feces or digetive tracts of animals (Biavati et al., 1988).
Methanothermaceae *Methanothermus*	Extreme thermophile from volcanic springs

Methanococcales   Isolated mostly from marine or coastal environments
Methanococcaceae
Methanococcus

Methanomicrobiales
Methanomicrobiaceae

Methanolacinia	Marine sediments
Methanospirillum	Mesophilic strains from various habitats
Methanoculleus *Methanocorpusculum*	Sewage sludge lacustrine sediments (Zhao et al., 1989).
Methanomicrobium	
Methanogenium	
Methanoplanaceae *Methanoplanus*	Symbiont of marine ciliate.
Methanosarcinaceae *Methanosarcina*	Freshwater and marine sediments, ricefields, lagoons, anaerobic sewage-sludge digestors, and rumen (Raimbault, 1981).

Genera not ascribed to a family (mostly isolated from salty biotopes):
Methanolobus Methanococcoides Methanohalophilus Halomethanococcus Methanohalobium Methanothrix and *Methanopyrus*

Methanogens can only metabolize a limited number of simple carbon compounds and hydrogen availability is a key factor for methanogenesis. As summarized by Garcia (1990):

Hydrogenotrophic methanogens (77% of the 68 described species) oxidize H_2 and reduce CO_2 to form methane. According to Conrad et al. (1985), H_2-dependent methanogenesis in sediments results mostly from H_2 transfer between microbial associations within flocks or consortia.

Methylotrophic methanogens (28% of the species) can use methyl compounds as methanol, methylamines, or dimethylsulfide; 10 species have been identified as obligate methylotrophs.

Acetotrophic methanogens (14% of the species) utilize acetate. The growth of virtually all methanogens is stimulated by acetate and its importance as a methane precursor in sediments has been documented (Cappenberg, 1974;, Cappenberg and Prins, 1974; Winfrey and Zeikus, 1977). Sixty percent of the hydrogenotrophic species also use formate. A few species use H_2 to reduce methanol to methane (hydrogeno-methylotrophic methanogens); others form methane in the presence of CO_2 and alcohols as hydrogen donors (alcoholotrophic methanogens). The importance of methanol and methylated amines as methane precursor in sediments varies with the abundance of decomposing plant materials such as algal mats (King, 1988). Methanol and methylated amines might be abundant in wetland ricefields after fertilizer application has induced the formation of large algal mats.

Methanogenesis in sediments is characterized by a complete degradation of organic matter while in the rumen of ruminants and the intestine of most animals, mineralization is incomplete since intermediate products are absorbed as food. The anaerobic degradation of organic matter to methane in sediments requires the cooperation of several types of bacteria within a substrate chain to provide the simple carbon compounds needed by methanogens. According to Conrad (1989), four types of bacteria are needed: a) hydrolytic and fermenting bacteria, b) H^+-reducing bacteria, c) homoacetogenic bacteria, and d) methanogenic bacteria. The first group hydrolyzes polymers and ferments the resulting monomers to smaller molecules such as alcohols, short chain fatty acids, H_2, and CO_2. Methanogens can immediately convert H_2/CO_2, formate, acetate, and a few other simple compounds including methanol, methylamines, and dimethylsulfide to CH_4 and CO_2. Fermentation products such as fatty acids, alcohols, aromates, and others cannot directly be utilized. They are oxidized by obligate H^+-producing bacteria to acetate and CO_2. Homoacetogens are very versatile bacteria that can use sugars, alcohols, fatty acids, purines, and aromatic compounds as well as methanol, formate, H_2, and CO_2 to produce acetate as the sole fermentation product (Dolfing, 1988).

All methanogens use NH_4^+ as a nitrogen source, and a few species are known to fix molecular nitrogen (Belay et al., 1984; Murray and Zinder, 1984).

3.2 Inhibitors of Methane Formation

Mineral terminal electron acceptors like nitrate or sulfate inhibit methanogenesis in sediments by channeling the electron flow to thermodynamically more efficient bacteria like denitrifiers or sulfate reducers (Balderston and Payne, 1976; Ward and Winfrey, 1985). Manganese and iron oxides should have the same effect. Methanogens, sulfate-reducers and homoacetogenic bacteria compete for H_2 produced by fermentative bacteria. Hydrogenotrophic homoacetogens do not significantly compete with methanogens for H_2 in sediments (Lovley and Klug, 1983). Since H_2 concentration is usually very low in such environments (Strayer and Tiedje, 1978), sulfate reducers are able to out-compete hydrogenotrophic methanogens in the presence of sulfate because of their higher affinity for H_2 and faster growth (Winfrey and Zeikus, 1977; Abram and Nedwell, 1978).

NaCl inhibits pure cultures of methanogens, though high concentrations (about 0.2 M) are required for several strains (Patel and Roth, 1977). In general, adding NaCl to a nonsaline soil inhibits methanogenesis (Koyama et al., 1970). Methanogenesis is inhibited by brackish water (Garcia et al., 1974, De Laune et al., 1983, Holzapfel-Pschorn et al., 1985, Bartlett et al., 1987). Inhibitory effects and interactions with sulfate-reducing bacteria are given as possible reasons (Mitsch and Gosselink, 1986). Competition for H_2 and toxicity of sulfide are the likely mechanisms. However, methanogenesis and sulfate reduction are not mutually exclusive when methane is produced from methanol or methylated amines for which sulfate reducers show little affinity (Oremland et al., 1982; Oremland and Polcin, 1982; Kiene and Visscher, 1987). Methanol is formed during anaerobic decomposition of plant pectins (Schink and Zeikus, 1980). In saline environments, degradation of osmoregulatory compounds such as glycinebetaine produces methylamines (King, 1984). Obligately methylotrophic methanogens constitute about half of the methanogenic population present in salt marsh sediments (Franklin et al., 1988), and methylotrophs are found in rice soils (Rajagopal et al., 1988).

Chemical substances inhibiting methanogenesis have been reviewed by Oremland and Capone (1988). The 2-bromoethane-sulfonic acid (BES), an analog to Coenzyme M, is a specific inhibitor of methanogenesis. Several chlorinated CH_4 analogues such as chloroform and methyl chloride have been identified to inhibit methanogenesis. Chloroform completely suppressed methane production in a paddy soil but did not hamper the turnover of glucose (Krumböck and Conrad, 1991) although evidence for glucose-utilizing H_2-syntrophic methanogenic bacterial associations has been found in glucose amended paddy soil (Conrad et al., 1989). Substances encountered in ricefields that inhibit methanogenesis and methane oxidation include DDT (McBride and Wolfe, 1971), acetylene (Raimbault, 1975), and nitrapyrin, an inhibitor of nitrification (Salvas and Taylor, 1980). Slow release of acetylene from calcium carbide, encapsulated in fertilizer granules highly reduced methane emission (Bronson and Mosier, 1991).

3.3 Methane-Oxidizing Bacteria

Methane oxidation may greatly limit the flux of CH_4 to the atmosphere (Bont et al., 1978). Holzapfel-Pschorn et al. (1986) reported that 67% of the CH_4 produced during a rice growing season in an Italian ricefield was oxidized. Sass et al. (1991) found that 58% was oxidized in a Texas ricefield. Schütz et al. (1989) reported that up to 90% of CH_4 generated at the late growth stage was oxidized.

Methane can be oxidized by aerobic and anaerobic bacteria. Several reviews have been published on methane-oxidizing bacteria and aerobic methane oxidation (Whittenbury et al., 1970 a,b; Higgins et al., 1981; Anthony, 1982; Crawford and Hanson, 1984). Aerobic methane-oxidizing (methanotrophic) bacteria constitute a group of eubacteria that grow only on methane or carbon compounds lacking carbon-carbon bonds such as methanol, formate, and methylated amines. One species

(*Methylobacterium organophilum*) can also grow on more complex organic compounds in combination with methane (Patt et al., 1974). All aerobic methanotrophs sequentially oxidize CH_4 to CO_2 via methanol, formaldehyde, and formate. Oxygen is essential for the growth of methane-oxidizing bacteria, but the required partial pressure may be low (Cicerone and Oremland, 1988), especially when methanotrophs fix nitrogen (Murrell and Dalton, 1983) or grow with nitrate as a nitrogen source (Toukdarian and Lidstrom, 1984). Aerobic methanotrophs, which require both methane and oxygen, are most active in ricefields at the interface of aerobic and anaerobic environments (floodwater-soil interface, rice rhizosphere).

Anaerobic oxidation of methane is poorly understood but appears to be an important methane sink in sulfate-containing environments, such as marine sediments or anoxic water (Alperin and Reeburgh, 1984; Iversen et al., 1987). The process has also been reported to occur in freshwater systems (Panganiban et al., 1979).

4. Flooded Rice Soils as Site for Methane Emission

In general, flooded rice soils provide an optimum environment for methane production and emission, especially in the tropics. Flooding a soil causes the essential low redox potential and anaerobic decomposition of organic matter and stabilizes the soil pH near neutral. In the tropics, the temperature of the reduced puddled layer becomes optimal for methanogenesis. Rice plants highly enhance the emission of methane. Variations in methane fluxes from rice paddies are caused by variations in soil properties, crop management, and related growth of rice.

4.1 Rice Soils

Neue (1989) characterized a typical soil profile of a flooded rice soil during the middle of a growing season as follows:

Horizon   Description
Ofw   A layer of standing water that becomes the habitat of bacteria, phytoplankton, macrophytes (submerged and floating weeds), zooplankton, and aquatic invertebrates and vertebrates. The chemical status of the floodwater depends on the water source, soil, nature, and biomass of aquatic fauna and flora, cultural practices, and rice growth. The pH of the standing water is determined by the alkalinity of the water source, soil pH, algal activity, and fertilization. Because of the growth of algae and aquatic weeds, the pH and oxygen content undergo marked diel fluctuations. During daytime, the pH may increase up to 11 and the standing water becomes oversaturated with O_2 due to photosynthesis of the aquatic biomass. Standing water stabilizes the soil water regime, moderates the soil temperature regime, prevents soil erosion, and enhances C and N supply.
Apox   The floodwater-soil interface that receives sufficient O_2 from the floodwater to maintain a pE + pH above the range where NH_4^+ becomes the most stable

form of N. The thickness of the layer may range from several mm to several cm depending on pedoturbation by soil fauna and the percolation rate of water.

Apg   The reduced puddled layer is characterized by the absence of free O_2 in the soil solution and a pE + pH low enough to reduce iron oxides.

Apx   This layer has increased bulk density, high mechanical strength, and low permeability. It is frequently referred to as plow pan or traffic pan.

B   The characteristics of the B horizon depend highly on water regime. In epiaquic moisture regimes the horizon generally remains oxidized, and mottling occurs along cracks and in wide pores. In aquic moisture regimes, the whole horizon or at least the interior of soil peds remain reduced during most years.

The chemistry and biology of rice soils have frequently been reviewed (Ponnamperuma, 1972, 1981, 1984a, 1985; Patrick and Reddy, 1978; De Datta, 1981; Watanabe and Roger, 1985; Yu, 1985; Patrick et al., 1985; Roger et al., 1987; Neue, 1988).

The duration and pattern of flooding and saturation are important criteria for methane formation. Saturation can be caused by groundwater (aquic moisture regime) or surface water (epiaquic moisture regime). Flooding an air-dried cultivated soil drastically changes the hydrosphere, atmosphere, and biosphere of that soil. Flooding highly limits diffusion of air into the soil. The O_2 supply cannot meet the demand of aerobic organisms, and facultative and anaerobic organisms proliferate using oxidized soil substrates as electron acceptors in their respiration. Consequently, the redox potential falls sharply according to a sequence predicted by thermodynamics and CO_2 and HCO_3^- concentrations increase to very high levels. As a result, the soil pH of acid soils increases while that of sodic and calcareous soils decreases, stabilizing between 6.5 and 7.2. Flooding and puddling render most soils an ideal growth medium for rice by supplying abundant water, buffering soil pH near neutral, enhancing N_2 fixation, and increasing diffusion rates, mass flow, and availability of most nutrients. In less favorable soils, flooding may result in toxicities of Fe, H_2S, or organic acids, or deficiencies of Zn or S.

The anaerobic fermentation produces an array of organic substances, many of them transitory and not found in aerobic soils. The major gaseous end products are CO_2, H_2S, and CH_4. The description of the paddy soil profile clearly indicates that methane formation mainly takes place in the reduced Apg horizon. In aquic moisture regimes, the B horizon may also become a source of methane. But in general carbon contents of B horizons are low and their organic matter is less degradable. In epiaquic moisture regimes, methane oxidation may predominate in the B horizon. The same holds true for the Apox layer. Harrison and Aiyer (1913) established early on that all methane diffusing into the aerobic surface layer is oxidized. This was reconfirmed by Bont et al. (1978). They found that 10 ml of a suspension of rice soil oxidized 2 ml of methane within 24 hours when incubated aerobically. Methane may also be oxidized in shallow floodwater since it is often oversaturated with O_2 due to assimilation of the aquatic flora.

In deepwater ricefields, the deeper layers of the floodwater may also become anoxic during the crop cycle (Whitton and Rother, 1988), permitting methanogenesis from the large quantity of organic material available from rice culms, nodal roots, and dead aquatic biomass.

4.2 Temperature Regimes of Rice Soils

Rice is grown under widely differing temperature regimes. The temperature of flooded soils at planting may range from 15°C in northern latitudes to 40°C in equatorial wetlands. Rice physiologists have studied extensively the effects of air and water temperature on rice growth characteristics (Matsushima et al., 1964 a,b; Yoshida, 1981), but there is only little information on the temperature regimes of flooded rice soils and their effects on the chemistry of the soils (Kondo, 1952; Cho and Ponnamperuma, 1971; Gupta, 1974; Sharma and De Datta, 1985). Seasonal and diel temperature changes likely influence methane formation and emission. Holzapfel-Pschorn and Seiler (1986) reported a marked influence of soil temperature on the methane flux with doubling of emission rates when temperature increased from 20 to 25°C. Diel variation of methane emission is correlated with temperature fluctuation (Schütz et al., 1989).

Most isolates of methanogenic bacteria are mesophilic with temperature optima of 30 to 40°C (Acharya, 1935; Vogels et al., 1988). Psychrophilic acetate-utilizing methanogens with a temperature optimum below 20°C seem to occur in acidic peat, which generally shows substantial rates of methane production (Svensson, 1984). The temperature optimum for the production of methanogenic substrates by fermenting bacteria may not concur with the optimum for methanogenesis. In subtropical regions or at high altitudes, the accumulation of intermediate metabolites may reach toxic levels, especially early in the rice-growing season, because of low temperatures. Specific drainage techniques with increased percolation rates and/or intermittent aeration periods are practiced to remedy such accumulations. In tropical lowlands, high temperature throughout the growing seasons stimulates degradation and methane production.

In flooded conditions, soil temperature varies in response to the meteorological regime acting upon the atmosphere-floodwater and floodwater-soil interfaces. The changing properties of soil and floodwater (i.e., temporal changes in reflectivity, heat capacity, thermal conductivity, incoming water temperature, and water flow) as well as vegetation interact with these external influences. Hackman (1979) reported that floodwater temperatures are above minimum air temperature but below maximum air temperature if daily amplitudes of air temperature are high, while water temperatures are above maximum air temperatures if daily fluctuations are low. Neue (1988) reported that floodwater temperature in Philippine ricefields always exceeded ambient air temperature and showed lower daily fluctuations. The temperature of the puddled layer closely followed the temperature of the floodwater and decreased with depth. The annual mean soil temperature at 2:00 p.m. was 33°C at 7 cm depth, its daily maximum equaled or exceeded the maximum air temperature on most days.

Floodwater transmits short-wave radiation to the soil while reducing the upward escape of emitted long-wave radiation. Thus, a "greenhouse effect" is produced, heating floodwater and soil. Diel temperature amplitudes of the floodwater are highly moderated because of the high heat capacity of water and because evaporation of water consumes energy from the floodwater but not directly from the soil. The high thermal conductivity of flooded and puddled soils, in which the bulk densities may be reduced to only 0.2-0.5 g/cm^3, enhances the downward conduction to the dense layer. Dissolved and suspended particles and aquatic biomass in the floodwater change the absorption of radiation, and depth of floodwater changes the heat capacity. The temperature of both floodwater and soils may rise above 40°C in unplanted soils with muddy floodwater of shallow depth. Floodwater temperature is lowered by canopy shading, flow of water, and through rainfall. In ricefields where floodwater has been drained for transplanting or seeding, soil temperatures may reach 50°C in the top centimeter because of increased heat absorption and reduced heat capacity, thermal conductivity, evaporation, and ventilation.

Aselmann and Crutzen (1990) computed monthly distributions of global methane emissions from linearly temperature-dependent methane fluxes in the range from 300 to 1000 mg/m^2 per day for temperatures from 20 to 30°C and constant emission of 300 mg/m^2 per day for temperatures below 20°C. Emissions of methane in the northern hemisphere reveal low monthly values (1.5 - 3 Tg) in December to April and a bell-shaped distribution between May and November with a clear peak of about 16 Tg in August. The southern hemisphere reveals highest emission rates (up to 2 Tg) in the months of February and March. The largest sources where computed between 20 and 30°N (South China, North India, Pakistan, Bangladesh, North Myanmar) with 37.6 Tg/yr, followed by 10 to 20°N (South India, South Myanmar, Thailand, Cambodia, Laos, Vietnam, North and Central Philippines, Brunei, Kalimantan) with 22 Tg/yr, 30 to 40°N (Central China, Japan, Korea) with 8 Tg/yr, 0 to 10°S with 6.5 Tg/yr (most of Indonesia) and 0-10°N (Sri Lanka, Malaysia, South Philippines, Brunei, Kalimantan) with 5 Tg/yr. Though rice ecologies have not been discriminated explicitly, the computed distribution of emission rates clearly reflects the importance of irrigated rice ecologies since double- and triple-cropped areas account for a major part of irrigated rice.

4.3 Organic Matter Accumulation and Decomposition

Readily mineralizable organic matter derived from primary production or organic amendments are the main source for methane formation in wetland rice soils. The net primary production of wetland rice soils (Table 4) can be deduced from yield statistics and estimates of aquatic biomass and weed growth during fallow periods.

In 1988, worldwide wetland rice production was 477 million tons (t), of which upland rice contributed about 28 million t. Based on a shoot/grain ratio of 3/2 (Ponnamperuma, 1984b) and a root/shoot ratio of 0.17 (Yoshida, 1981; Watanabe and Roger, 1985), the total dry matter production of wetland rice amounts to 1123 million t. Adding 74 million t dry matter of aquatic biomass [600 kg/ha season (Roger and Watanabe, 1984; Watanabe and Roger, 1985)] and 200 million t of weed dry matter [2 t/ha during fallow periods (Buresh and De Datta, 1991)] amounts to a total dry matter production of 1512 million t or 1500 g/m^2 per yr.

It is assumed that, on an average, 15% of the straw, 50% of the weeds, and all roots and aquatic biomass amounting to 390 million t dry matter or 156 million t carbon are returned to the soil. If a maximum of 30% of the returned carbon is transformed to methane as found by Neue (1985) in studies with ^{14}C-labeled straw in soils prone to methane formation, 60 Tg of methane would be globally produced in wetland ricelands annually. The input of degradable organic carbon is likely higher due to organic amendments. Reliable data on amounts of organic manures added are lacking. Based on long-term yield trials in the Philippines, a relation among soil C content, N fertilizer rates, and rice grain yields was established. The optimum C content in puddled and flooded soils was found to be 2 - 2.5%, corresponding to 0.20 - 0.25% total nitrogen (Neue, 1985; Smith et al., 1987). Since almost 90% of the tropical soils studied by Kawaguchi and Kyuma (1977) had less than the optimum total nitrogen content, moderate organic amendments seem to be essential to sustain or increase soil fertility and rice yields. In some instances, the returned net primary production of organic matter seems to be sufficient.

Table 4. Annual net primary production in wetland ricefields.

Source	Dry Weight	Returned to soil	
		%	(million t/yr)
Wetland rough rice	449	--	--
Wetland rice straw[a]	674	15	101
Wetland rice roots[b]	115	100	115
Aquatic biomass (algaeweeds)	74	100	74
Fallow weeds	200	50	100
TOTAL	1512	26	390

[a] Shoot:grain ratio = 1.50; [b] Root:shoot ratio = 0.17

The rate and pattern of organic matter addition and decomposition control the rate and pattern of methane formation. Anaerobic fermentation produces an array of organic substances, many of them transitory and not found in well-aerated soils. Ponnamperuma (1984a) listed various gases, hydrocarbons, alcohols, carbonyls, volatile fatty acids, nonvolatile fatty acids, phenolic acids, and volatile S compounds. Methanogens constitute the last step in the electron transfer chain generated by the anaerobic degradation of organic matter. Submergence of soils retards initial decomposition of rice straw in the field only slightly compared with upland soils (Neue and Scharpenseel, 1987; Neue, 1988). The rate of decomposition decreases with soil depth (Neue, 1985). Decomposition of the remaining, more resistant metabolites and residues is similar, with half-lives of about 2 years in all soils and water regimes if the following conditions for flooded soils are met:
- soil is intensively puddled each cropping season;
- soil temperature of the puddled layer is 30-35°C;
- neutral pH;
- low soil bulk density and wide soil/water ratio;
- shallow floodwater;
- high and balanced nutrient supply;
- no long-lasting accumulation of organic acids;
- permanent supply of energy-rich photosynthetic aquatic and benthic

biomass;
- high diversity of micro- and macroorganisms that provide successive fermentation down to CO_2, CH_4, H_2, and NH_3;
- supply of O_2 into the reduced layer by rice root excretion and oligochaete population; and
- diel oversaturation of the floodwater with O_2 due to photosynthetic aquatic biomass enhancing the aeration function of oligochaetes.

Decomposition and, accordingly, methane production are retarded in wetland rice soils with low and imbalanced nutrient supply, high bulk density, and low biological diversity and activity, as demonstrated in the Aeric Paleaquult of Northeast Thailand (Snitwongse et al., 1988). If the biological activity is restricted to bacteria, as in laboratory experiments, the decomposition of rice straw in flooded soils is highly retarded (Capistrano, 1988). Only 7-18% of the incorporated straw was decomposed after 100 days following the order San Manuel clay loam (pH 6.6) > Maahas clay loam (pH 5.5) > Louisiana clay (pH 4.9). These results clearly demonstrate the high limitations of laboratory incubation studies.

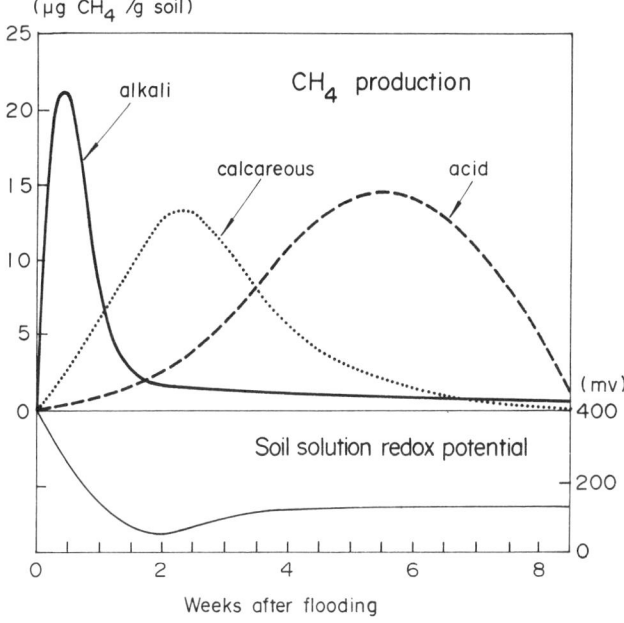

Fig. 3. Methane formation in alkaline, calcareous, and acid soils.

In calcareous and alkaline soils, methane production may occur within hours after flooding an air-dried soil, while in acid soils it may take weeks before methane is formed (Figure 3). In very acid soils, methane may not be formed at any time. The formation of methane is preceded by the production of volatile acids. Short-term H_2

evolution immediately follows the disappearance of O_2 after flooding. Thereafter, CO_2 production increases, and finally, with decreasing CO_2, CH_4 formation increases (Takai et al., 1956; Neue and Scharpenseel, 1984). The addition of organic substrates enhances the fermentation process. With increasing temperature up to 35°C, decomposition starts earlier and is more vigorous in every case. At high temperatures, the formation of CO_2 and CH_4 occurs earlier and is stronger (Yamane and Sato, 1961). The period of occurrence and the amount of the gaseous products and volatile acids depend largely on temperature and reducing conditions.

The ratio of CO_2 to CH_4 formation is controlled by the fermentation chain and the ratio of the oxidizing capacity (amount of reducible O_2, NO_3^-, Mn(IV), and Fe(III)) to the reducing capacity (Takai, 1961). The actual capacity is highly influenced by O_2 diffusion from the atmosphere, floodwater, and plant roots; the soil bulk density (soil-water ratio); and fertilization. Less CH_4 and higher accumulation of volatile acids are found in soils with higher bulk density (lower soil:water ratio). Digging tubificidae (earthworms) in the top soil decrease methane formation but increase methane emission by enhancing fluxes of gases.

Consecutive addition of organic substrates through plant growth and photosynthetic biomass production in the floodwater maintain the fermentation chain. The low specific activity of CH_4 produced after adding ^{14}C-labeled rice straw in field experiments (IRRI, 1981) was caused by degradation of newly produced photosynthetic biomass and root exudates. In permanently flooded soils, methane is only produced in significant amounts after soil-borne production or addition of readily mineralizable organic substrates. Humification of organic matter in wetland rice soils is less than that in aerobic soils. Humus of seasonal flooded soils has lower H_2 and N contents, its degree of unsaturation and its content of carboxyl and phenolic groups is lower, but its alcoholic and methoxyl groups are higher (Kuwatsuka et al., 1978; Tsutsuki and Kuwatsuka, 1978; Tsutsuki and Kumada, 1980). Humification indices in flooded soils, given as the ratio of nonhumified and humified materials (Sequi et al., 1986), are high (low humification) in topsoils and decrease with depth. Very acid rice soils have more humified materials.

Submergence is often equated with retarded decomposition and accumulation of organic matter. But wetland rice soils in the tropics fall into wet soils with high temperature in all seasons that show rapid mineralization and weak humification (Bonneau, 1982; Neue and Scharpenseel, 1987), both of which favor methane formation.

4.4 Redox Potential

The supply of biodegradable carbon and the activity of the edaphon are the key to most of the characteristic biochemical and chemical processes in flooded soils (Neue, 1988). These processes include soil reduction and associated electrochemical changes; N immobilization and fixation; production of an array of organic compounds, especially organic acids; and release of NH_4^+, CO_2, H_2S, and CH_4. Since methane is produced only by strictly anaerobic bacteria (methanogens), a sufficient low redox potential is required.

The magnitude of reduction is determined by the amount of easily degradable organic substrates; their rate of decomposition; and the amounts and kinds of reducible nitrates, iron and manganese oxides, sulfates, and organic compounds. A rapid initial decrease of Eh after flooding in most soils is caused by high decomposition rates of organic substrates and a low buffer of nitrates and Mn oxides. The most important redox buffer systems in rice soils are Fe(III) oxyhydroxides/Fe(II) and organic compounds stabilizing the Eh somewhat between +100 and −100 mV in most soil solutions. Measurements in the bulk soil may reveal Eh values as low as −300 mV because of direct contact with reduced surfaces of soil particles. The most important interacting chemical changes after flooding an air-dried acid soil are shown in Figure 4a,b.

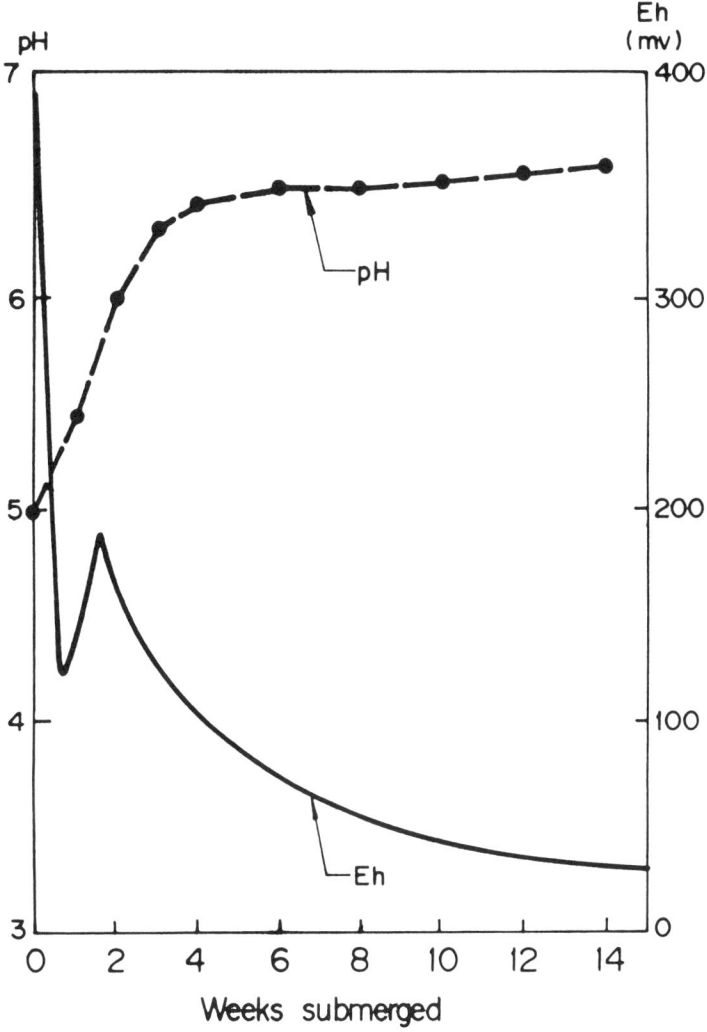

Fig. 4a. Kinetics of pH and Eh in the soil solution of a flooded Ultisol at 30°C (adapted from Ponnamperuma, 1985).

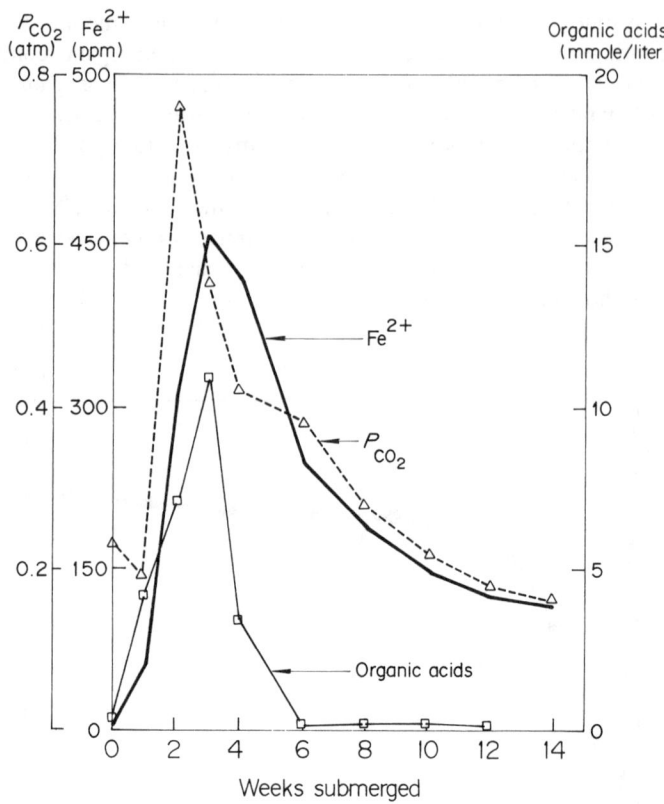

Fig. 4b. Kinetics of P_{CO_2}, water-soluble Fe^{2+} and organic acids in the soil solution of a flooded Ultisol at 30°C (adapted from Ponnamperuma, 1985).

Although the reduction of flooded soils proceeds stepwise in a thermodynamic sequence (Ponnamperuma, 1972; Patrick and Reddy, 1978), the given oxidation-reduction systems are only partially applicable to field conditions. The mineral phases present in soils are not pure and often unknown, and a large portion of reduced Fe^{2+} and Mn^{2+} ions are held at ion exchange sites (Tsuchiya et al., 1986). Changes in pH and activities of reactants and resultants can also alter the order of redox reactions. As a consequence, reduction potentials of a given redox reaction span a fairly wide range not only because of variations at microsites. Nevertheless, redox potentials of the bulk soil (corrected to pH 7) of at least –50 mV are needed for the formation of CH_4.

Chemical reactions that are favored thermodynamically are not necessarily favored kinetically. The lack of effective coupling and the slowness of redox reactions mean that catalysis is required if equilibrium is to be attained. In soils, the catalysis of redox reactions is mediated by microbial organisms. Equilibrium is dependent entirely on the growth and ecological behavior of the soil microbial population and the degree to which the reagents and products can diffuse and mix.

Soil organisms are important with regard to kinetic aspects of redox by affecting the rate of a redox reaction but not its standard free energy change (Sposito, 1981).

Soils low in active iron with high organic matter may attain Eh values of -200 to -300 mV within 2 weeks after submergence (Ponnamperuma, 1972). In soils high in both iron and organic matter, the Eh may rapidly fall to -50 mV and then slowly decline over weeks and level off. Soils where the redox potential is controlled by a ferritic, ferruginous, or oxidic mineralogy and/or the soil reaction is strong acidic or allic are less prone to methane formation (Neue et al., 1990).

4.5 Partial Pressure of CO_2

The partial pressure of CO_2 directly influences CH_4 production since CO_2 is a carbon source for methane. It also affects CH_4 production indirectly because the accumulation of CO_2 coupled with the formation of HCO_3^- buffers the pH near neutral in all flooded soils.

The increase in pH of acid soils is initially brought about by soil reduction of Fe-oxyhydroxides. The pH decrease of sodic and calcareous soil and the final regulation of the pH rise in acid soils are the result of CO_2 accumulation. The pH values at steady state of flooded alkaline, calcareous, and acid soils are highly sensitive to the partial pressure of CO_2. Carbon dioxide that accumulates in large amounts profoundly influences the chemical equilibria of almost all divalent cations (Ca^{2+}, Mg^{2+}, Fe^{2+}, Mn^{2+}, Zn^{2+}) in flooded soils as well as methane formation. Parashar et al. (1990) found the highest emission rates of CH_4 at a pH of 8.2. Acharya (1935) reported that the preliminary stage of acid formation is more tolerant to pH reactions, but gas formation is greatly impeded outside the range of pH 7.5 - 8.

Up to 2.6 t CO_2/ha is produced in the puddled layer during the first few weeks of flooding (IRRI, 1964). After the addition of organic substrates, the partial pressure of CO_2 in a flooded soil may reach a peak of almost 100 kPa (Ponnamperuma, 1985; Neue and Bloom, 1989). Typical values in flooded soils range from 5 to 20 kPa (Kundu, 1987; Patra, 1987). Carbon dioxide concentrations greater than 15 kPa retard root development, leading to wilting and reduced nutrient uptake (Dent, 1986).

At soil temperatures found in flooded tropical soils, CO_2 and CH_4 formation occur sooner and in larger amounts than in cooler climates (Tsutsuki and Ponnamperuma, 1987). The amount of CH_4 found in the soil solution and in gas bubbles of flooded soils may be up to 3 times higher than that of CO_2 after the initial stage of flooding (Martin et al., 1983). The change in favor of CH_4 is likely caused by assimilation of CO_2 and precipitation of carbonates rather than reduction of CO_2 to CH_4, but the controlling processes still need elucidation. According to Takai (1970), the bulk of CH_4 is formed through decarboxylation of acetic acid, which would result in a 1:1 ratio of CO_2 and CH_4 formation.

4.6 Correlations Between Soil Properties and Methane Formation

Neue et al. (1990) identified four crucial parameters for high methane production in wetland rice soils into four crucial parameters aside from carbon supply and water regime: temperature, texture and mineralogy, Eh/pH buffer, and salinity. He suggested that soils are not prone to high methane production if one or more of the following soil characteristics following limits of Soil Taxonomy (USDA, 1975) are met:

- EC > 4 dS/m while flooded,
- acidic or allic reaction,
- ferritic, gibbsitic, ferruginous, or oxidic mineralogy,
- > 40% of kaolinitic or halloysitic clays,
- < 18% clay in the fine earth fraction if the water regime is epiaquic, and
- drought-prone during cultivation period.

Table 5. Coefficient of correlation between soil characteristics and the number of methanogenic bacteria/methane production potential (calculated from data from 29 soils given by Garcia et al., 1974).

	Range	Mean	log no. of methanogens per g of soil	log methane production potential[a]
Clay (%)	2.8-66	28	− 0.488**	− 0.524 **
Silt (%)	7.9-58	23	− 0.227	− 0.145
Sand (%)	1.7-82	37	+ 0.491**	+ 0.486 **
ECs (dS/m)	0.04-5.3	0.95	− 0.543**	− 0.384 *
pH (7 d after flooding [DAF])	3.4-6.8	5.4	+ 0.589**	+ 0.522 **
Eh [7 (DAF) (mV)	+400-135	+116	− 0.661**	− 0.646 **
Carbon content (%)	0.4-9.0	2.1	− 0.067	− 0.071
Total nitrogen (%)	0.04-0.31	0.12	+ 0.044	+ 0.192
C/N	9.9-29	17	− 0.299	− 0.400 *
$S-SO_4^{2-}$[zero day] (mg/kg)	53-1690	380	− 0.437*	− 0.265
$N-NO_3^-$ [zero day] (mg/kg)	0-41.8	2.7	− 0.235	+ 0.005
Cl^- [zero day] meq/100g)	0-42.9	7.1	− 0.496**	− 0.358
No. of denitrifiers (log 10 No./g)	1.7-5.4	3.9	+ 0.305	+ 0.153
Denitrification potential[b]	40->1500	> 370	− 0.637**	− 0.509 **
No. of sulfate-reducing bacteria (log 10 No./g)	1.8-5.9	3.4	− 0.144	− 0.071
Sulfate-reducing index[c]	0-5	2.2	− 0.257	− 0.227
Rice growth index[d]	0-7.8	3.3	+ 0.321	+ 0.473 **

[a] Methane produced during anaerobic incubation of soil at 37°C during 8 to 12 days after flooding;
[b] Denitrification of 100 mg/kg $N-NO_3$ (KNO_3) at 30°C;
[c] Percentage of dead rice plants in standardized conditions of growth favoring sulphatoreduction;
[d] Weight of grain produced in pot experiment.

* and ** = significant at 1% and 5% levels, respectively.

Soils comprising these features are Oxisols, most of the Ultisols, and some of the Aridisols, Entisols, and Inceptisols. Rice soils that are prone to methane production mainly belong to the orders of Entisols, Inceptisols, Alfisols, Vertisols, and Mollisols. Correlations between methane formation and physicochemical features (Tables 5 and 6), calculated from data of 29 soils given by Garcia et al. (1974), support this concept:

Methane production potential is negatively correlated with Eh, ECs, chloride content, sulfate content, and C-N ratio but positively correlated with pH. The pH/Eh values 1 week after flooding, the denitrification potential, and the rice growth index show higher correlations. The high negative correlation with the clay content and positive correlation with the sand content indicate dominance of kaolinitic clays, and/or effect of the large number of salt affected mangrove soils in the sample. Considering only the 11 non-saline soils results in a highly significant correlation between soil carbon as well as soil nitrogen content and methane production potential but not with the number of methanogenic bacteria. This shows that the carbon content influences the activity but not necessarily the density of methanogenic bacteria. In the sample of non-saline soils there is no correlation between methane production potential and clay as well as sand. The positive correlation between methane production and the rice growth index, measured as rice grain yield, clearly indicates that increasing rice production enhances methane formation. Since the rice growth index is also an index for soil fertility of rice soils, it is evident that improving wetland soil fertility combined with current cropping technologies increases methane production and likely methane emission.

Table 6. Correlation between the number of methanogenic bacteria/methane production potential and pH/Eh at different days after flooding (DAF)(calculated from data of 29 soils given by Garcia et al., 1974).

	DAF methanogens	Log no. of production potential	Log methane
pH	0	+ 0.112	+ 0.030
	7	+ 0.589**	+ 0.522**
	14	+ 0.411*	+ 0.293
	21	+ 0.541**	+ 0.385*
	28	+ 0.539**	+ 0.268
Eh	0	− 0.107	− 0.197
	7	− 0.661**	− 0.646**
	14	− 0.438*	− 0.361*
	21	− 0.443**	− 0.303
	28	− 0.380*	− 0.097

* and ** = significant at 1% and 5% levels respectively.

5. Rice Cultivars and Methane Emission

Rice plants play an important role in the flux of methane. Up to 90% of the methane released from the rice soil to the atmosphere is emitted via the rice plant (Bont et al., 1978; Seiler, 1984; Holzapfel-Pschorn et al., 1986). The aerenchyma and intracellular space of rice plants mediate the transport of CH_4 from the reduced soil to the atmosphere (Raimbault et al., 1977). However, up to 80% of the methane produced is apparently oxidized in the rhizosphere (Holzapfel-Pschorn et al., 1985) and the oxidized soil floodwater interface (Figure 5).

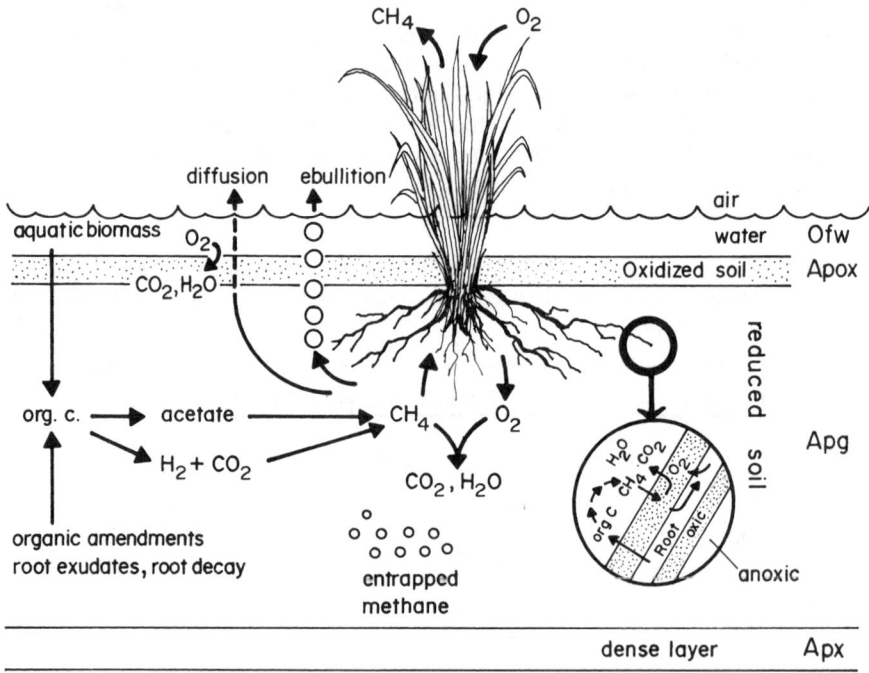

Fig. 5. Schematic of production, reoxidation, and emission of CH_4 in a paddy field. (Modified from Schütz et al., 1989.)

The aerenchyma of rice plants acts as a chimney but the transport mechanisms still have to be elucidated. The well-developed air spaces in leaf blades, leaf sheath, culm and roots provide an efficient gas exchange between the atmosphere and the anaerobic soil. Atmospheric O_2 is supplied via the aerenchyma to the roots for respiration. Oxygen diffusion from rice roots seems to constitute an important part of the root-oxidizing power aside from enzymatic oxidation due to hydrogen peroxide production. Because of the abundance of methane-oxidizing bacteria present in the rhizosphere, its potential for methane oxidation is very high. At tillering, Bont et al. (1978) counted in the rhizosphere 10 times more methane-oxidizing bacteria than in the bulk anaerobic soil and 1/3 more than in the oxidized soil-water interface. They found significant increases in CH_4 emission from cultivar IR36 when suppressing CH_4 oxidation with acetylene at the soil-water interface. However, acetylene had only a small effect on emission rates when applied to the rhizosphere. Bont et al. (1978) concluded that the utilization of O_2 by reduced substances and microbial activity other than methanotrophs in the root-soil interfacial region exceeds the supply of O_2 by the root. Consequently, the aerobic zone surrounding the root is too thin to get the diffusing CH_4 oxidized or the rhizosphere is, for the most part, anaerobic. Rice plants may not only mediate the flux of CH_4, they enhance biological activities in soils and their root exudates and degrading roots may be an important source of CH_4 formation. Sass et al. (1991) found that spatial variability of methane production coincided with spatial distribution of roots in wetland ricefields.

Large cultivar differences in root oxidation power (Figure 6) and in emission rates (Parashar et al., 1990) open up the possibility of breeding rices that emit less methane. The inheritance of underlying traits has still to be elucidated.

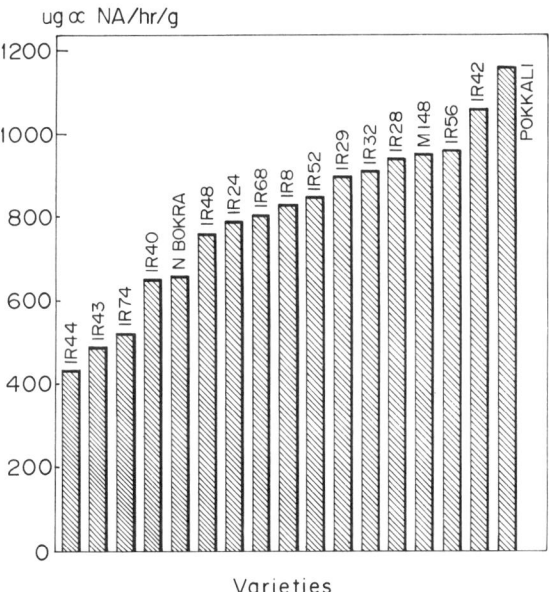

Fig. 6. Root oxidizing power of selected rices.

6. Agronomic Practices Affecting CH₄ Production

Various rice culture systems have been developed to suit the physical, biological, and socioeconomic conditions of different regions and environments. Little is known about the effect of agronomic practices on methane fluxes in wetland ricefields.

6.1 Water Control

Water control (irrigation and drainage) is one of the most important factors in rice production. In many ricefields, crops suffer from either too much or too little water because of rainfall pattern and topography.

During the monsoonal rainy season in tropical Asia, ricelands are naturally flooded. Excess water is a serious constraint in river, lacustrine, and coastal floodplains. Bunding (raising levees around the field) of ricefields and terracing and levelling of sloping land considerably change the water regime of that land. The overall effect is that uncontrolled runoff of water is minimized and more water, whether from natural sources or from irrigation is retained on or in the soil. Bunding and leveling of ricefields is a perfect measure of erosion control and allows efficient water harvesting and water conservation.

The following parameters of floodwater regimes are important for rice growth and should affect methane fluxes as well:
- duration and depth of flooding,
- regularity of flooding as determined by climatic relief and regime of rivers, and
- degree and pattern to which flooding is controlled by irrigation, flood protection, and drainage.

Single ricefields may have two distinct flooding regimes in a given year, especially in pronounced wet-dry monsoonal climates. Ricefields may be naturally shallow to moderately deep flooded during the rainy season and dry out or become shallow flooded by irrigation water during the dry season.

Floodwater control is the prerequisite for most technology changes in rice cultivation. Floodwater and soil water regime primarily determine methanogenesis and will likely be key issues for reducing methane fluxes. Aerating wetland soils to reduce methane fluxes without hampering rice production is a tempting mitigation technology.

But water stress at any growth stage may reduce yield. Moisture stress of 50 kPa (slightly above field capacity) may reduce grain yield to 20-25% of the yield of continually flooded treatments (De Datta, 1981). The rice plant is most sensitive to water deficit during the reproductive stage causing a high percentage of sterility (Yoshida, 1981). Water deficits during the vegetative stage may reduce plant height, tiller number, and leaf area that may also highly reduce yields if plants do not recover before flowering. The duration of a moisture stress is more important than the growth stage at which the stress occurred.

Short aeration periods at the end of the tillering stage and just before heading may improve wetland rice yields (Wang Zhaoqian, 1986) but only if it is followed by flooding. Intermittent irrigation or keeping soils only saturated considerably lowers rice yields (Borrell et al., 1991). Intermittent drying periods or percolation rates of up to 35 mm/day are associated with maximum rice yields in subtropical China and Japan (Wang Zhaoqian, 1986; Iwata et al., 1986). Percolation rates significantly increase yields only above production levels of 6 t/ha (Honya, 1966). Permanent year-round flooding or saturation, which may favor methane production, increases gleying and reduces soil fertility (Li Shi-jun and Li Xue-yuan, 1981), except in acid sulfate soils and some iron toxic soils. Although yields in a triple rice cropping system at IRRI have slightly declined over the past 22 years (Greenland, 1985), there is little evidence that high percolation rates, intermittent drying, or dry fallow periods are needed for high rice yields in most tropical wetland soils. The magnitude of aeration (oxidation) needed is likely dependent and interlinked with decomposition pattern and accumulation of organic and inorganic toxins, and nutrient imbalances involving Fe, Mn, Zn, S, and P (Neue, 1988, 1989). Much more information is needed to understand these interrelationships.

The high water demand for wetland rice becomes a constraint in areas and seasons of limited water resources. Wetland rice requires, on an average, 1240 mm of water (Yoshida, 1981), while upland crops may need less than half (Maesschalck et al., 1985). No methane is produced when upland crops are cultivated on riceland in seasons in which the land is naturally not flooded. This is commonly practiced in areas of limited water resources, especially since modern rice cultivars with short growth duration are available leaving sufficient water in the soil for a following upland crop. If water supply is assured, shifts to upland crops, which would reduce methane emission, are highly dependent on socioeconomic conditions.

6.2 Land Preparation

Tillage operations vary according to water availability, soil texture, topography, rice culture, and resources available. Kawaguchi and Kyuma (1977) found that 40% of the tropical rice soils they studied had at least 45% clay. Soils with such high clay content have a poor structure and are hard when dry. Since hand- and animal-powered tillage are still common in most Asian countries and the principal form of mechanization is only the 10-15 hp tiller (hand tractor), wet tillage is the preferred land preparation. Wet tillage comprises land soaking until the soil is saturated, then plowing, puddling and harrowing. One third of the total water required for a rice crop is needed for the wet field preparation. Two weeks are required to prepare the field for transplanting.

According to De Datta (1981), the advantages of wet tillage are:
• improved weed control;
• ease of transplanting;
• establishment of reduced soil condition which improves soil fertility and fertilizer management;
• reduced draft requirement;
• reduced water percolation;

- reliability of monsoon rains by the time land preparation is completed; and
- higher fertilizer efficiency, especially for N fertilizer.

Flooding and puddling a soil provide an ideal growth medium for rice by supplying abundant water, buffering soil pH near neutral, enhancing N fixation and carbon supply, and increasing diffusion rates, mass flow, and availability of most nutrients.   Standing water stabilizes the soil moisture regime, moderates soil temperature, and prevents soil erosion.

When initial crop growth at early monsoonal rainfall becomes essential because of subsequent floods, as in deepwater rice ecosystems, tillage and seeding are done in dry soils. Less than 200 mm of rainfall for the planting month leads to dryland preparation and seeding. Labor constraints associated with seedbed preparation, land preparation, and transplanting may also force farmers in other rainfed rice ecosystems to dryland tillage and seeding to ensure timely crop establishment. In most rice-growing countries where large power units can be employed because of available capital (as in the United States, Australia, most of Latin America, and Europe), dryland tillage is commonly practiced. In the United States and Australia rice is mostly also sown in dry soil, which is flooded after crop establishment. Upland ricelands are never flooded and tillage is, of course, the same as for other upland crops.

Information on the effect of land preparation on methane fluxes is lacking. Dryland tillage and dry seeding shorten the anaerobic phase and may slow down the decrease of the redox potential resulting in delayed and likely lower methane production. Minimum or zero tillage should have similar effects. However, these land preparation and seeding techniques require likely new rice cultural types, higher fertilizer rates, higher powered tillage implements, and alternative seeding and weeding techniques.

6.3 Seeding and Transplanting

Transplanted rice is the major practice of rice culture in most of tropical Asia. Direct seeding of pregerminated rice in wet prepared soils becomes popular in areas with good water control and if manpower is lacking or becoming expensive. Pregerminated seeds are mostly broadcast onto puddled fields without standing water (saturated soil moisture) and the field is flooded after crop establishment. The crop duration (vegetative phase) is shorter in direct seeded rice, avoiding delay due to seedbed preparation, transplanting, and reduced initial growth because of the transplanting shock. Weed control in broadcast seeded rice and possible moisture stresses because of insufficient water control are the main obstacles in direct wet seeded rice. The yield potential for direct wet seeded rice is similar to that of transplanted rice (De Datta, 1981).

The advantages of transplanting rice seedlings are:
- lower seed requirement (increased tillering),
- save seed establishment (control of pests and fertilization),
- variable schedule of field establishment is possible without risk,
- less requirements for floodwater control, and

- tolerance to biotic and abiotic stresses increases with seedling age.

Methane fluxes should vary between direct seeded and transplanted rice because
- crop duration is shorter in direct seeded rice,
- soil surface is aerated for 7-14 days after land preparation in direct seeded rice,
- growth pattern and canopy development differ, and
- transplanting causes additional soil disturbances.

6.4 Fertilization

The most deficient nutrient for high wetland rice yields is nitrogen, followed by P, K, and Zn. The choice of nitrogen source depends on the method and time of application. Most farmers apply nitrogen fertilizer in two or three parts. The first part is applied during final land preparation or shortly after planting and the remainder as topdressing at later growth stages, especially at the early panicle stage. The most common source of N-fertilizer in wetland rice is urea followed by ammonium-containing fertilizers like ammonium sulfate. The source of nitrogen used as topdressing at later growth stages is less critical because of rapid uptake. In general, K and P are basically applied during the final land preparation. Potassium chloride is the principal fertilizer source of K and superphosphate is the primary source of P fertilizer. On acid rice soils, phosphate rock may be applied. Zn may be added by seed treatments, dipping seedling roots in ZnO solution, or broadcasting Zn salts at the time Zn deficiency symptoms occur.

Studies on fertilizer use and crop management to minimize nitrogen losses (up to 60% due to volatilization of NH_3, nitrification denitrification) and to increase the efficiency of fertilizer have recently been reviewed (De Datta, 1981; De Datta and Patrick, 1986; De Datta, 1987). For basal application, ammonium-containing or ammonium producing (urea) N fertilizer are recommended (De Datta, 1981) to minimize denitrification losses. To reduce volatilization losses incorporation or deep placement of N fertilizer has to be done without standing water at final harrowing. Broadcasting basal N-fertilizer into floodwater results in extensive N losses (as ammonia) to the atmosphere due to high pH values as a result of the algal assimilation or alkaline irrigation water (Fillery and Vlek, 1986).

Reports on the influence of the mineral fertilizer application (source, mode, and rate) on CH_4 production and emission are inconsistent. The complex interrelationships of fertilization on the biochemistry (CH_4 production and oxidation) of flooded soils and on plant growth (plant-mediated emission) still have to be elucidated. Increasing the number of tiller/m^2 and enhancing root growth in methane enriched soil layers through fertilization will obviously increase methane emission. As discussed above, encapsulating methane inhibitors in fertilizer seems to be very promising.

It is evident that organic amendments of flooded soils increase CH_4 production and emission (Schütz et al., 1989) by lowering the Eh and providing carbon sources. Addition of plant residues accelerate and intensify Eh and pH changes (Katyal, 1977). The effect of vetch, which has a narrow C-N ratio, is greater than that of rice straw (Yu, 1985). Changes are more pronounced when organic substrates are added to soils low in organic matter (Nagarajah et al., 1989). Increasing the soil bulk density of flooded soils retards organic matter decomposition, increases the concentration and residue time of organic acids, and reduces the speed of Eh and pH changes as well as methane formation.

Though organic amendments are propagated to sustain soil resources, actual application of organic substrates into wetland ricefields seems on the decline (see Figure 7). In China the production of green manure increased sharply after 1960 and peaked sometime in the 1970s (13.2 million ha), followed by a steep decline to only 6.6 million ha in 1987 (Stone, 1990). In Japan, the decline of green manure cultivation started already in the 1950s. According to Kanazawa (1984), the total addition of organic substrates to ricefields in Japan decreased from 6 t/ha in 1965 to 2.7 t/ha in 1980.

Based on the content of readily mineralizable carbon, humified substrates like compost should produce less methane per unit carbon while rice straw or green manures likely produce more. Application of compost did not remarkably enhance methane emission while application of rice straw significantly increased methane emission irrespective of soil type (Yagi and Minami, 1990). Sound technologies have to consider both maintaining or increasing soil fertility and reducing methane emission. In a sustainable wetland rice system, it may be advisable to minimize rather than to maximize organic soil amendments.

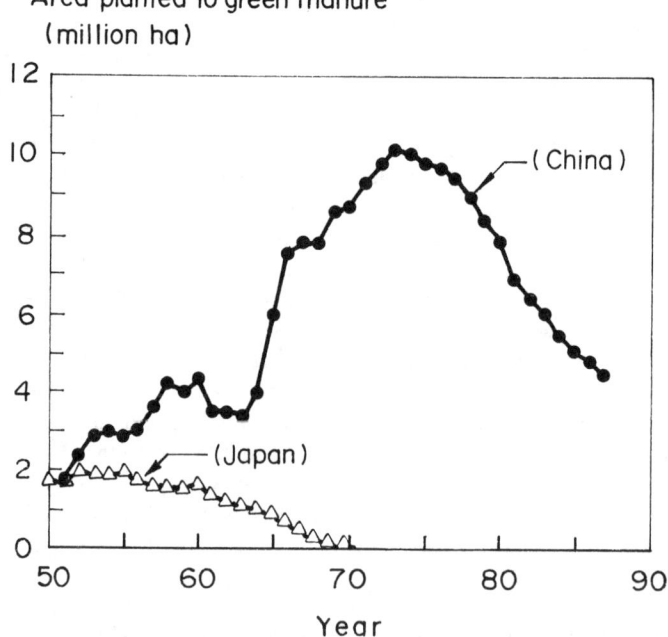

Fig. 7. Area planted to green manure in China (1952-1987) (Stone, 1990) and in Japan (Watanabe, 1984).

6.5 Pest Control

Control of pests in wetland rice ranges from varietal resistance through cultural control, biological control, and chemical control. Application of pesticides to the floodwater, soil surface, or into the soil may have significant effects on methane fluxes, especially if it adversely affects the aquatic and soil flora and fauna. Many cultural control measures, such as cropping pattern, crop residue management, tillage, water, fertility, weeding, or plant spacing and population, affect pests in rice but should affect methane fluxes at the same time. For example, the mechanical disturbance of the soil during weeding (2-3 times per season by hand or small implements) increases the release of gases trapped in the soil. Weeds become an additional source of methane since they are commonly returned to the reduced soil. On the other hand, aquatic weeds may provide a more efficient pathway for CH_4 to the atmosphere than rice plants, as indicated by Holzapfel-Pschorn et al. (1986).

7. Summary

Ricefields provide ideal environments for methanogenesis, especially irrigated ricefields in the tropics, because of anaerobic conditions at neutral pH, optimum temperature, and high easily degradable C inputs. Rice plants favor methane fluxes by supplying carbon and acting as chimney. The high C input in rice soils due to a high primary production both by the crop and the photosynthetic aquatic biomass and organic amendments favor methane emission from ricefields.

Irrigated rice ecologies seem to be the major source for increased global methane emissions from ricefields. The assured water supply and control, the intensive soil preparation, and the resultant improved growth of rice, mediating the methane flux to the atmosphere, favor methane production and emission. Methane emissions should be much lower and highly variable in space and time in rainfed rice because of drought periods during the growing season and the poorer growth of rice. In deepwater rice, methane production may be high but related emission rates are unknown. Upland rice is believed not to be a source of methane emission because upland rice is never flooded for a significant period of time.

Global extrapolations of emission rates are still highly uncertain and tentative. Accounting for variations of emission rates due to climate, soil properties, duration and pattern of flooding, organic amendments, fertilization and rice growth reveals that most published extrapolations are too high. Adjusting a basic emission rate of $0.5 \text{ g m}^{-2} \text{ day}^{-1}$ according to rice ecologies or soil types results in global emission rates of about 40 Tg year^{-1} (Neue, 1992; Bachelet and Neue, 1993). But reported measurements of CH_4 fluxes in rice fields do not account for ebullition induced by soil disturbance due to wet tillage, transplanting, fertilization, weeding, pest control and harvest. A large proportion of soil entrapped methane that may account for up to 90% of the methane generated and is oxidized in undisturbed rice fields likely escapes to the atmosphere during these cultural practices. Therefore, the global CH_4 emission rate from rice fields is likely higher than 40 Tg year^{-1}.

Methane from ricefields may contribute up to one third of the global anthropogenic methane emission, and mitigation technologies are required to stabilize atmospheric methane concentration in the long term. Possible mitigation technologies include reducing C inputs of easily degradable carbon, increasing soil and plant-mediated methane oxidation, reducing emission pathways through the selection and breeding of rice cultivars, and preventing or reducing methane formation through intermittent aeration, sources and mode of fertilizer, and application of chemical inhibitors. The most effective mitigation technology would be to shift from wetland rice to upland cropping whenever possible.

However, technologies that will be accepted by farmers have to be not only environmentally but also socio-economically sound. Because of the limited existing knowledge and the complexity of any methane-mitigating technology, a comprehensive interdisciplinary research approach of various biological and social sciences and common sense are needed to develop and implement feasible technologies.

Acknowledgement. We thank Dr. J. L. Garcia (ORSTOM, France) for providing us unpublished data and for his useful comments.

References

Abram, J.W., D.B. Nedwell. 1978. Inhibition of methanogenesis by sulfate reducing bacteria competing for transferred hydrogen. *Arch. Microbiol., 117*:89-92.

Acharya, C.N. 1935. Studies on the anaerobic decomposition of plant materials. II. Some factors influencing the anaerobic decomposition. *Biochem. J., 29*:953-960.

Alperin, M.J., W.S. Reeburgh. 1984. Geochemical observations supporting anaerobic methane oxidation. In: *Microbial Growth on C-1 Compounds* (R.L. Crawford and R.S. Hanson, eds.), American Society of Microbiology, Washington D.C. p. 282-289.

Anthony, C. 1982. *The Biochemistry of Methylotrophs.* Academic Press, San Diego California.

Aselmann, I., Crutzen, P.J. 1990. Global inventory of wetland distribution and seasonality net primary production and estimated methane emission. In: *Soils and the Greenhouse Effect* (A.F. Bouwman, ed.), John Wiley & Sons, Chichester, England, p 441-450.

Bachelet, D., H.U. Neue. 1993. Methane emissions from wetland rice areas of Asia. *Chemosphere, 26* (1-4):219-246.

Balderston, W.L., W.J. Payne. 1976. Inhibition of methanogenesis in salt marsh sediments and whole-cell suspensions of methanogenic bacteria by nitrogen oxides. *Appl. Environ. Microbiol., 32*:264-269.

Bartlett, K.B., D.S. Bartlett, R.C. Harriss, D.I. Sebacher. 1987. Methane emissions along a salt marsh salinity gradient. *Biogeochemistry, 4*:183-202.

Belay, N.R., R. Sparling, L. Daniels. 1984. Dinitrogen fixation by a thermophilic methanogenic bacterium. *Nature, 312*:286-288.

Biavati, B., M. Vasta, J.G. Ferry. 1988. Isolation and characterization of *Methanosphaera cuniculi* sp.nov. *Appl. Environ. Microbiol., 54*:786-771.

Blotevogel, K.H., U. Fischer, M. Mocha, S. Jannsen. 1985. *Methanobacterium thermoalcapiphilum* sp. nov. a new moderately alkaliphilic and thermophilic autotrophic methanogen. *Arch. Microbiol., 142*:211-217.

Bolle, H.J., W. Seiler, B. Bolin. 1986. Other greenhouse gases and aerosols, assessing their role for atmospheric radiative transfer. In: *The Greenhouse Effect, Climatic Change, and Ecosystems* (B. Bolin, B.R. Döös, J. Jäger, and R.A. Warrick, eds.), Chichester, New York,

Brisbane, Toronto, Singapore, Wiley and Sons; p 157-203.

Bonneau, M. 1982. Soil temperature. In: *Constituents and Properties of Soils* (M. Bonneau and B. Souchier, eds.), Academic Press, London, England, p 366-371.

Bont, J.A.M. de, K.K. Lee, D.F. Bouldin. 1978. Bacterial oxidation of methane in rice paddy. *Ecol. Bull., 26*:91-96.

Borrell, A.K., S. Fukai, A.L. Garside. 1991. Irrigation methods for rice in tropical Australia. *Int. Rice Res. Newsl., 16* (3):28.

Bouwman, A.F. 1990. *Soils and the Greenhouse Effect.* (A.F. Bouwman, ed.), John Wiley.

Bronson, K.F., A.R. Mosier. 1991. Effect of encapsulated calcium carbide on dinitrogen, nitrous oxide, methane and carbon dioxide emissions in flooded rice. *Biology and Fertility of Soils, 3*:116-120.

Buresh, R.J., S.K. De Datta. 1991. Nitrogen dynamics and management in rice-legume cropping systems. *Adv. Agron., 45*:1-59.

Capistrano, R.F. 1988. Decomposition of ^{14}C-labelled rice straw in 3 submerged soils under controlled laboratory conditions. M.S. thesis, University of the Philippines at Los Baños Laguna, Philippines.

Cappenberg, T.E. 1974. Interrelations between sulfate-reducing and methane-producing bacteria in bottom deposits of a fresh-water lake. I. Field observation. *Anton. Leeuwenhoek J. Microbiol. Serol., 40*:285-295.

Cappenberg, T.E., R.A. Prins. 1974. Interrelations between sulfate-reducing and methane-producing bacteria in bottom deposits of a fresh-water lake. III. Experiments with ^{14}C-labelled substrates. *Anton. Leeuwenhoek J. Microbiol. Serol., 40*:457-469.

Cho, D.Y., F.N. Ponnamperuma. 1971. Influence of soil temperature on the chemical kinetics of flooded soils and the growth of rice. *Soil Sci., 112*:184-194.

Cicerone, R.J., R.S. Oremland. 1988. Biogeochemical aspects of atmospheric methane. *Global Biogeochem. Cycles, 2*:299-327.

Conrad, R. 1989. Control of methane production in terrestrial ecosystems. In: *Exchange of Trace Gases Between Terrestrial Ecosystems and the Atmosphere* (M.O. Andreae and D.S. Schimel, eds.), S. Bernhard Dahlem Konferenzen. Wiley, New York, p 39-58.

Conrad, R., R. Bonjour, M. Aragno. 1985. Aerobic and anaerobic microbial consumption of hydrogen in geothermal spring water. *FEMS Microbiol. Lett., 29*:201-206.

Conrad, R., H.P. Mayer, M. Wüst. 1989. Temporal change of gas metabolism by hydrogen-syntrophic methanogenic bacterial association in anoxic paddy soil. *FEMS Microbiol. Ecol., 62*:265-274.

Crawford, R.L., R.S. Hanson (eds.) 1984. Microbial growth on C1 compounds. *Proceedings of the 4th International Symposium American Society for Microbiology*, Washington D.C.

De Datta, S.K. 1981. *Principles and Practices of Rice Production.* John Wiley and Sons New York USA.

De Datta, S.K. 1987. Advances in soil fertility research and nitrogen fertilizer management for lowland rice. In: *Efficiency of Nitrogen Fertilizer for Rice.* International Rice Research Institute, P.O. Box 933, Manila, Philippines, p 27-41.

De Datta, S.K., W.H. Patrick (eds.). 1986. Nitrogen economy of flooded rice soils. *Development in Plant and Soil Sciences.* Martin Nijhoff Publication, Dodrecht, The Netherlands.

De Laune, R.D., E.J. Smith, W.H. Patrick. 1983. Methane release from Gulf Coast wetlands. *Tellus, 35B*:8-15.

Dent, D. 1986. Acid sulfate soils: a baseline for research and development. *ILRI Publication 39.* Wageningen, The Netherlands.

Dickinson, R.E., R.J. Cicerone. 1986. Future global warming from atmospheric trace gases. *Nature, 319*:109-115.

Dolfing, J. 1988. Acetogenesis. In: *Biology of Anaerobic Microorganisms* (A.J.B. Zehnder, ed.) Wiley, New York, p 417-468.

FAO - Food and Agriculture Organization. 1988. *Quarterly Bulletin of Statistics. Vol. 1 No.*

4. FAO, Rome, Italy.

Fillery, I.R.P., P.L.G. Vlek. 1986. Reappraisal of the significance of ammonia volatilization as a N loss mechanism in flooded ricefields. In: *Development in Plant and Soil Sciences* (S.K. De Datta and W.H. Patrick, eds.), Martin Nijhoff Publ., Dodrecht, The Netherlands, p. 79-98.

Franklin, N.J., W.J. Wiebe, W.B. Whitman. 1988. Populations of methanogenic bacteria in Georgia salt marsh. *Appl. Environ. Microbiol., 54*:1,151-1,157.

Garcia, J.L. 1990. Taxonomy and ecology of methanogens. *FEMS Microbiol. Rev., 87*:297-308.

Garcia, J.L., M. Raimbault, V. Jacq, G. Rinaudo, P. Roger. 1974. Activities microbiennes dans les sols de rizieres du senegal: relations avec les proprietes physicochimiques et influence de la rhizosphere. *Rev. Ecol. Biol., 11* (2):169-185.

Greenland, D.J. 1985. Physical aspects of soil management for rice-based cropping systems. In: *Soil Physics and Rice.* International Rice Research Institute, P.O. Box 933, Manila, Philippines, p. 1-16.

Gupta, G.P. 1974. The influence of temperature on the chemical kinetics of submerged soils. Ph.D. thesis, Indian Agricultural Research Institute, New Delhi, India.

Hackman, Ch.W. 1979. Rice field ecology in Northeastern Thailand. The effect of wet and dry season on a cultivated aquatic ecosystem. In: *Monogr. Biol., 34* (J. Illies, ed.), W. Junk Publisher, 22 p.

Harrison, W.H., P.A.S. Aiyer. 1913. The gases of swamp rice soil. I. Their composition and relationship to the crop. Memoires, Department of Agriculture, India. *Chem. Ser., 5*(3): 65-104.

Higgins, I.J., D.J. Best, R.C. Hammond, D.C. Scott. 1981. Methane-oxidizing microorganisms. *Microbiol. Rev., 45*:556-590.

Holzapfel-Pschorn, A., W. Seiler. 1986. Methane emission during a cultivation period from an Italian rice paddy. *J. Geophys. Res., 91*:11,803-11,814.

Holzapfel-Pschorn, A., R. Conrad, W.W. Seiler. 1985. Production oxidation and emission of methane in rice paddies. *FEMS Microbiol. Ecol., 31*:343-351.

Holzapfel-Pschorn, A., R. Conrad R, W. Seiler. 1986. Effects of vegetation on the emission of methane from submerged paddy soil. *Plant Soil, 92*:223-233.

Honya, K. 1966. *Fundamental conditions for high yields of rice.* [in Japanese]. Nobunkyo Publishing, Tokyo.

IRRI - International Rice Research Institute. 1964. Annual report for 1963. P.O. Box 933, Manila, Philippines, 201 p.

IRRI - International Rice Research Institute. 1981. Annual report for 1980. P.O. Box 933, Manila, Philippines, 306 p.

IRRI - International Rice Research Institute. 1989. IRRI toward 2000 and beyond. P. O. Box 933, Manila, Philippines.

IRRI - International Rice Research Institute. 1991. World rice statistics 1990. P. O. Box 933, Manila, Philippines.

Iversen, N., R.S. Oremland, M.J. Klug. 1987. Big Soda Lake (Nevada) 3 pelagic methanogenesis and anaerobic methane oxidation. *Limnol. Oceanogr., 32*:804-814.

Iwata, S., S. Hasegawa, K. Adachi. 1986. Water flow balance and control in rice cultivation. In: *Wetlands and Rice in Subsaharan Africa* (A.S.R. Juo and J.A. Lowe, eds.), IITA Ibadan Nigeria, p. 69-86.

Kanazawa, N. 1984. Trends and economic factors affecting organic manures in Japan. In: *Organic Matter and Rice.* International Rice Research Institute, P.O. Box 933, Manila Philippines, p 557-568.

Katyal, J.C. 1977. Influence of organic matter on chemical and electrochemical properties of some flooded soils. *Soil Biol., 9*:259-266.

Kawaguchi, K., K. Kyuma. 1977. *Paddy Soils in Tropical Asia: Their Material Nature and Fertility.* The University Press of Hawaii, Honolulu, Hawaii, USA.

Khalil, M.A.K., R.A. Rasmussen. 1987. Atmospheric methane: trends over the last 10000 years. *Atmos. Environ., 21* (11):2,445-2,452.

Khalil, M.A.K., R.A. Rasmussen. 1989. Climate induced feedback for the global cycles of methane and nitrous oxide. *Tellus, 41B*:554-559.

Khalil, M.A.K., R.A. Rasmussen. 1990. Constraints on the global sources of methane and an analyses of recent budgets. *Tellus, 428*:229-236.

Kiene, R.P., P.T. Visscher. 1987. Production and fate of methylated sulfur compounds from methionine and dimethylsulfoniopropionate in anoxic marine sediments. *Appl. Environ. Microbiol., 53*:2,426-2,434.

King, G.M. 1984. Metabolism of trimethylamine choline and glycine betaine by sulfate-reducing and methanogenic bacteria in marine sediments. *Appl. Environ. Microbiol., 48*:719-725.

King, G.M. 1988. Methanogenesis from methylated amines in a hypersaline algal mat. *Appl. Environ. Microbiol., 54*:130-136.

Kondo, Y. 1952. Physiological studies on cool-weather resistance of rice varieties. Nogyo Gijutsi Kenkyusho Hokodu Di seiri Inde. Sakrimotsu Ippan (*National Institute of Agriculture Science Bulletin Japan Series) D 3*:113-228.

Koyama, T., M. Hishida, T. Tomino. 1970. Influence of sea salts on the soil metabolism. II. On the gaseous metabolism. *Soil Sci. Plant Nutr., 16*:81-86.

Krumböck, M., R. Conrad. 1991. Metabolism of position-labelled glucose in anoxic methanogenic paddy soil and lake sediment. *FEMS Microbiol. Ecol., 85*:247-256.

Kundu, D.K. 1987. Chemical kinetics of aerobic soils and rice growth. Ph.D. thesis, Indian Agricultural Research Institute, New Delhi, India.

Kuwatsuka, S., K. Tsutsuki, K. Kumada. 1978. Chemical studies on humic acids. I. Elementary composition of humic acid. *Soil Sci. Plant Nutr., 23*:337-347.

Li Shi-jun, Li Xue-yuan. 1981. Stagnancy of water in paddy soils under the triple cropping system and its improvement. In: *Proceedings of Symposium on Paddy Soil*. Institute of Soil Science. Academica Sinica, ed. Science Press, Beijing, and Springer-Verlag, Berlin, p. 509-516.

Lovley, D.R., M.J. Klug. 1983. Methanogenesis from methanol and from hydrogen and carbon dioxide in the sediments of a eutrophic lake. *Appl. Environ. Microbiol., 45*:1,310-1,315.

Maesschalck, G.H., M. Verplancke, M. De Boodt. 1985. Water use and wateruse efficiency under different management systems for upland crops. In: *Soil Physics and Rice*. International Rice Research Institute, P.O. Box 933, Manila, Philippines, p. 397-408.

Martin, U., H.U. Neue, H.W. Scharpenseel, P.M. Becker. 1983. Anaerobe Zersetzung von Reisstroh in einem gefluteten Reisboden auf den Philippinen. *Mitt. Dtsch. Bodenkdl. Gesellsch., 38*:245-250.

Mathrani, I.M., D.R. Boone, R.A. Mah, G.E. Fox, P.P. Lau. 1988. *Methanohalophilus zhilinae* sp.nov., an alkaliphilic halophilic methylotrophic methanogen. *Int. J. Sys. Bacteriol., 38*:139-142.

Matsushima, S., T. Tanaka, T. Hoshino. 1964a. Analysis of yield- determining process and its application to yield prediction and culture improvement of lowland rice. LXX combined effect of air temperature and water temperature at different stages of growth on the grain yield and its components of rice plants. *Proc. Crop Sci. Soc. Jpn., 33*:53-58.

Matsushima, S., T. Tanaka, T. Hoshino. 1964b. Analysis of yield- determining process and its application to yield prediction and culture improvement of lowland rice. LXX combined effect of air temperature and water temperature at different stages of growth on the growth and morphological characteristics of rice plants. *Proc. Crop Sci. Soc. Jpn., 33*:135-140.

McBride, B.C., R.S. Wolfe. 1971. Inhibition of methanogenesis by DDT. *Nature, 234*:551.

Miller, T.L., M.J. Wolin. 1985. *Methanosphaera stadtmaniae* gen.nov.sp.nov.: a species that forms methane by reducing methanol with hydrogen. *Arch. Microbiol., 141*:116-122.

Mitsch, W.J., J.G. Gosselink. 1986. *Wetlands.* Van Nostrand Reinhold Company New York, USA.

Moormann, F.R., N. Van Breemen. 1978. *Rice: Soil Water Land.* International Rice Research Institute, P.O. Box 933, Manila, Philippines.

Murray, P.A., S.H. Zinder. 1984. Nitrogen fixation by a methanogenic bacterium. *Nature, 312*:284-286.

Murrell, J.C., H. Dalton. 1983. Nitrogen fixation in obligate methanotrophs. *J. Gen. Microbiol., 129*:3,481-3,486.

Nagarajah, S., H.U. Neue, M.C.R. Alberto. 1989. Effect of Sesbania Azolla and rice straw incorporation on the kinetics of NH_4, K, Fe, Mn, Zn, and P in some flooded rice soils. *Plant Soil, 116*:37-48.

Neue, H.U. 1985. Organic matter dynamics in wetland soils. In: *Wetland Soils: Characterization, Classification and Utilization.* International Rice Research Institute, P.O. Box 933, Manila, Philippines, p. 109-122.

Neue, H.U. 1988. Holistic view of chemistry of flooded soil. In: *Proceedings of the First International Symposium on Paddy Soil Fertility,* 6-13 December 1988. International Board for Soil Research and Management, Bangkok, p. 21-56.

Neue, H.U. 1989. Rice growing soils: Constraints utilization and research needs. Pages 1-14 *in* Classification and management of rice growing soils. *FFFTC Book Series No. 39.* Food and Fertilizer Technology Center for the ASPAC Region, Taiwan, R.O.C.

Neue, H.U. 1992. Agronomic practices affecting methane fluxes from rice cultivation. In: *Trace Gas Exchange in a Global Perspective, Ecol. Bull. (Copenhagen), 42*:174-182 (D.S. Ojima and B.H. Svensson, eds.).

Neue, H.U., H.W. Scharpenseel. 1984. Gaseous products of the decomposition of organic matter in submerged soils. In: *Organic Matter and Rice.* International Rice Research Institute, P.O. Box 933, Manila, Philippines, p. 311-328.

Neue, H.U., H.W. Scharpenseel. 1987. Decomposition pattern of ^{14}C-labelled rice straw in aerobic and submerged rice soils of the Philippines. *Science Total Environ., 62*:431-434.

Neue, H.U., P.R. Bloom. 1989. Nutrient kinetics and availability in flooded soils. In: *Progress in Irrigated Rice Research.* International Rice Research Institute, P.O. Box 933, Manila, Philippines, p 173-190.

Neue, H.U., P. Becker-Heidmann, H.W. Scharpenseel. 1990. Organic matter dynamics soil properties and cultural practices in ricelands and their relationship to methane production. In: *Soils and the Greenhouse Effect* (A.F. Bouwman, ed.), John Wiley & Sons, Chichester, England, p. 457-466.

Oremland, R.S., S. Polcin. 1982. Methanogenesis and sulfate-reduction:competitive and noncompetitive substrate in estuarine sediments. *Appl. Environ. Microbiol., 44*:1,270-1,276.

Oremland, R.S., D.G. Capone. 1988. Use of "specific" inhibitors in biogeochemistry and microbial ecology. *Adv. Microbiol. Ecol., 10*:285-383.

Oremland, R.S., L.M. Marsh, S. Polcin. 1982. Methane production and simultaneous sulfate reduction in anoxic salt marsh sediments. *Nature (London), 296*:143-145.

Panganiban, A.T., T.E. Patt, W. Hart, R.S. Hanson. 1979. Oxidation of methane in the absence of oxygen in lake water samples. *Appl. Environ. Microbiol., 37*:303-309.

Parashar, D., C.J. Rai, P.K. Gupta, N. Singh. 1990. Parameters affecting methane emission from paddy fields. *Indian J. Radio Space Physics, 20*:12-17.

Patel, G.B., L.A. Roth. 1977. Effect of sodium chloride on growth and methane production of methanogens. *Can. J. Microbiol., 6*:893.

Patra, P.K. 1987. Influence of water regime on the chemical kinetics of soils and rice growth. Ph.D. thesis, Indian Agricultural Research Institute, New Delhi, India.

Patrick, W.H., Jr., C.N. Reddy. 1978. Chemical changes in rice soils. In: *Soils and Rice.* International Rice Research Institute, P.O. Box 933, Manila, Philippines, p 361-380.

Patrick, W.H., Jr., D.S. Mikkelsen, B.R. Wells. 1985. Plant nutrient behavior in flooded soil.

In: *Fertilizer Technology and Use, 3d Ed.* Soil Science Society of America Madison Wisconsin.

Patt, T.E., G.C. Cole, J. Bland, R.S. Hanson. 1974. Isolation and characterisation of bacteria that grow on methane and organic compounds as sole source of carbon and energy. *J. Bacteriol., 120*:955-964.

Ponnamperuma, F.N. 1972. The chemistry of submerged soils. *Adv. Agron., 24*:29-96.

Ponnamperuma, F.N. 1981. Some aspects of the physical chemistry of paddy soils. In: *Proceedings of the Symposium on Paddy Soils.* Science Press, Beijing People's Republic of China, p 59-94.

Ponnamperuma, F.N. 1984a. Effects of flooding on soils. In: *Flooding and Plant Growth* (T.T. Kozlowski, ed.), Academic Press, New York, USA, p 9-45.

Ponnamperuma, F.N. 1984b. Straw as a source of nutrients for wetland rice. In: *Organic Matter and Rice.* International Rice Research Institute, P.O. Box 933, Manila, Philippines, p 117-136.

Ponnamperuma, F.N. 1985. Chemical kinetics of wetland rice soils relative to soil fertility. In: *Wetland Soils: Characterization Classification and Utilization.* International Rice Research Institute, P.O. Box 933, Manila, Philippines, p 71-89.

Raimbault, M. 1975. Étude de l'influence inhibitrice de l'acétylene sur la formation biologique du méthane dans un sol de riziére. *Ann. Microbiol. (Inst. Pasteur), 126a*:217-258.

Raimbault, M. 1981. Inhibition de la formation de methane par l'acétylene chez Methananosarcina bakerii. *Cah. ORSTOM, Ser. Biol., 43*:45-51.

Raimbault, M., G. Rinaudo, J.L. Garcia, M. Boureau. 1977. A device to study metabolic gases in the rice rhizosphere. *Biol. Biochem., 9*:193-196.

Rajagopal, B.S., N. Belay, L. Daniels. 1988. Isolation and characterization of methanogenic bacteria from rice paddies. *FEMS Microbiol. Ecol., 53*:153-158.

Roger, P.A., I. Watanabe. 1984. Algae and aquatic weeds as source of organic matter and plant nutrients for wetland rice. In: *Organic Matter and Rice.* International Rice Research Institute, P.O. Box 933, Manila, Philippines, p. 147-168.

Roger, P.A., I.F. Grant, P.N. Reddy, I. Watanabe. 1987. The photosynthetic aquatic biomass in wetland ricefields and its effect on nitrogen dynamics. In: *Efficiency of N Fertilizers for Rice.* International Rice Research Institute, P.O. Box 933, Manila, Philippines, p 43-68.

Salvas, P.L., B.F. Taylor. 1980. Blockage of methanogenesis in marine sediments by the nitrification inhibitor 2-chloro-6-(trichloromethyl) pyridine (Nitrapin or N-serve) *Curr. Microbiol., 4*:305.

Sass, R.L., F.M. Fischer, P.A. Harcombe, F.T. Turner. 1991. Methane production and emission in a Texas rice field. *Global Biogeochem. Cycles, 4*:47-68.

Schink, B., J.G. Zeikus. 1980. Microbial methanol formation: a major end product of protein metabolism. *Curr. Microbiol., 4*:387-389.

Schönheit, P., H. Keweloh, R.K. Thauer. 1981. Factor F_{420} degradation in *Methanobacterium thermoautotrophicum* during exposure to oxygen. *FEMS Microbiol. Lett., 12*:347-349.

Schütz, H., A. Holzapfel-Pschorn, R. Conrad, H. Rennenberg, W. Seiler. 1989. A three-year continuous record on the influence of daytime season and fertilizer treatment on methane emission rates from an Italian rice paddy field. *J. Geophys. Res., 94*:16,405-16,416.

Seiler, W. 1984. Contribution of biological processes to the global budget of CH_4 in the atmosphere. In: *Current Perspectives in Microbial Ecology* (M.J. Kleig and C.A. Reddy, eds.), American Society of Microbiology, Washington D.C., p 468-477.

Seiler, W., R. Conrad. 1987. Contribution of tropical ecosystems to the global budget of trace gases especially CH_4 H_2 CO and N_2O. In: *The Geography of Amazonia: Vegetation and Climate Interactions* (R.E. Dickinson, ed.), Wiley, N.Y., p 133-162.

Sequi, P., M. De Nobili, L. Leita L, G. Cerciguani. 1986. A new index of humification. *Agrochemical, 30*:175-179.

Sharma, P.K., S.K. De Datta. 1985. Effects of puddling on soil physical properties and processes. In: *Soil Physics and Rice.* International Rice Research Institute, P.O. Box 933, Manila, Philippines, p 217-234.

Smith, J., H.U. Neue, G. Umali. 1987. Soil nitrogen and fertilizer recommendations for irrigated rice in the Philippines. *Agric. Sys., 24*:165-181.

Smith, P.H., R.E. Hungate. 1958. Isolation and characterization of *Methanobacterium ruminantium* n.sp. *J. Bacteriol., 75*:713-718.

Snitwongse, P., S. Pongpan, H.U. Neue. 1988. Decomposition of ^{14}C-labelled rice straw in a submerged and aerated rice soil in Northeastern Thailand. In: *Proceedings of the First International Symposium on Paddy Soil Fertility,* 6-13 December 1988. International Board for Soil Research and Management, Bangkok, p 461-480.

Sposito, G. 1981. *The Thermodynamics of Soil Solutions.* Clarendon Press, Oxford.

Stone, B. 1990. Evolution and diffusion of agricultural technology in China. In: *Sharing Innovation Global Perspectives on Food Agriculture and Rural Development* (N.G. Kotler, (ed.), International Rice Research Institute, P.O. Box 933, Manila, Philippines, p 35-93.

Strayer, R.F., J.M. Tiedje. 1978. Kinetic parameters of the conversion of methane precursors to methane in hypereutrophic lake sediment. *Appl. Environ. Microbiol., 36*:330-340.

Svensson, B.H. 1984. Different temperature optima for methane formation when enrichments from acid peat are supplemented with acetate or hydrogen. *Appl. Environ. Microbiol., 48*:389-394.

Takai, Y. 1961. Reduction and microbial metabolism in paddy soils (3) [in Japanese English summary]. *Nogyo Gijitsu (Agro. Technol.), 19*:122-126.

Takai, Y. 1970. The mechanism of methane fermentation in flooded soils. *Soil Sci. Plant Nutr., 16*:238.

Takai, Y., T. Koyama, T. Kamura. 1956. Microbial metabolism in reduction process of paddy soil. Part I. *Soil Plant Food, 2*:63-66.

Toukdarian, A.E., M.E. Lidstrom. 1984. Nitrogen metabolism in a new obligate methanotroph *Methylosinus* strain 6. *J. Gen. Microbiol., 130*:1,827-1,837.

Tsuchiya, K., H. Wada, Y. Takai. 1986. Leaching of substances from paddy soils. 4. Water solubilization of inorganic components in submerged soils. *Jpn. J. Soil Sci. Plant Nutr., 57* (6):593-597.

Tsutsuki, K., S. Kuwatsuka. 1978. Chemical studies on soil humic acids. II. Composition of oxygen-containing functional groups of humic acids. *Soil Sci. Plant Nutr., 24*:547-560.

Tsutsuki, K., K. Kumada. 1980. Chemistry of humic acids [in Japanese; English summary]. *Fert. Sci., 3*:93-171.

Tsutsuki, K., F.N. Ponnamperuma. 1987. Behavior of anaerobic decomposition products in submerged soils. Effects of organic material amendment soil properties and temperature. *Soil Sci. Plant Nutr., 33* (1):13-33.

USDA -- United States Department of Agriculture, Soil Conservation Service Soil Survey Staff (1975) Soil taxonomy: a basic system of soil classification for making and interpreting soil surveys. *USDA Agric. Handb. 436.* U.S. Government Printing Office, Washington, D.C.

Vogels, G.D., J.T. Keltjens, C. Van der Drift. 1988. Biochemistry of methane production. In: *Biology of Anaerobic Microorganisms* (A.J.B. Zehnder, ed.), Wiley New York, p 707-770.

Wang, Zhaoqian. 1986. Rice-based systems in subtropical China. In: *Wetlands and Rice in Subsaharan Africa* (A.S.R. Juo and J.A. Lowe, eds.), IITA Ibadan Nigeria, p 195-206.

Ward, D.M., M.R. Winfrey. 1985. Interactions between methanogenic and sulfate-reducing bacteria in sediments. *Adv. Aquatic Microbiol., 3*:141-179.

Watanabe, I. 1984. Use of green manures in Northeast Asia. In: *Organic Matter and Rice.* International Rice Research Institute, P.O. Box 933, Manila, Philippines, p 229-233.

Watanabe, I., P.A. Roger. 1985. Ecology of flooded ricefields. In: *Wetland Soils: Characterization Classification and Utilization.* International Rice Research Institute, P.O.

Box 933, Manila, Philippines, p 229-246.

Whittenbury, R., K.A. Phillips, J.K. Wilkinson. 1970a. Enrichment isolation and some properties of methane-utilizing bacteria. *J. Gen. Microbiol., 61*:205-218.

Whittenbury, R., S.L. Davies, J.F. Davey. 1970b. Exospores and cysts formed by methane-utilizing bacteria. *J. Gen. Microbiol., 61*:219-226.

Whitton, B.A., J.A. Rother. 1988. Environmental features of deepwater ricefields in Bangladesh during the flood season. In: *1987 International Deepwater Rice Workshop.* International Rice Research Institute, P.O. Box 933, Manila, Philippines, p 47-54.

Williams, R.T, R.L. Crawford. 1984. Methane production in Minnesota peatlands. *Appl. Environ. Microbiol., 47*:1,266-1,271.

Williams, R.T., R.L. Crawford. 1985. Methanogenic bacteria including an acid tolerant strain from peatlands. *Appl. Environ. Microbiol., 50*:1,542-1,544.

Winfrey, M.R, J.G. Zeikus. 1977. Effect of sulfate on carbon and electron flow during microbial methanogenesis in freshwater sediments. *Appl. Environ. Microbiol., 33*:275-281.

Worakit, S., D.R. Boone, R.A. Mah, M.E. Abdel-Samie, M.M. El-Halwagi. 1986. *Methanobacterium alcaliphilum* sp. nov. an H_2-utilizing methanogen that grows at high pH values. *Int. J. Syst. Bacteriol., 36*:380-382.

Yagi, K., K. Minami. 1990. Effects of organic matter application on methane emission from Japanese paddy fields. In: *Soil and the Greenhouse Effects* (A.F. Bouwman, ed.), John Wiley, p 467-473.

Yamane, I., S. Sato. 1961. Effect of temperature on the formation of gases and ammonium nitrogen in the waterlogged soils. *Rep. Inst. Agric. Res. Tokoku Univ., 12*:1-10.

Yoshida, S. 1981. *Fundamentals of Rice Crop Science.* International Rice Research Institute, P.O. Box 933, Manila, Philippines. 269 p.

Yu, T. 1985. *Physical Chemistry of Paddy Soils.* Springer-Verlag, Berlin.

Zeikus, J.G., D.L. Henning. 1975. *Methanobacterium arboriphilus* sp.nov. an obligate anaerobe isolated from wetwood of living trees. Antonie van Leeuwenhoek. *J. Microbiol. Serol., 41*:543-552.

Zhao, Y., D.R. Boone, R.A. Mah, J.E. Boone, L. Xun. 1989. Isolation and characterization of *Methanocorpusculum labreanum* sp.nov. from the LaBrea Tar Pits. *Int. J. Syst. Bacteriol., 39*:10-13.

10. Rice Agriculture: Emissions

M.J. Shearer[1] and M.A.K. Khalil[2]
Environmental Sciences & Resources Program[1]
Department of Physics[2]
Portland State University
Portland, Oregon 97207-0751 U.S.A.

1. Introduction

Rice agriculture has long been recognized as a major source of methane (CH_4). Global budgets of methane have generally included emissions of about 100 Tg/yr (Tg = Teragrams, 10^{12} grams) from rice agriculture (with a range of 50-300 Tg/yr), constituting about 20% of emissions from all sources (range 14%-40%) (Ehhalt and Schmidt, 1978; Donahue, 1979; Khalil and Rasmussen, 1983; Blake, 1984; Bolle et al., 1986; Bingemer and Crutzen, 1987; Cicerone and Oremland, 1988; Warneck, 1988; Khalil and Rasmussen, 1990). The Inter-governmental Panel on Climate Change (IPCC, 1995) estimates a source of 60 Tg/yr (range 20-100 Tg/yr). We estimate that rice agriculture contributes some 330 ppbv to the present atmospheric burden of CH_4 in the atmosphere, and may be responsible for some 20-30% of the increase of methane during the last century (Khalil et al., 1996).

Estimating the flux of methane from rice fields in various parts of the world requires knowledge of two factors: the emission rates and the regional or global extrapolant. The emission rate or flux depends on different internal and external variables (x_i). Internal variables include: soil characteristics; rice variety; and soil microbiology. External factors include: soil temperature driven by solar radiation; meteorological conditions; water level, which is affected by rainfall and availability of irrigation; and treatments such as the type and amount of fertilizers applied. The global flux, F_G, is usually calculated by equation (1):

$$F_G = \sum_R \overline{\phi}_R \, T_R \, A_R \qquad (1)$$

where $\overline{\phi}_R$ (g/m^2/day) is the measured, seasonally averaged emission rate from a region R; A_R (m^2) is the area of the region, presumably with similar characteristics, so that $\overline{\phi}_R$ is an accurate representation of flux for the entire area; and T_R is the growing season (days/year). The global extrapolant is $A_R \, T_R$. Any region, such as a country, may also be subdivided into similar regions, and the flux of CH_4 from rice agriculture from the country can be calculated based on an equation analogous to equation (1). The nature of the extrapolation process is discussed in more detail by

Mohammad Aslam Khan Khalil (Ed.)
Atmospheric Methane
© *Springer-Verlag Berlin Heidelberg 2000*

Khalil and Shearer (this volume). The following sections cover estimation of seasonal flux, area, and methane emission season.

2. Methane Emission Rates from Rice Fields

Direct flux measurements over the entire growing period provide precise and accurate values for ϕ_R . Usually, A_R and T_R are determined from agricultural statistics and are also well known. Problems with the extrapolation arise in associating a measured flux with an appropriate area and growing season and in the assumption that the measured flux is representative of the associated area and season.

This empirical approach requires whole season measurements of CH_4 emission rates from as many regions as possible so that the large variations of emissions from one region to another can be properly included in the estimate of global or regional emissions. Alternative approaches to global extrapolation also exist and are based on the knowledge of the processes or factors that control emission rates (Cao et al., 1995a). At present there are insufficient data for such approaches to produce better estimates of global or regional emissions.

2.1 Flux Measurements

During the last decade a number of systematic experiments have been reported on methane emissions from rice fields. All are based on static chamber methods. While there are many variants, the method consists of enclosing a small part ($0.1 - 100$ m^2) of the rice fields within a chamber and taking periodic samples. Methane, emitted from the enclosed area of the rice field, builds up in the chamber. The rate of accumulation is directly proportional to the flux or the rate of emission from the area covered by the chamber. The relationship is:

$$\phi = \frac{\rho \, M \, V}{N_A \, A} \times 10^{-6} \frac{dC}{dt} \qquad (2)$$

where ϕ is the flux, ρ is the density of air (molecules/m^3), M is the molecular weight of CH_4, V is the volume of the chamber (m^3), A is the area covered by the chamber (m^2), N_A is Avogadro's number, C is the concentration of methane (ppbv) and dC/dt is in (ppbv/hr). The resulting units for CH_4 fluxes are mg/m^2/hr.

The advantages of chamber methods are that they are inexpensive, easy to use in remote locations and can be coupled to a highly sensitive and precise measurement method such as gas chromatography. Since the fluxes of methane are quite large, the plants need not be exposed to the unnatural conditions of the chamber environment for very long; often 10 minute exposures are sufficient. This fact also makes chamber methods suitable for methane measurements even though they may not be convenient for other gases.

The disadvantages are that placing the chamber can disturb the soil and release abnormal amounts of methane. Several methods have been devised to reduce, if not overcome this problem. In the studies of Khalil et al. (1991, 1998a) a permanent base is installed in the soil at the time rice is planted and chambers fit into grooves in this base. In studies reported by Schütz et al. (1989a, 1990) a large permanent chamber is used that has a lid that opens and closes, but the chamber itself is not removed until the rice is harvested. Both these methods create some feedbacks that may affect flux estimates. The chambers also affect the immediate environment of the rice plant by causing heating and a buildup of CO_2, which may affect emissions of methane. (See Livingston and Hutchinson, 1995, for a full review.) Secondly, since most chambers are small, the number of plants in the chamber must be representative of the overall planting density, or the flux measurements are affected. For example: in experiments at TuZu, China (Khalil et al., 1998a, 1998b) one, two, or four plants could be placed in the chamber, and still be close to the prevailing planting density of the rice fields studied. The number of plants inside the chambers had a complex effect on the measured emission rates varying throughout the growing season (Figure 1). Finally, the extrapolation of direct flux measurements from small areas to large regions may be unreliable because of heterogeneities within fields, within local regions, and within different parts of the same country.

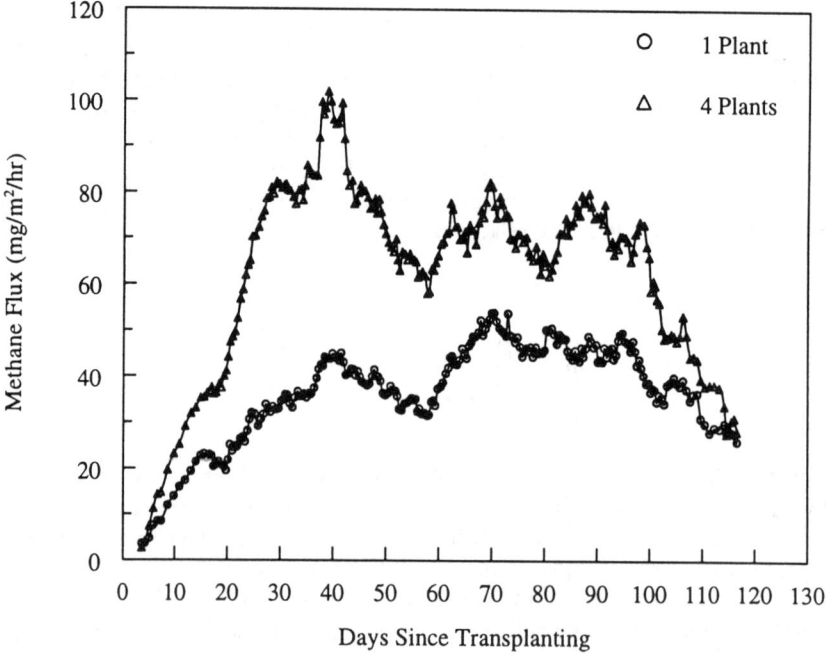

Fig. 1. The effect of planting density on methane emissions from rice fields during the growing season. The data are used for years when one plant per plot and four plants per plot measurements were made.

In recent years experiments have been designed to measure the flux throughout the growing season. The earliest experiments of this type made it clear that there are large systematic changes in methane emissions during the growing cycle. Such changes are driven by several factors including the growth of root mass, maturation process for the plants, availability of nutrients and fertilization, and the seasonal change of temperature and length of day (Schütz et al., 1989a; Yagi and Minami, 1990; Khalil et al., 1991).

2.2 Factors Affecting Methane Emissions

Methane emissions from rice cultivation result from a combination of three processes: methane production in the paddy soil, methane oxidation at the soil surface and in the root zone, and methane transport from the soil to the atmosphere by diffusion through the flood water, ebullition, and plant-mediated transport.

Methane is produced in saturated soils by anaerobic bacteria (methanogens), which use chemicals from the decay of organic matter as their food source and produce methane as a by-product. This process takes place only where oxygen is not available, such as in flooded paddy fields and wetlands. While the life cycle of methanogens is understood, predicting the population fluctuations in rice fields is not yet possible. Methanogens do not directly break down organic matter but require other bacteria to produce the substrates they need for food. The population dynamics of the various interdependent strains of bacteria may account for year to year variation of methane fluxes found in field studies where all other variables are nearly the same. Chemicals exuded from the root system of the plants may also influence the growth of methanogens or competing bacteria populations (Boone, this volume; Neue and Roger, this volume; and references therein, Wassmann et al., 1993a).

Experiments to measure the production and emission of methane from paddy soils showed that not all methane produced in the soil is emitted to the atmosphere (Holzapfel-Pschorn et al., 1985; Schütz et al., 1989b, Sass et al., 1992). Laboratory studies of paddy soil have shown that 50% to 90% of the methane produced in the soil can be destroyed before it reaches the surface (summarized in Neue and Roger, this volume). Strains of bacteria that consume methane (methanotrophs) are believed to live in both soil and water. They oxidize the carbon in CH_4 and produce CO_2 as a by-product. This reaction can take place even in the low oxygen environment of a paddy soil. Air transported to the roots of the rice plant creates an oxygenated zone around the roots. There the methanotrophs can break down methane before it is pumped out of the soil by the air circulation system of the plant.

In the same experiments, the pathways of methane emission to the atmosphere throughout the growing season were also measured. Most of the methane (up to 90%) was found to be transported through the intercellular spaces in the plant. About 10% of the methane was transported in bubbles from the soil, while less than 1% of the methane diffused through the soil and water to the atmosphere. At the beginning of the growing season all the methane was emitted through ebullition; as the rice plants developed, an increasing amount was transported through the plant. Different rice cultivars may affect the transport of methane by different root mass and tiller density, and affect methanogen populations by root exudates.

Water regime, nutrient availability, temperature, and soil factors such as texture, organic matter content, pH and redox potential all may affect the growth of the methanogen populations and the rice plants, and ultimately the methane flux from the paddy fields. Agricultural management, particularly irrigation, addition of fertilizers and other soil amendments, or planting high yield cultivars, affects these factors. Over the last 15 years, many experiments have been done to identify the most important factors and to quantify their effects.

2.2.1. Water Regimes. Rice agriculture may be divided into wetland and dryland (or upland) culture (Grist, 1986; Neue et al., 1990). In wetland culture, the soil is prepared by puddling to reduce water loss, and dikes or levees are built to contain the water. Wetland culture may be separated into irrigated and rainfed rice. Irrigation gives more control over the water regime, which may vary from flood water maintained over the field throughout the growing season, to one or more planned drainages, to flushing the field with water periodically to keep the soil saturated. Rainfed rice is planted during the local wet season; since all water is provided by rainfall the rice fields are subject to drought. Dry spells in any wetland water regime may decrease methane fluxes by increasing oxidation in the soil. Allowing the soils to dry during some stages of plant growth, especially heading and flowering, leads to drastically reduced yields (see summaries in Grist, 1986; Bouwman, 1991).

Dryland rice culture is usually a low-yield, subsistence agriculture, highly susceptible to drought, where no special preparations are made to retain water in the fields. Dryland rice may not be a source of methane at all, as the soil is not saturated long enough for methanogen populations to build up. It is usually ignored in global inventories of methane emissions for this reason, despite the large area of dryland rice in some parts of the world.

Agricultural reasons for mid-season drainage of rice fields are: weeding and fertilization (Husin et al., 1995); to enhance root growth of the rice plant (Chen et al., 1993); to decrease toxic by-products of soil reduction (Yagi et al., 1996); and to conserve irrigation water (Grist, 1986). Intermittent irrigation is practiced only where a high degree of control over water management is possible. Any irrigation management scheme using controlled drainage is practiced only when water is available to re-flood the field when needed (Grist, 1986).

Controlled field comparisons of water regimes have been carried out in Texas, USA (Sass et al., 1992; Sigren et al., 1997b); near Beijing, P.R. China (Chen et al., 1993; Yao and Chen, 1994b); in Indonesia (Husin et al., 1995); and Japan (Yagi et al., 1996). Because of the expense and manpower requirements for a full experimental design, most of these experiments have one field per irrigation treatment with multiple plots. None of the studies completed thus far can give an estimate of both within and between field variation in a single growing season. Also, since a "flooded" control plot may actually have 0 to 3 drained periods, the results of the experiments may not be exactly comparable.

Sass et al. experimented with four water management regimes: the control, permanent flooding after a seeded rice crop was established, maintained until about 10 days before harvest; one mid-season drainage (similar to practice in Japan, see e.g. Yagi and Minami, 1990); multiple (3) drainages; and late season flood. The mid-season drainage reduced methane flux by about 48%; multiple drainages reduced fluxes by nearly 90%. The late season flood had higher methane emissions and lower rice yields. The experiment was continued at a nearby site (Sigren et al., 1997b) with unexpectedly larger fluxes throughout the growing season from the fields with 2 drainages compared to the flooded control for the first year of the study, but about a 40% overall reduction in the field with one mid-season drainage compared to the control for the next year.

In Chen et al. (1993), one experiment provided a direct comparison between two common water management techniques: intermittent irrigation, i.e., after flooding, the field was allowed to dry naturally, and re-flooded 3 to 5 days later; and flooded irrigation, i.e., fields immersed from transplant to rice maturity. The seasonal average methane fluxes from intermittent irrigation were nearly 60% lower than fluxes from the flooded irrigation plot. Yao and Chen (1994b) compared intermittent irrigation to "local" flooding in fields near Beijing. Local flooding consists of "field baking" where the field is allowed to drain for a few days twice during the growing season to enhance root development. Intermittent irrigation fluxes were 12% lower than the flooded field over the growing season.

Husin et al. (1995) tested three water management techniques: flooded (5 cm water depth), intermittent irrigation (field allowed to drain then reflooded); and saturated soil (field flushed to keep soil at field capacity). All fields were maintained at saturated conditions until 8 days after transplant, when the first two irrigation treatments were flooded. All fields were allowed to go to saturated soil conditions twice during the growing season for fertilization and weeding. They remained saturated for 3-4 days before re-flooding. The intermittently irrigated plots had 40-55% lower emissions than flooded plots, and saturated soil plot emissions were 60-85% lower than flooded plots. The saturated soil plots had 4 times the weeds of the other water management treatments; rice yield was lower than the other treatments but the difference was not statistically significant.

Yagi et al. (1996) found a reduction in methane fluxes of 40-45% between flooded irrigation and intermittent irrigation for two growing seasons at Ryugasaki, Japan. The flooded plots in both seasons were continuously flooded until three to four weeks before harvest. In the first year, two mid-season drainages of 4 to 6 days were compared to the flooded plot. In the second experiment, eleven short drainages from four hours to three days in length took place throughout the growing season. The latter form of intermittent irrigation simulated irrigation practice by Japanese farmers. The authors used continuous sampling to measure fluxes, and captured a burst of emitted methane just as the water table dropped below the soil. They suggest that methane emissions from intermittent irrigation would be 7-10% higher if these bursts were measured.

In summary, reducing the duration of flooding on a rice field reduced the seasonal emission of methane by about 50% (range: ± 40%). The brief periods without standing water in the rice fields appear to increase oxidation of methane at the soil surface, and may also lead to decreases in the methanogen population. In global inventories, fluxes from intermittent irrigation may be used to represent rainfed rice areas with periodic droughts, and some areas of irrigated rice.

2.2.2. Soil Amendments. Some fertilizer treatments have also been found to affect methane flux. The addition of rice straw, manure, compost, and other organic amendments appears to enhance methane production, probably by providing a food supply for the methanogens (Schütz et al., 1989a; Yagi and Minami, 1990; Sass et al., 1991a; Cicerone et al., 1992; Chen et al., 1993; Lindau and Bollich, 1993; Jermsawatdipong et al., 1994; Denier van der Gon and Neue, 1995a; Debnath et al., 1996; Khalil et al., 1998c). Organic amendments appear to augment methane emission most during the early part of the growing season, before decaying litter and roots from the rice plant are added to the soil. Seasonal fluxes increase 100 to 200% with addition of organic amendments, though in some extreme cases the flux increased by greater than 10 times the mineral fertilizer control (Cicerone et al., 1992; Jermsawatdipong et al., 1994: Pathumthani, Thailand). In a few cases where fluxes were already high and organic fertilizer application was low (Wassmann et al., 1993b), or manure had been applied heavily in all fields in years prior to the test (Khalil et al., 1998b) the effect of additional organic matter application was not significant.

Nitrogen fertilizers, especially ammonium sulfate, may inhibit methane production, possibly by chemical competition (reviewed in Bouwman, 1991; Wang et al., 1992). In field trials, most nitrogen fertilized plots have seasonal emissions ± 20% of the control, suggesting that a positive influence on plant growth, and hence methane transport, may offset expected negative effects due to soil chemistry. Cicerone et al. (1992) and Lindau (1994) found nitrogen fertilization increased methane fluxes from 20 to 270% compared to non-nitrogen fertilized control plots. Near Beijing, Yao and Chen (1994a) compared alternative nitrogen fertilizers to a fertilization with ammonium bicarbonate (the nitrogen fertilizer used locally). They found equivalent seasonal fluxes for all fertilizers the first year of the experiment, but fluxes from urea and ammonium sulfate were about 60% lower than the control the second year. Schütz et al. (1989a) found seasonal methane fluxes for nitrogen fertilized plots to be within ± 20% of the unfertilized control for the first two years of measurements, with an average of 50% reduced fluxes for incorporated urea and ammonium sulfate for the third year. Surface applied fertilizers had fluxes equivalent to previous years. At present, the net effect of nitrogen fertilizers on flux is difficult to predict, particularly since two or more years of applications may be needed for the differences to be expressed.

Use of mineral fertilizers in rice fields has been increasing globally (Neue et al., 1990); however, there are still large regions where they are unavailable or not economic. In particular, organic manures are still widely used in some parts of Asia and may result in higher fluxes for the largest rice growing areas.

2.2.3. Rice Cultivars. Most of the methane emitted from rice paddies passes through the stems of the rice plant (Schütz et al., 1989b). Root biomass has been shown to be related to methane production (Sass et al., 1990), and possibly soil oxidation as well (Denier van der Gon and Neue, 1996). Different rice cultivars may have different capacity for gas transport, oxidation, or root exudation.

In recent years several comparisons of methane emission through different cultivars were made in both field and greenhouse experiments. Lindau et al. (1995) found semi-dwarf rice varieties emitted 36% less methane than tall rice varieties over early and ratoon rice crops in Louisiana, USA. Sigren et al. (1997a) found 30-50% lower emissions from a semi-dwarf species compared to a tall species over three growing seasons. Further tests indicated that neither transport nor biomass was significantly different between the two cultivars; however, soil acetate concentrations were higher for the first 40 days after flooding in soil cores taken from fields planted with the tall cultivar. Shalini-Singh et al. (1997) found methane emission roughly correlated with aboveground biomass from pot experiments with five rice varieties; Husin et al. (1995) suggested that greater tiller density led to higher methane emissions from one of two cultivars field tested in Indonesia. In greenhouse experiments, Wang et al. (1997) found high correlation between methane emission and root dry weight, and between root dry weight and total carbon released from roots; and between dry weight of plant and root weight.

In summary, tests in the United States suggest that semi-dwarf rice varieties emit less methane than tall varieties, possibly because of lower methane production. Cultivars with lower above-ground biomass and fewer tillers also tend to have lower methane emissions.

2.2.4. Soil Properties. Neue et al. (1990) identified soil texture, mineralogy, and Eh/pH buffer systems as the soil properties most important in affecting methane production. Experiments by Yagi and Minami (1990) in Japan, and Sass et al. (1990, 1991a, 1994) in the U.S., have directly measured the effect of different soil textures during the same growing season. The differences in fluxes between the soils studied in Japan may reflect the ability of the soils to sustain a reducing environment. The study in Texas found lower methane fluxes from the soils with a higher clay content, which perhaps favored a buildup of soil toxins. In laboratory incubations of methane from twenty different rice soils of the Philippines, Neue et al. (1994) found the highest methane emissions from soils with organic carbon content greater than 2% and low clay fraction.

CH_4 is formed only after the soil environment has become completely reduced in the series: NO_3^-, Mn^{4+}, Fe^{3+}, SO_4^{2-}. The pH of acid soils increases during the reduction of Fe-oxyhydroxides and Eh decreases (see Neue and Roger, this volume). Masscheleyn et al. (1993) found in laboratory tests of paddy soils that a soil redox level (Eh) of -150 mV was necessary for methane production to begin. Further testing found optimum pH for methane production to be 7 ± 0.1, with methane production increasing greatly as Eh dropped below -200 mV and pH approached 7 (Wang et al., 1993). Addition of gypsum ($CaSO_4$) to rice fields reduced methane emissions by over 50%, probably by increasing SO_4^{2-} concentration in the soil to favor sulfate-reducing bacteria over methanogens (Denier van der Gon and Neue, 1994). Aluminum in soils, and soil salinity may also reduce methanogenesis (Chattopadhyay et al., 1994; Denier van der Gon and Neue, 1995b). Naturally saline soils have not yet been tested, though an incursion of brackish water was implicated in lower fluxes in Spain (Seiler et al., 1984).

2.2.5. Weather Factors. Local weather factors such as solar radiation, cloud cover, and rainfall, influence both plant growth and the environment for methanogenic bacteria. All of the weather factors affect soil temperature. Since most biological processes respond to warmer temperatures up to a maximum, soil temperature was compared in early studies to methane flux. Schütz et al. (1989a), Sass et al. (1991b) and Khalil et al. (1991) all found a positive relationship between soil temperature and daily CH_4 fluxes, which changed slightly throughout the growing season. Parashar et al. (1993) tested the methane emission response to temperature by placing heating coils in a rice field. They found that methane emission increased rapidly at temperatures above 24° C, reaching a maximum at 34.5° C. Higher temperatures inhibited production. Khalil et al. (1998b) found a good correlation between soil temperature and methane flux, with Q_{10}s of about 2-3 for diurnal temperature variations, using data from six growing seasons.

Temperature relationships may also change from season to season or from site to site within a season. Changes in flux over the growing season have been found in every season-long field experiment (see e.g. Schütz et al., 1989a; Yagi and Minami, 1990; Khalil et al., 1998a). Much of the seasonal cycle may be attributed to the growth stage of the plant; therefore the effect of the plant growing season must be removed to calculate large scale temperature effects. Figure 2 shows the relationship between flux and temperature from a data set with seasonal and diurnal cycles removed.

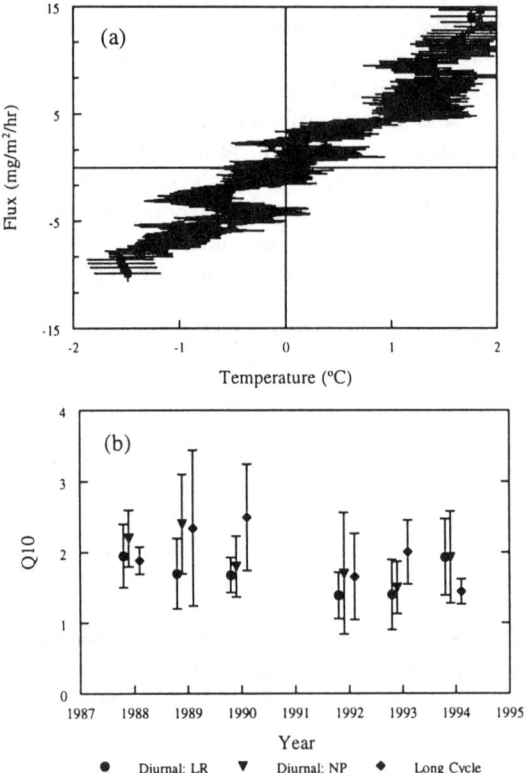

Fig. 2. (a) Relationship between soil temperature and methane flux after subtracting the seasonal cycle from the data. Horizontal bars are the standard errors of the temperature. **(b)** An estimate of the temperature response of rice field methane emission; the Q_{10} and associated 90% confidence limits are shown. Diurnal data was analyzed using linear regression (LR) and nonparametric (NP) statistical methods. Long cycles are identified weather patterns longer than one day and less than 30 days. All data are from rice fields in Sichuan, China (Khalil et al., 1998b).

2.3. Estimates of Areas Planted to Rice

Estimates of rice growing area are usually taken from published agricultural statistics supplied by the United Nations, the International Rice Research Institute [IRRI], or the agricultural agencies of individual countries (see, for example: U.N. Food and Agriculture Organization [FAO] Yearbook; IRRI World Rice Statistics; China Agricultural Yearbook; all various years). For convenience, agricultural statistics are reported by political or administrative subdivisions. The FAO statistics count areas twice if the same field grows two crops of rice per year.

Aselmann and Crutzen (1989) and Matthews et al. (1991) did careful global rice area allocations to provide data for global transport-chemistry models for methane; they concentrated on different variables, and the authors made different types of information available. Aselmann and Crutzen provided detailed tables of the percent of the area in 2.5° latitude by 5° longitude boxes. The percentages reflect the area in rainfed and irrigated rice only for South Asia (no upland / dryland rice was included) but show total area for Africa, Central and South America. Matthews et al. allocated areas by 1° latitude by 1° longitude cells using total rice areas by country; this database is available electronically at present.

Estimating the areas under different types of water management, and the amounts and location of organic amendments applied to rice fields, is more difficult. Huke (1982) and Huke and Huke (1988) have published statistics on water management areas in south and east Asia. Other areas of the world are not as well documented. Estimating the areas with organic amendment applications requires information about availability of rice straw, manures, and green manures, and the economics of applying them to rice fields, all of which are currently not available.

2.4. Season Length

The estimate of the season length from the literature may be confused by whether "season" refers to the season of methane emission, the rice growing season (planting or transplanting to harvest), the total growing season (the frost free period), or the total growing period, which in the case of transplanted rice includes the time in the seedling beds. For example, in the study of Sass et al. (1991a) the total growing season is about 245 days, the rice growing season was 140 days from planting to harvest, while methane was emitted during only 85 days (flood irrigated period). Because the number of areas where the methane emission season has actually been measured is quite limited, some other variable must be used to approximate it. Whether the crop is directly seeded or is transplanted may affect the length of time the crop is kept flooded and thus affect the total CH_4 emissions over the growing season. The Khalil et al. (1998a) data were obtained from a region where the rice was transplanted to the fields, then standing water was maintained in the fields until harvest about 120 days later. In the study by Sass et al. (1991a) the crop was seeded, and permanently flooded after the young plants were established. Out of a growing season of 140 days, the crop was flooded for only 85 days. Planting date affects the stage of the crop during the longest and warmest days. Sass et al. (1991b) compared the differences in flux over a planting season by planting a crop at one-month intervals. The later plantings ripened more quickly and were irrigated for a shorter period. Recent measurements made following rice crop harvest indicate that about 10% of the seasonal methane emission is flushed from the soil as it dries (Denier van der Gon et al., 1996), and may be emitted during wet periods in the fallow season (Bronson et al., 1997); effectively lengthening the methane emission season by a few days.

Matthews et al. (1991) used crop calendars to estimate rice growing seasonality by country; autumn, winter and summer crops by Indian state; and early rice, double/late crop rice, and single (mid-season) crop rice by Chinese province. The crop calenders tend to overestimate the methane emission season in single crop, mid-season rice by about 40 days, while multiple crop growing seasons usually are within 15 days of the emission period (Table 1).

Table 1. Season Lengths for Methane Emissions from Rice Fields

References	Planting Season Days	CH$_4$ Emission Period Days	Growing Season days[†]
Holzapfel-Pschorn and Seiler (1986): Italy	147	126	122
Cicerone et al. (1983): California, USA	145	100	153
Sass et al. (1991a): Texas, USA	140	85	153
Lindau and Bollich. (1993): Louisiana, USA[‡]	188	143	153
Yagi and Minami (1990): Japan	140	115	153
Denier van der Gon et al. (1996): Philippines, Dry Season	105	105	90
Chen et al. (1993): Beijing, PRC	100	87	137
Khalil et al. (1998a): Sichuan, PRC	120	120	168
Khalil et al. (1998c): Guangzhou, PRC			
Early Season	106	106	122
Late Season	95	95	138

[†] Estimated from Matthews et al. (1991). See text.
[‡] Season length includes first (main) and ratoon rice crops.

3. Estimates of Regional and Global Emissions

Every methane budget includes an estimate of emissions from rice paddies. As more information becomes available, the global estimates have been continually modified, starting at an estimate of almost 300 Tg/year and presently about 60 (20 - 100) Tg/year (Khalil and Rasmussen, 1990, and references therein).

3.1. Global Estimates

Recent estimates of the global source range from 50 to 100 Tg (10^{12} grams) per year. Estimates of global emissions of methane from rice agriculture usually concentrate on regions in south and east Asia, where over 85% of the area planted to rice is located (United Nations [U.N.], 1996), and where agricultural practices are most likely to favor methane production in rice paddies. Nine of the top ten rice growing countries are in Asia (by 1996 area: India, China, Indonesia, Bangladesh, Thailand, Vietnam, Burma, Philippines, Brazil, and Pakistan). Of the non-Asian countries in the top twenty rice planting countries, three (Brazil, Madagascar, and Nigeria) have 60% to 80% of their total area planted in dryland rice (Grist, 1986); the fourth country (USA) uses different cultivation practices than most other large rice growing countries.

Source estimates vary greatly with the assumptions made on the importance of different factors affecting the methane flux, and the information on the factors currently available. Ideally, each rice growing region should have an average methane flux factor associated with it which uniquely reflects the soil, climate and cultivation practices of the area. Practically, estimators must rely on the information that is available and their best judgement to apply measured fluxes from one region to another where data is not available. Recent global estimates are summarized in Table 2 (Matthews et al. (1991) assumed a global source of 100 Tg).

Table 2. Estimates of global methane emissions from rice agriculture.

Study	Tg/year	Year of Estimate
Aselmann and Crutzen (1989)	92 (53[†])	1985
Khalil and Shearer (1994)	66 (58 [‡])	1990
IPCC (1995)	60 (20 - 100)	Not given
Cao et al. (1996)	53	1993

[†] Number in parentheses assumes a global average flux of 13 mg/m^2/hr
[‡] Revised using fluxes from Khalil et al. (1998a, 1998c)

The major difference between these estimates is in the way a weighted average flux rate is calculated. Aselmann and Crutzen (1989) calculated fluxes from the work of Holzapfel-Pschorn and Seiler (1986) in Italy, with a base rate of 300 mg/m²/day (12.5 mg/m²/hr) for average soil temperatures of 20°C and below, and assuming a linear relationship between flux and temperature up to 1000 mg/m²/day (42 mg/m²/hr) for soil temperatures of 30°C. Khalil and Shearer (1994) used averaged fluxes from the studies available at that time. Flux rates were reduced by 40% for areas of rainfed rice, and Matthews et al.'s growing seasons of 140 to 170 days were reduced by 30 days; growing seasons of 110 to 140 days were reduced by 20 days; and growing seasons of fewer than 110 days were reduced by 10 days. The IPCC/OECD estimate is the midpoint in the range of estimated global emissions from various sources. Cao et al. (1996) estimated fluxes using a process-based model, with soil texture, carbon flow, and temperature as part of the input data.

3.2. Regional Emissions

Newer estimates of methane emission usually are produced by some form of geographical information system (GIS), with information on area of rice, irrigation, soil type and organic matter, weather, etc., apportioned to some arbitrary area scheme; presently 0.5° latitude × 0.5° longitude is the smallest subdivision. These estimates are used as model input. However, much of the agricultural information is published by political boundary, since local governments compile information on crop growth for their own use. The methane source, apportioned to 10 degree latitude bands, is shown in Figure 3. The overall similarity between the estimates is the result of the allocation of rice area; all three studies use the FAO Agricultural Yearbook data. The Khalil and Shearer estimate and Aselmann and Crutzen (1989) differ from Matthews et al. (1991) in the 20 to 30° N latitude band because the former studies reduce the total area by the estimated area in dryland cultivation. The Khalil and Shearer estimate also reduces rice area in Africa and Central and South America by the percent in dryland cultivation.

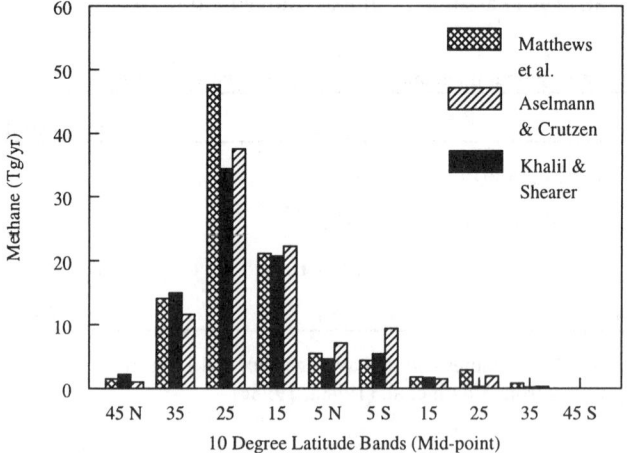

Fig. 3. A comparison of latitudinal allocation of methane emissions from rice agriculture for three global studies.

Estimates of methane emission by country differ by the calculation of base flux and the way it is associated with rice growing areas, and can vary widely. For example, India and China are the two largest rice growing countries in the world, and have some of the widest range of emissions estimates (Table 3). Estimates of emissions by country are difficult to constrain, so that nearly any estimate may be justified.

Table 3. Comparison of different estimates of methane emissions from two countries.

Study		India	China	Year
Matthews et al. (1991)		27.6	21.6	1984
Bachelet and Neue (1993)[†]	Yield	18.5	21.3	1984
	Soil	17.5	14.7	
	NPP	18.4	13.5	
Khalil and Shearer (1994)		15.3	23.0	1990
Parashar et al. (1996)		4.0		1991
Cao et al. (1995b)			16.2	1993
Yao et al. (1996)			15.3	1991

[†] Yield: estimates made using IRRI rice yield data and regional estimates of incorporated organic matter; Soil: estimate modified using the FAO soil map, and an estimate on the methane potential of different soil types (Neue et al., 1994); NPP: Net Primary Production. Area allocation was from Matthews et al.

3.3. Trends: The Role of Rice Agriculture in the Budget of Atmospheric Methane

Analysis of the role of rice paddies in the increase of atmospheric methane suggests that it was an important factor, particularly from the middle of this century to the present. However, the rate of increase has slowed in the last decade. Predicting the rice agriculture source of methane using either the trend of past emissions or the population will probably lead to large over estimates of the future source (Khalil and Rasmussen, 1990; Khalil et al., 1993).

The increase in rice agriculture was likely one of the main contributors to the increase of methane during the last century. We estimate that rice agriculture contributes some 300 ppbv of methane to the present atmosphere and may be responsible for some 20% of the increase of methane during the last century (Khalil et al., 1996). However, factors that were important causes of the increase in the past are changing and probably will not be as important in the future (see Figure 4). The time series of global methane emissions from rice estimated by Khalil and Shearer is largely influenced by the tremendous increase in area planted to rice in the last four decades. By the early 1980s this growth had slowed significantly.

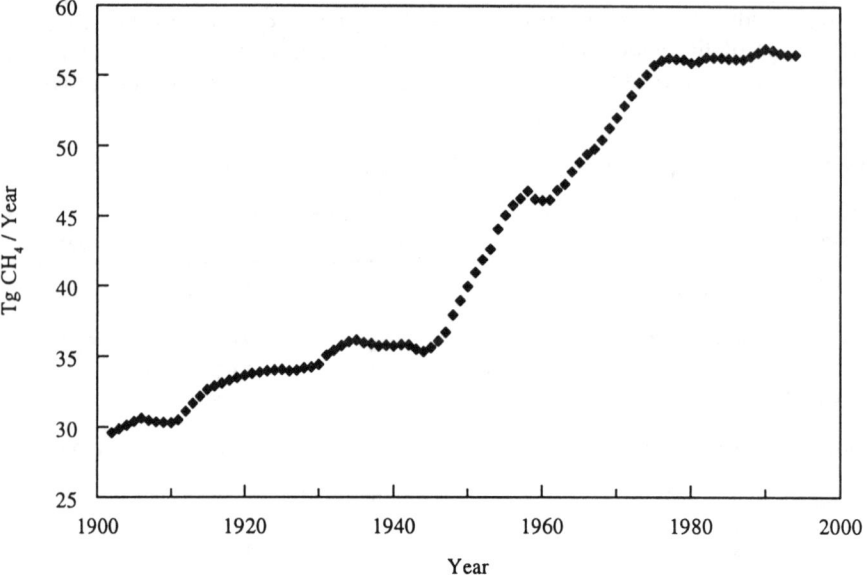

Fig. 4. The time series of global CH$_4$ emissions from rice agriculture.

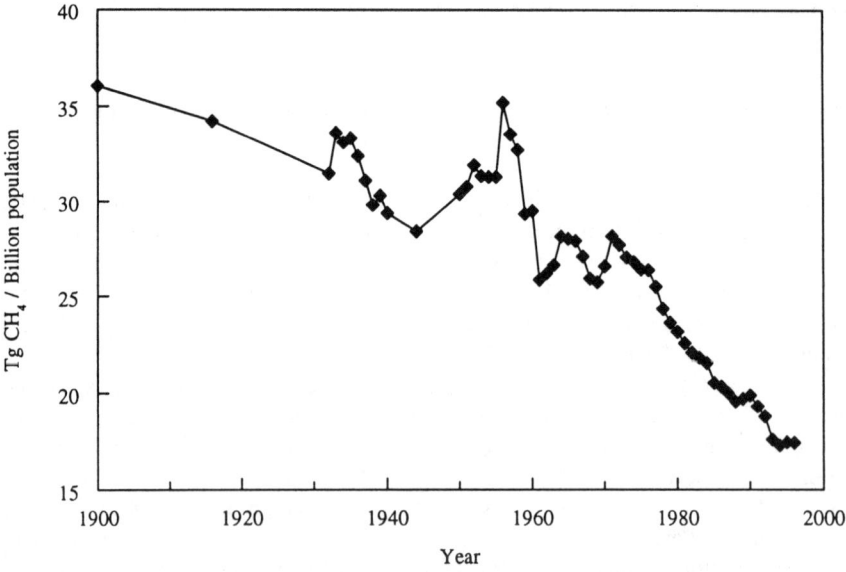

Fig. 5. The per capita emissions of methane from rice fields in China during the last century. The per capita emissions have declined by 35% from the 1930s to the 1980s (Khalil et al., 1993).

We constructed time series of harvested rice area for all rice growing countries listed in the 1995 FAO Production Yearbook (U.N., 1996). Historical statistics compiled by Mitchell (1980, 1982 and 1983), and more recent data available from the United Nations (FAO Yearbook, various years between 1956 and 1996) were the primary sources of information. Additional statistics for China came from the China Agriculture Yearbook (various years) and USDA (1984).

Figure 5 shows the time series of per capita rice field methane emissions for China. Except for a brief period in the 1950s, per capita emissions have steadily declined, despite the increase in area planted (nearly 10 million hectares). We believe this makes the future trends of population of little use in predicting methane emissions from rice growing countries. The efforts to increase rice yield without increasing the area planted, such as irrigation, fertilizers, and high yield rice cultivars, will affect the future methane emissions from rice paddies. Economic, political and sociological factors may also affect rice consumption and hence the methane emitted.

4. Summary

Global estimates of methane emission from rice agriculture are derived by summing up regional estimates, which are calculated by multiplying flux, area and seasonal factors (Eq. 1). Any part of the equation may introduce uncertainty. It is difficult to assess whether a seasonal average flux measurement is correctly associated with area and growing season extrapolants, since seasonal fluxes can vary widely even in nearby areas.

A decade ago, the most comprehensive controlled studies of methane emission were from the United States, Europe, and Japan. Since then, seasonal measurements have been made in China, India, Thailand, Indonesia and the Philippines. No seasonal measurements have been reported from the continents of Africa or South America, each with about 5% of global rice area. The assumption that dryland rice is not a significant source of methane uses the best information we have available, but it has never been tested. The difficulties in estimating the season of methane emission from the growing season for rice was demonstrated in Table 2. However, except for the limited number of areas where sampling has been done, the season of methane emission must be estimated from the local growing season. Using the crop calendar growing season without adjustments will probably overestimate the methane emission season.

Global estimates from three different sources have good general agreement on the latitudinal emissions from rice agriculture. When the regions of interest are individual countries, different assumptions lead to a much larger disagreements.

Rice agriculture is an important component of the methane budget, and in the past century was important in the increase of atmospheric methane. However, in the 1980s, emissions from rice agriculture appear to have nearly stabilized. Future emissions from rice agriculture will depend on efforts to increase production without increasing the area planted, particularly through fertilization, irrigation, and planting high productivity hybrid rice cultivars.

Acknowledgments. We thank Dr. R.A. Rasmussen, R. Dalluge, and D. Stearns for their contributions. Financial support for this project was provided in part by the U.S. Department of Energy (grant number DE-FG06-85ER60313). Additional support was provided by the Andarz Co.

References

Aselmann, I., P.J. Crutzen. 1989. Global distribution of natural freshwater wetlands and rice paddies, their net primary productivity, seasonality and possible wetland emissions. *J. of Atmos. Chem., 8*:307-358.

Bachelet, D., H.U. Neue. 1993. Methane emissions from wetland rice areas of Asia. *Chemosphere, 26(1-4)*:219-238.

Bingemer, H.G., P.J.Crutzen. 1987. The production of methane from solid wastes. *J. Geophys. Res., 92*:2181-2187.

Blake, D.R. 1984. Increasing concentrations of atmospheric methane. Ph.D. dissertation, Univ. of California at Irvine.

Bolle, H.-J., W. Seiler, B. Bolin. 1986. Other greenhouse gases and aerosols. In: *The Greenhouse Effect, Climate Change and Ecosystems (SCOPE 29)*, Chapter 4 (J. Wiley and Sons, N.Y., 1986) 157-198.

Bouwman, A.F. 1991. Agronomic aspects of wetland rice cultivation and associated methane emissions. *Biogeochemistry, 15*:65-88.

Bronson, K.F., U. Singh, H.-U. Neue, E.B. Abato, Jr. 1997. Automated chamber measurements of methane and nitrous oxide flux in a flooded rice soil: II. Fallow period emissions. *Soil Sci. Soc. Am. J. 61*: 988-993.

Cao, M., J.B. Dent, O.W. Heal. 1995a. Modeling methane emissions from rice paddies. *Global Biogeochemical Cycles, 9*:183-195.

Cao, M., J.B. Dent, O.W. Heal. 1995b. Methane emissions from China's paddyland. *Agriculture Ecosystems & Environment 55*: 129-137.

Cao, M., K. Gregson, S. Marshall, J.B. Dent, O.W. Heal. 1996. Global methane emissions from rice paddies. *Chemosphere 33*:879-897.

Chattopadhyay, B.D., A.R. Thakur, P.C. Mangal, N.N. Purkait, S.K. Saha. 1994. Inhibitory role of aluminum in methane emission in rice fields. *Indian J. of Exp. Biol.* 32: 495-500.

Chen, Z., L. Debo, K. Shao, B. Wang. 1993. Features of CH_4 emission from rice paddy fields in Beijing and Nanjing, China. *Chemosphere, 26(1-4)*:239-246.

China Agriculture Yearbook 1985, 1986, 1987, 1989. Agribookstore, Hampton, VA.

Cicerone, R.J., J.D. Shetter, C.C. Delwiche. 1983. Seasonal variation of methane flux from a California rice paddy. *J. Geophys. Res., 88*(C15):11,022-11,024.

Cicerone, R.J., R.S. Oremland. 1988. Biogeochemical aspects of atmospheric methane. *Global Biogeochemical Cycles, 2*:299-327.

Cicerone, R.J., C.C. Delwiche, S.C. Tyler, P.R. Zimmerman. 1992. Methane emissions from California rice paddies with varied treatments. *Global Biogeochemical Cycles, 6(3)*:233-248.

Debnath, G., M.C. Jain, S. Kumar, K. Sarkar, S.K. Sinha. 1996. Methane emissions from rice fields amended with biogas slurry and farm yard manure. *Climatic Change 6*: 97-109.

Denier van der Gon, H.A.C., H.-U. Neue. 1994. Impact of gypsum application on the methane emission from a wetland rice field. *Global Biogeochemical Cycles*, 8: 127-134.

Denier van der Gon, H.A.C., H.-U. Neue. 1995a. Influence of organic matter incorporation on the methane emission from a wetland rice field. *Global Biogeochemical Cycles*, 9: 11-22.

Denier van der Gon, H.A.C., H.-U. Neue. 1995b. Methane emission from a wetland rice field as affected by salinity. *Plant and Soil* 170: 307-313.

Denier van der Gon, H.A.C., H.-U. Neue. 1996. Oxidation of methane in the rhizosphere of plants. *Biol. and Fert. of Soils* 22: 359-366.

Denier van der Gon, H.A.C., N. van Breeman, H.-U. Neue, R.S. Lantin, J.B. Aduna, M.C.R. Alberto, R. Wassmann. 1996. Release of entrapped methane from wetland rice fields upon soil drying. *Global Biogeochemical Cycles 10*: 1-7.

Donahue, T.M. 1979. The atmospheric methane budget. In: *Proceedings of the NATO Advanced Study Institute on Atmospheric Ozone: Its Variation and Human Influences*, A.C. Aikin, Ed. (U.S. Department of Transportation, Washington D.C).

Ehhalt, D., U. Schmidt. 1978. Sources and sinks of atmospheric methane. *PAGEOPH, 116*:452-464.

Grist, D.H. 1986. *Rice*, 6th Edition. Longman, Inc., New York, U.S.A.

Holzapfel-Pschorn, A., R. Conrad, W. Seiler. 1985. Production, oxidation and emission of methane in rice paddies. *FEMS Micro. Ecol., 31*:343-351.

Holzapfel-Pschorn, A., W. Seiler. 1986. Methane emission during a cultivation period from an Italian rice paddy. *J. Geophys. Res., 91*(D11):11,803-11,814.

Huke, R.E. 1982. *Rice Area by Type of Culture: South, Southeast, and East Asia.* International Rice Research Institute (IRRI), Los Baños, Philippines.

Huke, R.E., E.H. Huke. 1988. *Human Geography of Rice in South Asia.* International Rice Research Institute (IRRI), Los Baños, Philippines.

Husin, Y.A., D. Murdiyarso, M.A.K. Khalil, R.A. Rasmussen, M.J. Shearer, S. Sabiham, A. Sunar, H. Adijuwana. 1995. Methane flux from Indonesian wetland rice: The effects of water management and rice variety. *Chemosphere, 31*: 3153-3180.

IPCC. 1995. *Climate Change 1994, Radiative Forcing of Climate Change and An Evaluation of the IPCC IS92 Emission Scenarios.* J.T. Houghton, L.G. Meira Filho, J. Bruce, Hoesung Lee, B.A. Callander, E. Haites, N. Harris and K. Maskell, eds., Cambridge University Press, G.B.

IRRI. 1991. *World Rice Statistics 1990.* IRRI Dept. of Agricultural Economics, Los Baños, Philippines.

Jermsawatdipong, P., J. Murase, P. Prabuddham, Y. Hasathon, N. Khomthong, K. Naklang, A. Watanabe, H. Haraguchi, M. Kimura. 1994. Methane emission from plots with differences in fertilizer application in Thai paddy fields. *Soil Sci. Plant Nutr.*, 40: 63-71.

Khalil, M.A.K., R.A.Rasmussen. 1983. Sources, sinks and seasonal cycles of atmospheric methane. *J. Geophys. Res., 88*:5131-5144.

Khalil, M.A.K., R.A. Rasmussen. 1990. Constraints on the global sources of methane and an analysis of recent budgets. *Tellus, 42B*:229-236.

Khalil, M.A.K., R.A. Rasmussen, M.-X. Wang, L. Ren. 1991. Methane emissions from rice fields in China. *Environ. Sci. Tech., 25*:979-981.

Khalil, M.A.K., M.J. Shearer, R.A. Rasmussen. 1993. Methane sources in China: Historical and current emissions. *Chemosphere, 26(1-4)*:127-142.

Khalil, M.A.K., M.J. Shearer. 1994. Methane emissions from rice agriculture. In: *International Anthropogenic Methane Emissions: Estimates for 1990*, Report to Congress, M.J. Adler, ed. U.S. Environmental Protection Agency, EPA 230-R-93-010.

Khalil, M.A.K., M.J. Shearer, R.A. Rasmussen. 1996. Atmospheric methane over the last century. *World Resource Review 8*: 481-492.

Khalil, M.A.K., R.A. Rasmussen, M.J. Shearer, R.W. Dalluge, Ren L.-X., Duan C.-L. 1998a. Measurements of methane emissions from rice fields in China. *J. Geophys. Res.*, 103(D19): 25,181-25,210.

Khalil, M.A.K., R.A. Rasmussen, M.J. Shearer, R.W. Dalluge, Ren L.-X., Duan C.-L. 1998b. Factors affecting methane emissions from rice fields. *J. Geophys. Res.*, 103(D19): 25,219-25,231.

Khalil, M.A.K., R.A. Rasmussen, M.J. Shearer, Z.-L. Chen, H. Yao, J. Yang. 1998c. Emissions of methane, nitrous oxide, and other trace gases from rice fields in China. *J. Geophys. Res.*,103(D19): 25,241-25,250.

Lindau, C.W., P.K. Bollich. 1993. Methane emissions from Louisiana first and ratoon crop rice. *Soil Science*, 156: 42-48.

Lindau, C.W. 1994. Methane emissions from Louisiana rice fields amended with nitrogen fertilizers. *Soil Biol. Biochem.*, 26: 353-359.

Lindau, C.W., P.K. Bollich, R.D. DeLaune. 1995. Effect of rice variety on methane emission from Louisiana rice. *Agri. Ecosys. and Environ.* 54: 109-114.

Livingston, G.P. and G.L. Hutchinson. 1995. Enclosure-based measurement of trace gas exchange: applications and sources of error. In *Biogenic Trace Gases: Measuring Emissions from Soil and Water*. P.A. Matson and R.C. Harriss, eds. Blackwell Science Ltd., Oxford, England.

Masscheleyn, P.H., R.D. DeLaune, W.H. Patrick, Jr. 1993. Methane and nitrous oxide emissions from laboratory measurements of rice soil suspension: effect of soil oxidation-reduction status. *Chemosphere, 26(1-4)*:251-260.

Matthews, E., I. Fung, J. Lerner. 1991. Methane emission from rice cultivation: geographic and seasonal distribution of cultivated areas and emissions. *Global Biogeochemical Cycles*, 5(1):3-24.

Mitchell, B.R. 1980. *European Historical Statistics*, 2nd Rev. Ed. Facts on File, New York, U.S.A.

Mitchell, B.R. 1982. *International Historical Statistics, Africa and Asia*. New York University Press.

Mitchell, B.R. 1983. *International Historical Statistics, The Americas and Australasia*. Gale Research Co., Detroit, Michigan.

Neue, H.-U., P. Becker-Heidmann, H.W. Scharpenseel. 1990. Organic matter dynamics, soil properties, and cultural practices in rice lands and their relationship to methane production. In: *Soils and the Greenhouse Effect*. A.F. Bouwman, ed. J. Wiley and Sons, N.Y., USA.

Neue, H.-U., R.S. Lantin, R. Wassmann, J.B. Aduna, Ma. C.R. Alberto, M.J.F. Andales. 1994. Methane emission from rice soils in the Philippines. In: *CH₄ and N₂O: Global emissions and controls from rice fields and other agricultural and industrial sources*. K. Minami, A. Mosier, R.Sass., ed. NIAS Series 2, Yokendo Publ., Tokyo, Japan.

Parashar, D.C., P.K. Gupta, J. Rai, R.C. Sharma, N. Singh. 1993. Effect of soil temperature on methane emission from paddy fields. *Chemosphere, 26*:247-250.

Parashar, D.C., A.P. Mitra, P.K. Gupta, J. Rai, R.C. Sharma, N. Singh, et al. 1996. Methane budget from paddy fields in India. *Chemosphere 33*: 737-757.

Sass, R.L., F.M. Fisher, P.A. Harcombe, F.T. Turner. 1990. Methane production and emission in a Texas rice field. *Global Biogeochemical Cycles, 4*(1):47-68.

Sass, R.L., F.M. Fisher, P.A. Harcombe, F.T. Turner. 1991a. Mitigation of methane emissions from rice fields: possible adverse effects of incorporated rice straw. *Global Biogeochemical Cycles, 5*(3):275-287.

Sass, R.L., F.M. Fisher, F.T. Turner, M.F. Jund. 1991b. Methane emission from rice fields as influenced by solar radiation, temperature, and straw incorporation. *Global Biogeochemical Cycles, 5*(4):335-350.

Sass, R.L., F.M. Fisher, Y.B. Wang, F.T. Turner, M.F. Jund. 1992. Methane emission from rice fields: the effect of floodwater management. *Global Biogeochemical Cycles 6(3)*:249-262.

Sass, R.L., F.M. Fisher, S.T. Lewis, M.F. Jund, F.T. Turner. 1994. Methane emission from rice fields: Effect of soil properties. *Global Biogeochemical Cycles 8(2)*:135-140.

Schütz, H., A. Holzapfel-Pschorn, R. Conrad, H. Rennenberg, W. Seiler. 1989a. A 3-year continuous record on the influence of daytime, season, and fertilizer treatment on methane emission rates from an Italian rice paddy. *J. Geophys. Res.*, *94*(D13):16405-16416.

Schütz, H., W. Seiler, R. Conrad. 1989b. Processes involved in formation and emission of methane in rice paddies. *Biogeochemistry, 7*:33-53.

Schütz, H., W. Seiler, H. Rennenberg. 1990. Soil and land use related sources and sinks of methane (CH₄) in the context of the global methane budget. In: *Soils and the Greenhouse*

Effect. A.F. Bouwman, ed. J. Wiley and Sons, N.Y.

Seiler, W., A. Holzapfel-Pschorn, R. Conrad, D. Scharffe. 1984. Methane emission from rice paddies. *J. of Atmos. Chem., 1*:241-268.

Shalini-Singh, S. Kumar, M.C. Jain. 1997. Methane emission from two Indian soils planted with different rice cultivars. *Biol. Fertil. Soils* 25: 285-289.

Sigren, L. K., G.T. Byrd, F.M. Fisher, and R.L. Sass. 1997a. Comparison of soil acetate concentrations and methane production, transport, and emission in two rice cultivars. *Global Biogeochem. Cycles 11*(1):1-14.

Sigren, L. K., S.T. Lewis, F.M. Fisher, and R.L. Sass. 1997b. Effects of field drainage on soil parameters related to methane production and emission from rice paddies. *Global Biogeochem. Cycles 11*(2):151-162.

United Nations. 1977, ..., 1997. *1976, ..., 1996 FAO Production Yearbook, vols. 30, ..., 50.* Food and Agriculture Organization of the United Nations, Rome.

U.S. Dept. of Agriculture. 1984. *Agricultural Statistics of People's Republic of China, 1949-1982.* International Economics Division, Economic Research Service, Statistical Bulletin No. 714. U.S. Govt. Printing Office, Washington, D.C.

Wang, Z., R.D. DeLaune, C.W. Lindau, W.H. Patrick, Jr. 1992. Methane production from anaerobic soil amended with rice straw and nitrogen fertilizers. *Fertilizer Research* 33: 115-121.

Wang, Z.P., R.D. DeLaune, P.H. Masscheleyn, W.H. Patrick, Jr. 1993. Soil redox and pH effects on methane production in a flooded rice soil. *Soil Sci. Soc. Am. J.* 57: 382-385.

Wang, B., H.U. Neue, H.P. Samonte. 1997. Effect of cultivar difference ('IR72', 'IR65598' and 'Dular') on methane emission. *Agric. Ecosys. and Environ.* 62:31-40.

Warneck, P. 1988. *Chemistry of the Natural Atmosphere* (Academic Press, N.Y.)

Wassmann, R., H. Papen, H. Rennenberg. 1993a. Methane emission from rice paddies and possible mitigation strategies. *Chemosphere, 26(1-4)*:201-218.

Wassmann, R., H. Schütz, H. Papen, H. Rennenberg, W. Seiler, Dai A., Shen R., Shangguan X. Wang M. 1993b. Quantification of methane emissions from Chinese rice fields (Zhejiang Province) as influenced by fertilizer treatment. *Biogeochemistry 20*: 83-101.

Yagi, K., K. Minami. 1990. Effect of organic matter application on methane emission from some Japanese paddy fields. *Soil Sci. Plant Nutr., 36*(4):599-610.

Yagi, K., H. Tsuruta, K. Kanda, K. Minami. 1996. Effects of water management on methane emission from a Japanese rice paddy field: Automated methane monitoring. *Global Biogeochem. Cycles 10*(2): 255-267.

Yao H., Chen Z.L. 1994a. Effect of chemical fertilization on methane emission from rice paddies. *J. of Geophys. Res., 99*(D8): 16,463-16,470.

Yao H., Chen Z.L. 1994b. Seasonal variation of methane flux from a Chinese rice paddy in a semi arid, temperate region. *J. of Geophys. Res., 99*(D8): 16,471-16,477.

Yao H., Zhuang Y.-h., Chen Z.L. 1996. Estimation of methane emission from rice paddies in mainland China. *Global Biogeochem. Cycles 10*: 641-649.

11. Biomass Burning

Joel S. Levine,[1] Wesley R. Cofer III[2], and Joseph P. Pinto[1]
[1]Atmospheric Sciences Division, NASA Langley Research Center
Hampton, Virginia 23681-2199
[2] NCEA, MD 52, U.S. Environmental Protection Agency
Research Triangle Park, North Carolina 27711

1. Introduction

Our planet is a unique object in the solar system due to the presence of a biosphere with its accompanying biomass and the occurrence of fire (Levine, 1991a). The burning of living and dead biomass is a very significant global source of atmospheric gases and particulates. Crutzen and colleagues were the first to assess biomass burning as a source of gases and particulates to the atmosphere (Crutzen et al., 1979; and Seiler and Crutzen, 1980). However, in a recent paper, Crutzen and Andreae (1990) point out that "Studies on the environmental effects of biomass burning have been much neglected until rather recently but are now attracting increased attention." The "increased attention" includes the Chapman Conference on Global Biomass Burning. Much of the information presented here is based on material from this conference (Levine, 1991b). Biomass burning and its environmental implications have also become important research elements of the International Geosphere-Biosphere Program (IGBP) and the International Global Atmospheric Chemistry (IGAC) Project (Prinn, 1991).

The production of atmospheric methane (CH_4) by biomass burning will be assessed. Field measurements and laboratory studies to quantify the emission ratio of methane and other carbon species will be reviewed. The historic database suggests that global biomass burning is increasing with time and is controlled by human activities. Present estimates indicate that biomass burning contributes between about 27 and 80 Teragrams per year (Tg/yr; Tg = 10^{12} grams) of methane to the atmosphere. This represents 5 to 15% of the global annual emissions of methane. Measurements do indicate that biomass burning is the overwhelming source of CH_4 in tropical Africa. However, if the rate of global biomass burning increases at the rate that it has been over the last few decades, then the production of methane from biomass burning may become much more important on a global scale.

Mohammad Aslam Khan Khalil   (Ed.)
Atmospheric Methane
© *Springer-Verlag Berlin Heidelberg 2000*

2. Gaseous Emissions Due to Biomass Burning

Biomass burning includes the combustion of living and dead material in forests, savannas, and agricultural wastes, and the burning of fuel wood. Under the ideal conditions of complete combustion, the burning of biomass material produces carbon dioxide (CO_2) and water vapor (H_2O), according to the reaction:

$$CH_2O + O_2 \rightarrow CO_2 + H_2O \tag{1}$$

where CH_2O represents the average composition of biomass material. Since complete combustion is not achieved under any conditions of biomass burning, other carbon species, including carbon monoxide (CO), methane (CH_4), nonmethane hydrocarbons (NMHCs), and particulate carbon, result by the incomplete combustion of biomass material. In addition, nitrogen and sulfur species are produced from the combustion of nitrogen and sulfur in the biomass material.

While CO_2 is the carbon species overwhelmingly produced by biomass burning, its emissions into the atmosphere resulting from the burning of savannas and agricultural wastes are largely balanced by its reincorporation back into biomass via photosynthetic activity within weeks to years after burning. However, CO_2 emissions resulting from the clearing of forests and other carbon combustion products from all biomass sources including CH_4, CO, NMHCs, and particulate carbon, are largely "net" fluxes into the atmosphere since these products are not reincorporated into the biosphere when the land is converted to another use.

Biomass material contains about 40% carbon by weight, with the remainder hydrogen (6.7%) and oxygen (53.3%) (Bowen, 1979). Nitrogen accounts for between 0.3 and 3.8% and sulfur for between 0.1 and 0.9%, depending on the nature of the biomass material (Bowen, 1979). The nature and amount of the combustion products depend on the characteristics of both the fire and the biomass material burned. Hot, dry fires with a good supply of oxygen produce mostly carbon dioxide with little CO, CH_4, and NMHCs. The flaming phase of the fire approximates complete combustion, while the smoldering phase approximates incomplete combustion, resulting in greater production of CO, CH_4, and NMHCs. The percentage production of CO_2, CO, CH_4, NMHCs, and carbon ash during the flaming and smoldering phases of burning based on laboratory studies is summarized in Table 1 (Lobert et al., 1991). Typically for forest fires, the flaming phase lasts on the order of an hour or less, while the smoldering phase may last up to a day or more, depending on the type of fuel, the fuel moisture content, wind velocity, topography, etc. For savanna grassland and agricultural waste fires, the flaming phase lasts a few minutes and the smoldering phase lasts up to an hour.

Table 1. Percentage of production of CO_2, CO, CH_4, and NMHCs during flaming and smoldering phases of burning based on laboratory experiments (Lobert et al., 1991).

	Percentage in burning stage	
	Flaming	Smoldering
CO_2	63	37
CO	16	84
CH_4	27	73
NMHCs	33	67

3. Emission Ratios

The total mass of the carbon species (CO_2 + CO + CH_4 + NMHCs + particulate carbon) M_C is related to the mass of the burned biomass (m) by $M_C = f \times m$, where f = mass fraction of carbon in the biomass material, i.e., 40-45%. To quantify the production of gases other than CO_2, we must determine the emission ratio (ER) for each species. The emission ratio for each species is defined as:

$$ER = \frac{\Delta X}{\Delta CO_2} \qquad (2)$$

where ΔX is the concentration of the species X produced by biomass burning, and ΔCO_2 is the concentration of CO_2 produced by biomass burning. $\Delta X = X^* - X$ where X^* is the measured concentration of X in the biomass burn smoke plume, and X is the background (out of plume) atmospheric concentration of the species. Similarly, $\Delta CO_2 = CO_2^* - CO_2$, where CO_2^* is the measured concentration in the biomass burn plume, and CO_2 is the background (out of plume) atmospheric concentration of CO_2.

In general, all species emission factors are normalized with respect to CO_2, as the concentration of CO_2 produced by biomass burning may be directly related to the amount of biomass material burned by simple stoichiometric considerations as discussed earlier. Furthermore, the measurement of CO_2 in the background atmosphere and in the smoke plume is relatively simple.

For the reasons outlined above, it is most convenient to quantify the combustion products of biomass burning in terms of the species emission ratio (ER). Measurements of the emission ratio for CH_4 and CO normalized with respect to CO_2 for diverse ecosystems (for example, wetlands, chaparral, and boreal; for different phases of burning, flaming, smoldering phases and combined flaming and smoldering phases, called "mixed") are summarized in Table 2. Measurements of the emission ratio for CH_4 normalized with respect to CO_2 for various burning sources in tropical Africa are summarized in Table 3.

Table 2. Emission ratios for CO, CH_4, and NMHCs for diverse ecosystems (in units of $\Delta X/\Delta CO_2$, in percent; ± = standard deviation) (Cofer et al., 1991).

Wetlands	CO	CH_4	NMHCs
Flaming	4.7 ± 0.8	0.27 ± 0.11	0.39 ± 0.17
Mixed	5.0 ± 1.1	0.28 ± 0.13	0.45 ± 0.16
Smoldering	5.4 ± 1.0	0.34 ± 0.12	0.40 ± 0.15
Chaparral			
Flaming	5.7 ± 11.6	0.55 ± 0.23	0.52 ± 0.21
Mixed	5.8 ± 2.4	0.47 ± 0.24	0.46 ± 0.15
Smoldering	8.2 ± 1.4	0.87 ± 0.23	1.17 ± 0.33
Boreal			
Flaming	6.7 ± 1.2	0.64 ± 0.20	0.66 ± 0.26
Mixed	11.5 ± 2.1	1.12 ± 0.31	1.14 ± 0.27
Smoldering	12.1 ± 1.9	1.21 ± 0.32	1.08 ± 0.18

Table 3. Emission ratio for CH_4 for different fires in tropical Africa (in units of $\Delta CH_4/\Delta CO_2$ in percent; ± = standard deviation) (Delmas et al., 1991).

Type of Combustion	Emission Ratio = $\Delta CH_4/\Delta CO_2$	
	Mean	Range
Natural savanna bushfire	0.28 ± 0.04	0.23 - 0.34
Forest fire	1.23 ± 0.60	0.56 - 2.22
Emissions from traditional charcoal	2.06 ± 2.86	6.7 - 14.2
oven	1.79 ± 0.81	1.04 - 3.2
Firewood	0.14	
Charcoal		

Table 4. Average emission factors for CO_2, CO, and CH_4 for diverse ecosystems (in units of grams of combustion product carbon to kilograms of fuel carbon; ± = standard deviation) (Radke et al., 1991).

	CO_2	CO	CH_4
Chaparral-1	1644 ± 44	74 ± 16	2.4 ± 0.15
Chaparral-2	1650 ± 31	75 ± 14	3.6 ± 0.25
Pine, Douglas fir and brush	1626 ± 39	106 ± 20	3.0 ± 0.8
Douglas fir, true fir and hemlock	1637 ± 103	89 ± 50	2.6 ± 1.6
Aspen, paper birch and debris from jack pine	1664 ± 62	82 ± 36	1.9 ± 0.5
Black sage, sumac, and chamise	1748 ± 11	34 ± 6	0.9 ± 0.2
Jack pine, white and black spruce	1508 ± 16	175 ± 91	5.6 ± 1.7
"Chained" and herbicidal paper birch and poplar	1646 ± 50	90 ± 21	4.2 ± 1.3
"Chained" and herbicidal birch, polar and mixed hardwoods	1700 ± 82	55 ± 41	3.8 ± 2.8
Debris from hemlock, deciduous and Douglas fir	1600 ± 70	83 ± 37	3.5 ± 1.9
Overall average	1650 ± 29	83 ± 16	3.2 ± 0.5

Table 5. Emission factors for CO_2, CO, CH_4, and NMHC and ash based on laboratory experiments (in % of fuel carbon; Lobert et al., 1991).

	Mean	Range
CO_2	82.58	49.17 – 98.95
CO	5.73	2.83 – 11.19
CH_4	0.42	0.14 – 0.94
NMHC (as C)	1.18	0.14 – 3.19
Ash (as C)	5.00	0.66 – 22.28
Total sum C	94.91	

Some researchers present their biomass burn emission measurement in the ratio of grams of carbon in the gaseous and particle combustion products to the mass of the carbon in the biomass fuel in kilograms. Average emission factors for CO_2, CO, and CH_4 in these units for diverse ecosystems are summarized in Table 4 and emission factors for CO_2, CO, CH_4, NMHCs, and carbon ash in terms of percentage of fuel carbon based on laboratory experiments are summarized in Table 5. Inspection of Tables 2-5 indicates that there is considerable variability in both the emission ratio and the emission factor for carbon species as a function of ecosystem burning and the phase of burning (flaming or smoldering).

A recent compilation of CO_2-normalized emission ratios for carbon species is listed in Table 6. This table gives the range for both field measurements and laboratory studies and provides a "best guess."

Table 6. CO_2-Normalized emission ratios for carbon species: summary of field measurements and laboratory studies (in units of grams C in each species per kilograms of C in CO_2) (Andreae, 1991).

	Field Measurements	Laboratory Studies	"Best Guess"
CO	6.5 – 140	59 – 105	100
CH_4	6.2 – 16	11 – 16	11
NMHCs	6.6 – 11.0	3.4 – 6.8	7
Particulate organic carbon (including elemental carbon)	7.9 – 54		20
Element carbon (black soot)	2.2 – 16		5.4

4. Emission of Methane

Once the mass of the burned biomass (M) and the species emission ratios (ER) are known, the gaseous and particulate species produced by biomass combustion may be calculated. The mass of the burned biomass (M) is related to the area (A) burned in a particular ecosystem by the following relationship (Seiler and Crutzen, 1980):

$$M = A \times B \times \alpha \times \beta \qquad (3)$$

where B is the average biomass material per unit area in the particular ecosystem (g/m^2), α is the fraction of the average above-ground biomass relative to the total average biomass B, and β is the burning efficiency of the above-ground biomass. Parameters B, α, and β vary with the particular ecosystem under study and are determined by assessing the total biomass before and after burning.

The total area burned during a fire may be assessed using satellite data. Recent reviews have considered the extent and geographical distribution of biomass burning from a variety of space platforms: astronaut photography (Wood and Nelson, 1991), the NOAA polar orbiting Advanced Very High Resolution Radiometer (AVHRR) (Brustet et al., 1991a; Cahoon et al., 1991; Robinson, 1991a, 1991b), the Geostationary Operational Environmental Satellite (GOES) Visible Infrared Spin Scan Radiometer Atmospheric Sounder (VAS) (Menzel et al., 1991); and the Landsat Thematic Mapper (TM) (Brustet et al., 1991b).

Hence, the contribution of biomass burning to the total global budget of methane or any other species depends on a variety of ecosystem and fire parameters, including the particular ecosystem that is burning (which determines the parameters B, α and β), the mass consumed during burning, the nature of combustion (complete vs. incomplete), the phase of combustion (flaming vs. smoldering), and knowledge of how the species emission factors (EF) vary with changing fire conditions in various ecosystems. The contribution of biomass burning to the global budgets of any particular species depends on precise knowledge of all these parameters. While all these parameters are known imprecisely, the largest uncertainty is probably associated with the total mass (M) consumed during biomass burning on an annual basis (and there are large year-to-year variations in this parameter!). The total biomass burned annually according to source of burning is summarized in Table 7 (Seiler and Crutzen, 1980; Hao et al., 1990; Crutzen and Andreae, 1990; Andreae, 1991). The estimate for carbon released of 3940 Tg/yr includes all carbon species produced by biomass combustion (CO_2 + CO + CH_4 + NMHCs + particulate carbon). About 90% of the released carbon is in the form of CO_2 (about 3550 Tg/yr).

Table 7. Global estimates of annual amount of biomass burning and the resulting release of carbon to the atmosphere (Seiler and Crutzen, 1980; Crutzen and Andreae, 1990; Hao et al., 1990; and Andreae, 1991).

Source of burning	Biomass burned (Tg/yr)[1]	Carbon released (Tg C/yr)[2]	CH_4 released (Tg CH_4/yr)[3]
Savanna	3690	1660	21.9
Agricultural waste	2020	910	12.0
Fuel wood	1430	640	8.4
Tropical forests	1260	570	7.5
Temperate and boreal forests	280	130	1.7
Charcoal	21	30	0.4
World total	8700	3940	51.9

[1] 1 Tg (teragram) = 10^6 metric tons = 10^{12} grams.

[2] Based on a carbon content of 45% in the biomass material. In the case of charcoal, the rate of burning has been multiplied by 1.4.

[3] Assuming that 90% of the carbon released is in the form of CO_2 and that the "best guess" emission ratio of C:CH_4 to C:CO_2 is 1.1% (see Table 5), and CH_4 (Tg) = C:CH_4 (Tg) * 16/12.

Knowledge of the CO_2-normalized emission ratio for CH_4 coupled with information on the total production of CO_2 due to biomass burning allows us to estimate the total annual global production of CH_4 due to biomass burning. Field measurements and laboratory studies indicate that the emission ratio for CH_4 is in the range of 6.2 to 16 grams of carbon in the form of CH_4 (C:CH_4) per kilogram of carbon in the form of CO_2 (C:CO_2) (see Table 6), which corresponds to a C:CH_4 to C:CO_2 emission ratio in the range of 0.62 to 1.6%. Using a "best guess" of 1.1% and assuming that biomass burning produces about 3550 Tg/yr of C:CO_2, then the global annual production of C:CH_4 due to biomass burning is in the range of 21.7 to 56 Tg/yr, which converts to 29 to 75 Tg/yr of CH_4, with a "best guess" of 52 Tg/yr of CH_4. The production of CH_4 by different burning sources on a global scale is summarized in Table 7. A detailed study using a chemical transport model with a 1° x 1° spatial grid yielded an annual average CH_4 production due to biomass burning of 63.4 Tg (Taylor and Zimmerman, 1991), which is somewhat smaller than the maximum CH_4 production value calculated here of 74.7 Tg CH_4/yr. Assuming that the total annual global production of CH_4 from all sources is about 500 Tg CH_4 (Cicerone and Oremland, 1988), then the range of CH_4 we calculate corresponds to between 6% and 15% of the global emissions of CH_4, while the calculations of Taylor and Zimmerman (1991) suggest that biomass burning produces about 14% of the global emissions of CH_4. Considering all of the uncertainties in these calculations, there is very good agreement between these two estimates. While an upper limit range of about 15% for the production of methane due to biomass burning may not seem very significant, the importance of this source is enhanced when we consider that the largest single global sources of methane do not produce much more than about 20% of the total.

Table 8. Total emissions of CO_2 and CH_4 from the burning of biomass in tropical Africa.

	Source				
	Savanna bushfires	Forest fires	Firewood burning	Charcoal production	Total
Biomass burned	2.52	0.13	0.12	0.11	2.88
CO_2 emission factor	1370	957	957	641	
CH_4 emission factor	1.65	6.94	5.42	21.0	
CO_2 emissions	3.45	0.12	0.11	0.07	3.75
CH_4 emissions	4.14	0.90	0.65	2..31	9.22

[1] Biomass burned in units of Gigatons dry matter = 10^9 metric tons = 10^3 Tg = 10^{15} grams
[2] Emission factors units of g gas/kg dry matter
[3] CO_2 emissions in units of Gigatons/yr
[4] CH_4 emissions in units of Teragram/yr

Delmas et al. (1991) have studied the CH_4 budget of tropical Africa. They considered the emission of CH_4 from biogenic processes in the soil and from biomass burning. They found that the dry African savanna soil is always a net sink for CH_4. They measured an average soil uptake rate for atmospheric CH_4 of 2×10^{10} CH_4 molecules cm^{-2} s^{-1}. They calculated the production of CH_4 (and CO_2) due to biomass burning and found that biomass burning supplies about 9.22 Tg CH_4 yr (and 3750 Tg CO_2 yr) (see Table 8). Hence, in tropical Africa, biomass burning, not biogenic emissions from the soil, controls the CH_4 budget.

In addition to the direct production of CH_4 by the combustion of biomass material, there is recent evidence to suggest that burning stimulates biogenic emissions of CH_4 from wetlands. Flux chamber measurements indicate higher fluxes of CH_4 from wetlands following burning. It has been suggested that combustion products, carbon dioxide, carbon monoxide, acetate, and formate entering the wetlands following burning are used by methanogenic bacteria in the metabolic production of CH_4

(Levine et al., 1990).

At present, biomass burning is a significant global source of several important radiatively and chemically active species. Biomass burning may supply 40% of the world's annual gross production of CO_2 or 26% of the world's annual net production of CO_2 (due to the burning of the world's forests) (Seiler and Crutzen, 1980; Crutzen and Andreae, 1990; Hao et al., 1990; Levine, 1990; Andreae, 1991; Houghton, 1991). Biomass burning supplies 10% of the world's annual production of CH_4, 32% of the CO; 24% of the NMHCs, excluding isoprene and terpenes; 21% of the oxides of nitrogen (nitric oxide and nitrogen dioxide); 25% of the molecular hydrogen (H_2); 22% of the methyl chloride (CH_3Cl); 38% of the precursors that lead to the photochemical production of tropospheric ozone; 39% of the particulate organic carbon (including elemental carbon); and more than 86% of the elemental carbon (Levine, 1990; Andreae, 1991).

5. Historic Changes in Biomass Burning

It is generally accepted that the emissions from biomass burning have increased in recent decades, largely as a result of increasing rates of deforestation in the tropics. Houghton (1991) estimates that gaseous and particulate emissions to the atmosphere due to deforestation have increased by a factor of 3 to 6 over the last 135 years. He also believes that the burning of grasslands, savannas, and agricultural lands has increased over the last century because rarely burned ecosystems, such as forests, have been converted to frequently burned ecosystems, such as grasslands, savannas, and agricultural lands.   In Latin America, the area of grasslands, pastures, and agricultural lands increased by about 50% between 1850 and 1985. The same trend is true for South and Southeast Asia. In summary, Houghton (1991) estimates that total biomass burning may have increased by about 50% since 1850. Most of the increase results from the ever-increasing rates of forest burning, with other contributions of burning (grasslands, savannas, and agricultural lands) having increased by 15% to 40%. The increase in biomass burning is not limited to the tropics.   In analyzing 50 years of fire data from the boreal forests of Canada, the U.S.S.R., the Scandinavian countries, and Alaska, Stocks (1991) has reported a dramatic increase in area burned in the 1980s. The largest fire in the recent past destroyed more than 12 million acres of boreal forest in the People's Republic of China and Russia in a period of less than a month in May 1987 (Cahoon et al., 1991).

The historic data indicate that biomass burning has increased with time and that the production of greenhouse gases from biomass burning has increased with time. Furthermore, the bulk of biomass burning is human-initiated. As greenhouse gases build up in the atmosphere and the Earth becomes warmer, there may be an enhanced frequency of fires. The enhanced frequency of fires may prove to be an important positive feedback in a warming Earth. However, it has been suggested that the bulk of biomass burning worldwide may be significantly reduced (Andrasko et al., 1991). Policy options for mitigating biomass burning have been developed by Andrasko et al. (1991). For mitigating burning in the tropical forests, where much of the burning is aimed at land clearing and conversion to agricultural lands, policy options include the marketing of timber as a resource and improved productivity of existing agricultural lands to reduce the need for conversions of forests to agricultural lands. Improved productivity will result from the application of new agricultural technology, e.g., fertilizers. For mitigating burning in tropical savanna grasslands, animal grazing could be replaced by stall feeding since savanna burning results from the need to replace nutrient-poor tall grass with nutrient-rich short grass. For mitigating burning on agricultural lands and croplands, incorporate crop wastes into the soil, instead of burning, as is the present practice throughout the world. The crop wastes could also be used as fuel for household heating and cooking rather than cutting down and destroying forests for fuel as is presently done.

6. Uncertainties and Future Research

The construction of a global emissions inventory for methane from biomass burning must account for the high degree of variability of these emissions in both space and time. Biomass burning exhibits strong seasonal and geographic variations. As shown earlier, methane emissions from biomass burning are highly dependent on the type of ecosystem being burned, which determines the total amount of biomass consumed and the extent of flaming and smoldering phases during combustion. The calculations by Taylor and Zimmerman (1991) go a long way towards deriving a global inventory in that they have simulated the variability of biomass burning. They scaled the burning rate inversely with precipitation as global data sets are currently not available. Satellite techniques, when they are developed, offer a promising way to obtain global coverage.

Taylor and Zimmerman also used a constant emission ratio in their calculations since measurements of the emission ratio for methane are lacking for many different ecosystems. While some data exist for mid-latitude ecosystems, measurements are needed to better define the contributions from burning tropical forests and savannas. In addition, airborne measurements are limited to the outer edges of biomass burn plumes so little is known about variability across the plume. The use of long path remote measurements across plumes is also planned for the future.

7. Summary

Biomass burning may be the overwhelming regional or continental-scale source of CH_4 as in tropical Africa and a significant global source of CH_4. Our best estimate of present methane emissions from biomass burning is about 51.9 Tg/yr, or 10% of the annual methane emissions to the atmosphere. Increased frequency of fires that may result as the Earth warms up may result in increases in this source of atmospheric methane.

It is appropriate to conclude this chapter with an observation of fire historian, Stephen Pyne (1991):

"We are uniquely fire creatures on a uniquely fire planet, and through fire the destiny of humans has bound itself to the destiny of the planet."

References

Andrasko, K.J., D.R. Ahuja, S.M. Winnett, D.A. Tirpak. 1991. Policy options for managing biomass burning to mitigate global climate change. In: *Global Biomass Burning: Atmospheric, Climatic, and Biospheric Implications* (J.S. Levine, ed.), The MIT Press, Cambridge, Massachusetts, 445-456.

Andreae, M.O. 1991. Biomass burning: Its history, use, and distribution and its impact on environmental quality and global climate. In: *Global Biomass Burning: Atmospheric, Climatic, and Biospheric Implications* (J.S. Levine, ed.), The MIT Press, Cambridge, Massachusetts, 3-21.

Bowen, H.J.M. 1979. *Environmental Chemistry of the Elements.* Academic Press: London, England.

Brustet, J.M., J.B. Vickos, J. Fontan, K. Manissadjan, A. Podaire, F. Lavenue. 1991a. Remote sensing of biomass burning in West Africa with NOAA-AVHRR. In: *Global Biomass Burning: Atmospheric, Climatic, and Biospheric Implications* (J.S. Levine, ed.), The MIT Press, Cambridge, Massachusetts, 47-52.

Brustet, J.M., J.B. Vickos, J. Fontan, A. Podaire, F. Lavenue. 1991b. Characterization of active fires in West African savannas by analysis of satellite data: Landsat thematic mapper. In: *Global Biomass Burning: Atmospheric, Climatic, and Biospheric Implications* (J.S. Levine, ed.), The MIT Press, Cambridge, Massachusetts, 53-60.

Cahoon, D.R., Jr., J.S. Levine, W.R. Cofer, III, J.E. Miller, P. Minnis, G.M. Tennille, T.W. Yip, B.J. Stocks, P.W. Heck. 1991. The great Chinese fire of 1987: A view from space. In: *Global Biomass Burning: Atmospheric, Climatic, and Biospheric Implications* (J.S. Levine, ed.), The MIT Press, Cambridge, Massachusetts, 61-66.

Cicerone, R.J., R.S. Oremland. 1988. Biogeochemical aspects of atmospheric methane. *Global Biogeochem. Cycles*, 2:299-327.

Cofer, W.R. III, J.S. Levine, E.L. Winstead, B.J. Stocks. 1991. Trace gas and particulate emissions from biomass burning in temperate ecosystems. In: *Global Biomass Burning: Atmospheric, Climatic, and Biospheric Implications* (J.S. Levine, ed.), The MIT Press, Cambridge, Massachusetts, 203-208.

Crutzen, P.J., L.E. Heidt, J.P. Krasnec, W.H. Pollock, W. Seiler. 1979. Biomass burning as a source of atmospheric gases CO, H_2, N_2O, NO, CH_3Cl, and COS. *Nature, 282*:253-256.

Crutzen, P.J., M.O. Andreae. 1990. Biomass burning in the tropics: Impact on atmospheric chemistry and biogeochemical cycles. *Science, 250*:1,669-1,678.

Delmas, R.A., A. Marenco, J.P. Tathy, B. Cros, J.G.R. Baudet. 1991. Sources and sinks of methane in the African savanna. CH_4 emissions from biomass burning. *J. Geophys. Res., 96*:7,287-7,299.

Hao, W.M., M.H. Liu, P.J. Crutzen. 1990. Estimates of annual and regional release of CO_2 and other trace gases to the atmosphere from fires in the tropics, based on the FAO statistics for the period 1975-1980. In: *Fire in the Tropical Biota: Ecosystem Processes and Global Challenges* (J.G. Goldammer, ed.), Springer-Verlag, Berlin-Heidelberg, Germany, 440-462.

Houghton, R.A. 1991. Biomass burning from the perspective of the global carbon cycle. In: *Global Biomass Burning: Atmospheric, Climatic, and Biospheric Implications* (J.S. Levine, ed.), The MIT Press, Cambridge, Massachusetts, 321-325.

Levine, J.S. 1990. Global biomass burning: Atmospheric, climatic, and biospheric implications. *EOS Transactions (AGU), 71:*1,075-1,077.

Levine, J.S. 1991a. The biosphere as a driver for global atmospheric change. In: *Scientists on Gaia* (S.H. Schneider and P.J. Boston, eds.), The MIT Press, Cambridge, Massachusetts, 353-361.

Levine J.S. (ed.). 1991b. *Global Biomass Burning: Atmospheric, Climatic, and Biospheric Implications*, The MIT Press, Cambridge, Massachusetts, 569 pages.

Levine, J.S., W.R. Cofer, III, D.I. Sebacher, R.P. Rhinehart, E.L. Winstead, S. Sebacher, C.R. Hinkle, P.A. Schmalzer, A.J. Koller, Jr. 1990. The effects of fire on biogenic emissions of methane and nitric oxide from wetlands. *J. Geophys. Res., 95*:1,853-1,864.

Lobert, J.M., D.H. Scharffe, W.-M. Hao, T.A. Kuhlbusch, R. Seuwen, P. Warneck, P.J. Crutzen. 1991. Experimental evaluation of biomass burning emissions: Nitrogen and carbon containing compounds. In: *Global Biomass Burning: Atmospheric, Climatic, and Biospheric Implications* (J.S. Levine, ed.), The MIT Press, Cambridge, Massachusetts, 289-304.

Menzel, W.P., E.C. Cutrim, E.M. Prins. 1991. Geostationary satellite estimation of biomass burning in Amazonia during BASE-A. In: *Global Biomass Burning: Atmospheric, Climatic, and Biospheric Implications* (J.S. Levine, ed.), The MIT Press, Cambridge, Massachusetts, 41-46.

Prinn, R.G. 1991. Biomass burning studies and the International Global Atmospheric Chemistry (IGAC) Project. In: *Global Biomass Burning: Atmospheric, Climatic, and Biospheric Implications* (J.S. Levine, ed.), The MIT Press, Cambridge, Massachusetts, 22--28.

Pyne, S.J. 1991. Sky of ash, earth of ash: A brief history of fire in the United States. In: *Global Biomass Burning: Atmospheric, Climatic, and Biospheric Implications* (J.S. Levine, ed.), The MIT Press, Cambridge, Massachusetts, 504-511.

Radke, L.F.; D.A. Hegg, P.V. Hobbs, J.D. Nance, J.H. Lyons, K.K. Laursen, R.E. Weiss, P.J. Riggan, D.E. Ward. 1991. Particulate and trace gas emissions from large biomass fires in North America. In: *Global Biomass Burning: Atmospheric, Climatic, and Biospheric Implications* (J.S. Levine, ed.), The MIT Press, Cambridge, Massachusetts, 209-224.

Robinson, J.M. 1991a. Fire from space: Global fire evaluation using IR remote sensing. *Intern. Journ. of Remote Sensing, 12*:3-24.

Robinson, J.M. 1991b. Problems in global fire evaluation: Is remote sensing the solution? In: *Global Biomass Burning: Atmospheric, Climatic, and Biospheric Implications* (J.S. Levine, ed.), The MIT Press, Cambridge, Massachusetts, 67-73.

Seiler, W., P.J. Crutzen. 1980. Estimates of grass and net fluxes of carbon between the biosphere and the atmosphere from biomass burning. *Climatic Change, 2*:207-247.

Stocks, B.J. 1991. The extent and impact of forest fires in north circumpolar countries. In: *Global Biomass Burning: Atmospheric, Climatic, and Biospheric Implications* (J.S. Levine, ed.), The MIT Press, Cambridge, Massachusetts, 197-202.

Taylor, J.A., P.R. Zimmerman. 1991. Modeling trace gas emissions from biomass burning. In: *Global Biomass Burning: Atmospheric, Climatic, and Biospheric Implications* (J.S. Levine, ed.), The MIT Press, Cambridge, Massachusetts, 345-350.

Wood, C.A., R. Nelson. 1991. Astronaut observations of global biomass burning. In: *Global Biomass Burning: Atmospheric, Climatic, and Biospheric Implications* (J.S. Levine, ed.), The MIT Press, Cambridge, Massachusetts, 29-40.

12. Wetlands

Elaine Matthews
Columbia University Center for Climate Systems Research
NASA GISS, 2880 Broadway, New York, NY 10025

1. Introduction

Wetlands are most likely the largest natural source of methane to the atmosphere accounting for ~20% of the current global annual emission of ~450-550 Tg (10^{12} g) (Khalil and Rasmussen, 1983; Cicerone and Oremland, 1988; Fung et al., 1991; Crutzen, 1991; Houghton et al., 1996). Measurements of methane from Greenland and Antarctic ice cores indicate atmospheric concentrations of ~350 ppbv during the Last Glacial Maximum rising to 650 ppbv during the pre-industrial Holocene (Stauffer et al., 1988; Chappellaz et al., 1990). Preindustrial source strengths of methane, consistent with historical concentrations and estimates based on isotopes, have been estimated at ~180-380 Tg methane annually (Khalil and Rasmussen, 1987; Stauffer et al., 1985; Chappellaz et al., 1993). Wetlands were the dominant preindustrial source with smaller contributions from wild fires, animals and oceans. During the last two hundred years, atmospheric methane concentrations have more than doubled to ~1750 ppbv (Etheridge et al., 1992). Annual increases of ~0.6% yr^{-1} in the 1980s declined to ~0.1% yr^{-1} in the 1990s (Rasmussen and Khalil, 1981; Craig and Chou, 1982; Khalil and Rasmussen, 1982, 1983, 1985; Ehhalt et al., 1983; Rasmussen and Khalil, 1984; Stauffer et al., 1985; Blake and Rowland, 1986, 1988; Pearman et al., 1986; Steele et al., 1987; Blake et al., 1988; Khalil et al., 1989; Lang et al., 1990a,b; Dlugokencky et al., 1994a,b, 1997). Currently, the total annual emission of methane is about twice that estimated for the preindustrial period, but both the relative and absolute contribution of wetlands is smaller than in the past due to increases in anthropogenic sources and reductions in wetland areas. However, climate and related biological interactions that presently control the distribution of wetlands and their methane emissions are expected to change during the next 50 to 100 years.

The role of wetlands in the cycle of methane has been studied for several decades. Early estimates of emissions were based on very few measurements and highly uncertain information about the wetland areal extent (e.g., Koyama, 1964; Ehhalt, 1974; Ehhalt and Schmidt, 1978; Blake, 1984; Seiler, 1984; Holzapfel-Pschorn and Seiler, 1986; Seiler and Conrad, 1987). While all these studies relied on the same wetland area of 2.6 x $10^{12} m^2$ (from Twenhofel 1926, 1951), the estimated global methane emission ranged from 11 to 300 Tg/year although the more recent estimates lie in the lower half of that range. This wide range of emission figures reflects

Mohammad Aslam Khan Khalil   (Ed.)
Atmospheric Methane
© *Springer-Verlag Berlin Heidelberg 2000*

differing assumptions about the magnitude and annual duration of methane fluxes. Sebacher et al. (1986) suggested a potential wetland area of 4.5-9.0 x 10^{12} m^2 for northern peatlands and estimated an annual methane emission from them of 45 - 106 Tg.

Using newer information about wetland distributions and environmental characteristics, estimates have converged around 100 Tg CH$_4$ yr^{-1} (Matthews and Fung, 1987; Aselmann and Crutzen, 1989; Bartlett et al., 1990) using global wetland areas of 5-6 x 10^{12}m^2. Recent modeling studies are consistent with these values. However, the similarities mask some major uncertainties about seasonal methane-production periods as well as important differences in the relative importance of the role of climatically and ecologically distinct wetland ecosystems, i.e., tropical/subtropical wetlands whose methane emissions are governed generally by large scale precipitation and flood cycles, and high-latitude wetlands whose highly seasonal emissions are controlled via interactions between temperature and water cycles. A complex suite of environmental parameters including soil chemistry, substrate quality and soil water status influence emissions in all these environments, further complicating the task of evaluating emissions.

Several developments in understanding the role of wetlands in the global methane cycle took place beginning in the late 1980s. Data on measured methane fluxes in wetland ecosystems expanded substantially. Measurements of concentrations and isotopic composition of atmospheric methane continued, providing more comprehensive information about sources, trends and seasonal cycles of methane. Modeling and synthesis techniques for analyzing the terrestrial and atmospheric information improved. Most recently, models of large-scale methane emissions, as well as the distribution of wetlands themselves, have made their appearance.

With respect to wetland ecosystems in particular, data expansion occurred in the following areas: (1) measurement studies increased to cover all ecosystems representative of global wetland areas (e.g., Africa, South America, Siberia); (2) periodic or single-date warm-season measurements characteristic of early field studies are augmented with measurements spanning complete growing seasons or full years in high-latitude environments (Alaska, Minnesota, Canada), measurements conducted during wet and dry seasons in tropical ecosystems (South America, Africa) and with time series of several years in Alaskan and Canadian ecosystems; (3) large-scale interdisciplinary field campaigns designed to characterize chemistry and dynamics of the regional troposphere have been conducted in Alaska, Canada, Central Africa and the Amazon Basin and integrate ground-based measurements (chambers, balloons, towers, floating platforms) with aircraft flights, and satellite overpasses. Finally, researchers have made progress toward modeling methane emissions from wetlands as well as the more difficult problem of modeling distributions of wetlands themselves.

This chapter provides an overview of the role of natural wetlands in the global cycle of methane. Table 1 summarizes aspects of wetlands important for assessing their role in the global methane cycle and indicates the complementary strengths of various approaches to understanding those features. These topics are covered in more detail in the following sections. Section 2 discusses wetland distributions and characteristics on a global scale including remote-sensing approaches to characterizing wetland extent and seasonality. Flux measurements and emission estimates at regional and global scales are discussed in Section 3. Section 4 focuses on modeling wetlands and their methane emissions. Section 5 summarizes research on assessing wetland and methane responses to climate change. The final section summarizes current understanding the role of natural wetlands in the methane budget, and outlines some remaining research questions.

Table 1. Ranking of strengths of wetland information: H = high/strong, M = medium, L = low/weak.

	Current Traditional [1]	Remote Sensing	Process Modeling	Hydrological Modeling
Area	M	H	L	M/H
Inundation status	M	H	L/M	M
Inundation seasonality				
Vegetation cover	L	M/H	L/M	M
Methane flux	H	H	L	L
Processes	L/M	M	H	L
	H	L	H	L

[1] includes measurements, surveys, local reports, vegetation, soil and inundation data

2. Distribution and Characteristics of Natural Wetlands

2.1 Areal Distribution and Seasonality

Early global estimates of methane emission from wetlands relied on very general information about wetland areas. Most of these estimates used a global wetland area of 2.6×10^{12} m^2 put forward in the 1920s (Twenhofel, 1926). In the late 1980s, several groups have compiled data sets specifically designed to evaluate methane emissions from natural wetlands, focusing on reducing uncertainties in wetland distributions and environmental characteristics. Matthews and Fung (1987) derived a data set by integrating three global digital data bases on vegetation, ponded soils and fractional inundation at 1° latitude by 1° longitude resolution, while that of Aselmann and Crutzen (1989) was compiled at 2.5° resolution from regional and local wetland reports. Global wetland areas derived from these studies are 5.3 and 5.7×10^{12} m^2, respectively. Relative regional distributions of areas from the two works are very similar.

About one-half of the total area lies between 50°-70°N. This high-latitude region is characterized by peat-rich ecosystems (bogs and fens) and a temperature-restricted thaw season resulting in highly seasonal emissions of methane. Approximately 35% of the global wetland area is broadly distributed in the latitude zone extending from 20°N to 30°S. This region is co-dominated by forested and nonforested swamps and marshes, with a smaller contribution from alluvial or floodplain formations. Most of these tropical wetlands, particularly floodplain habitats, undergo large seasonal expansion and contraction in response to precipitation cycles. Since many of them lie along river courses with exceptionally level topography over large distances, rapid and substantial changes in inundation during the year are common.

Differences between the two studies are discussed in Aselmann and Crutzen (1989). In brief, Aselmann and Crutzen's areas are slightly lower in the northern subtropics (10°-30°N), and in the southern zone between 20°S and 40°S which accounts for 9% of the Matthews and Fung total and 2% of the Aselmann and Crutzen total. The southern subtropical differences come primarily from the inclusion by Matthews and Fung of what are probably infrequently-flooded ephemeral wetlands in arid regions of Australia. The two studies exhibit a larger areal discrepancy in the southern tropics from the equator to 10°S. Aselmann and Crutzen indicate that ~20% of the global total occurs in this narrow tropical zone whereas Matthews and Fung arrived at a value equal to only ~10% of the global total. Causes for these tropical discrepancies are not clear although the studies relied on different sources and compilation methodologies. A sizable portion of the locations that disagree coincide with river systems which are associated with the largest areal uncertainties.

2.2 Wetland Classification and Characterization

Both wetland data sets discussed above used simple groupings of detailed wetland information primarily because such generalizations matched the ecosystem classes represented in the methane-flux measurements available in the late 1980s. For instance, Matthews and Fung (1987) classified 28 wetland vegetation types as described in the UNESCO (1973) vegetation classification system, in addition to ~100 other vegetation types occupying locations identified as wetlands using information on ponded soils and inundation, into five major groups: forested and nonforested bog, forested and nonforested swamp, and alluvial formations. Following regional wetland classifications of the local sources used in data collection, Aselmann and Crutzen (1989) identified 45 freshwater wetland types globally which were grouped into six broad categories following Level I of the hierarchical wetland classification devised for Canada by Zoltai and Pollett (1983). To accommodate tropical environments, they added floodplains to Zoltai and Pollett's (1983) system which classes wetlands according to physiognomic features such as peat structure, vegetation cover, and inundation depth and seasonality. The resulting groups are bog, fen, swamp, marsh, floodplain, and shallow lakes. With the exception of shallow lakes, the generalized classes of Aselmann and Crutzen (1989) and Matthews and Fung (1987) are comparable.

Field measurements characterizing methane production and emission from varied wetland ecosystems over extended time periods have increased substantially since the development of these data sets. This broadening of habitat coverage now justifies using more of the information contained in the wetland data sets to either reclassify or expand the classification of methane-significant ecosystems (Gore, 1983a,b; Glaser, 1987; National Wetlands Working Group, 1988; Reeburgh et al., 1998), although this has not yet been done.

A parallel approach is advocated by Sahagian and Melack (1996) who report a new functional classification of wetlands, paralleling recent functional classifications of vegetation (Running et al., 1994; DeFries et al., 1994). The System was developed to characterize the following wetland functions: (1) methane production, (2) carbon accumulation or export, (3) denitrification/N burial, and 4) sulfur cycling (DMS and H_2S production). Wetlands are therefore described with respect to the following suite of parameters most of which have been identified as controllers of methane production: (21) net primary production, (2) temperature, (30 water table and hydrology, (40 transport of organics and sediment, (50 vegetation type and morphology (6) chemical characteristics of organic materials (lignin, N content DOC quantity, chlorophyll), (7) salinity, (8) soil nutrient status, and (9) topography/geomorphology. This system represents a bridge between using existing data sets to characterize wetlands, modeling methane emissions from wetlands, and ultimately modeling wetlands and their response to changing climates.

2.3 Remote Sensing

Currently, the largest remaining uncertainties in estimating methane emissions from natural wetlands arise in part from difficulties in determining areal coverage of various wetland habitats, and seasonal and inter-annual variations in hydroperiod. Aselmann and Crutzen (1989) attempted to estimate maximum and minimum extent of wetland areas, and months of methane production, using hydrological and meteorological observations and wetland descriptions while Fung et al. (1991) applied a simple model based on temperature (for high latitude systems) and precipitation (for low latitude systems) to estimate seasonality and magnitude of emissions. They and others (see regional case studies included in Gore (1983b)) confirm that considerable uncertainties remain with respect to the seasonality of both areas, inundation, and emission despite the fact that the global distribution of wetlands is well understood. Seasonal expansion and contraction of tropical/subtropical swamps in response to precipitation and flood cycles is particularly difficult to estimate. For example, Bartlett et al. (1990) note that estimates of total Amazon floodplain are a range over an order of magnitude and areas covered by the three main categories of Amazon floodplain habitats are poorly known. A particular problem in characterizing the areal dimension of hydroperiod is the very local variations in topography that characterize many tropical wetlands such as the Sudd, the Amazon floodplain, and the Pantanal. In contrast, the methane production season for high latitude ecosystems is regulated largely by temperature, and habitats undergo less extreme areal changes over seasons although complex terrain complicates the identification of habitats associated with different methane fluxes and controllers.

Recently, Roulet et al. (1994) and Reeburgh et al. (1998) applied measured methane fluxes for wetland environments in Canada and Alaska, respectively, to high-resolution land-cover maps of the regions derived from LANDSAT data. In these well-studied northern regions with strongly seasonal inundation periods, identification of vegetation types with distinctive CH_4 fluxes reduces uncertainties in emissions. The rationale underlying remote sensing approaches is that more precise measurements of inundation and of areal coverage of wetland habitats with different methane emission characteristics will improve extrapolation of ecosystem flux measurements to wetland habitats (Roulet et al., 1992a; Bubier et al., 1995b; Roulet et al., 1994; Reeburgh et al., 1998) (Table 1). However, remote sensing techniques employing visible and near-infrared, thermal, microwave, and radar data offer varying degrees of success in improving areal estimates of wetland habitats on seasonal and inter-annual timescales, particularly with respect to open-water and saturated-soil environments. (Refer to reviews by Hess et al. (1990), Melack et al. (1994), Morrissey et al. (1994, 1996), and Sahagian and Melack (1996).)

Early attempts to characterize wetland environments using optical satellite data (e.g., Morrissey and Ennis, 1981; Walker et al., 1982; Rose and Rosendahl, 1983; Ormsby et al., 1985; Bartlett et al., 1989) proved these data useful primarily for assessing herbaceous environments characterized by standing water, for determining extent of open water in pond-dominated northern wetlands (e.g., Hudson Bay Lowlands, Alaska), and for distinguishing wetland habitats in boreal and arctic environments albeit without determining variations in water status within soils. However, the seasonal series of LANDSAT or AVHRR data required to improve areal estimates of wetland habitats and inundated conditions are typically unattainable due to persistently cloudy conditions, especially in the tropics. Furthermore, optical instruments do not penetrate canopies, a feature that severely limits the use of these data in tree-dominated wetlands which account for about two-thirds of the world's wetlands (Matthews and Fung, 1987).

In an overview of remote sensing of lakes and floodplains in the Amazon, Melack et al. (1994) assessed data availability, spatial resolution, and classification accuracy of a series of remotely-sensed data including aerial photography, satellite-borne optical and thermal sensors, satellite-borne passive microwave sensors (e.g., Scanning Multichannel Microwave Radiometer (SMMR) and the Special Sensor Microwave/Imager (SMM/I)), and synthetic aperture radars (SARs). They concluded that while SAR sensors are theoretically optimal for mapping inundation in the tropical Amazon because of their insensitivity to cloud cover and penetration of tree canopies, their fine resolution and narrow swaths prevent instantaneous synoptic evaluation of inundation over large areas which is a crucial element in environments with rapid and large seasonal variations. In addition, the limited wavelengths and polarizations available from currently operating SAR instruments requires combining data from several satellites to distinguish among wetland habitats, which in turn introduces problems of spatial coregistration and temporal coherence. Therefore, passive microwave instruments such as SMMR and SSM/I are likely the best suited for large-scale wetland characterization despite their coarse resolution relative to SARs.

Two passive microwave radiometers have provided global coverage since 1979. SMMR operated from 1979-1987 with 6 day global coverage and SMM/I has operated from 1987 to the present with 3-day global coverage, vertical and horizontal polarizations, and four frequencies. The highest frequencies, 37 GHz (both instruments) and 85.5 GHz (SMM/I only), offer 30 and 15 km resolution, respectively. Although SMMR provides synoptic views over large wetland regions, the coarse spatial resolution has limited its use in terrestrial studies to seasonal or inter-annual assessments (Choudhury, 1988, 1989; Giddings and Choudhury, 1989). In an attempt to minimize the problems associated with coarse resolution, Sippel et al. (1994) devised linear mixing models to incorporate observed microwave signatures of major end members into estimates of fractional inundation area.

The higher resolution, and multiple frequencies and polarizations available from SSM/I, appear more promising for assessing wetland extent particularly if combined with ancillary data for interpretation although additional problems exist (Neale et al., 1990; Achutuni et al., 1996; Prigent et al., 1997). Until recently, most land surface studies using microwave satellite observations focused on simple indices derived from linear combinations of microwave satellite measurements. However, atmospheric effects, especially cloud cover, may be responsible for a substantial part of the microwave signal, thus casting doubt on the interpretation of these indices solely in terms of surface properties (Tucker, 1989; Justice et al., 1989; Kerr and Njoku, 1993). This is particularly severe in wetland areas where clouds can persist for long periods of the year. Because flooding is typically associated with periods of heavy cloudiness and therefore maximum contamination of microwave radiances by clouds and rainfall, extreme caution is required when interpreting simple microwave indices. Even a cloud-free atmosphere can account for up to 15% of the microwave signal at 19 and 37 GHz, on average over a month, accompanied by large temporal and spatial variation (Prigent et al., 1997). Most importantly, even if cloudy conditions are avoided, and contributions of the cloud-free atmosphere are low, microwave radiation is directly influenced by variations in surface temperature making it impossible to directly compare microwave brightness temperatures over different time periods or different areas without accounting for surface temperature variations. Such effects therefore preclude direct comparisons of microwave brightness temperatures with surface parameters, although techniques exist to account for the effects in these data of surface temperature, atmosphere, and rainfall and clouds.

3. Distribution and Characteristics of Methane Emission from Wetlands

3.1 Flux Measurements

Several extensive syntheses of methane measurements and of techniques for estimating methane emission from wetlands are provided by Bartlett et al. (1990) for temperate, subtropical and tropical wetlands, and by Harriss et al. (1993) for northern high-latitude wetlands. These works were integrated and expanded in Bartlett and Harriss (1993). Refer to those publications for a comprehensive discussion of the measurements.

Methane emissions from natural wetlands show large variability resulting from the complex suite of environmental factors that affect the production of methane via anaerobic decomposition of organic material and the factors that affect transport, consumption and release of methane to the atmosphere. Fluxes measured at field sites and from soil samples have been independently correlated with local environmental and ecological factors that include temperature (Koyama, 1963; Baker-Blocker et al., 1977; King and Wiebe, 1978; Harriss et al., 1982; Mayer, 1982; Moore and Knowles, 1987, 1990; Crill et al., 1988a; Wilson et al., 1989; Klinger et al., 1994; Bubier et al., 1995a), water table position (Svensson, 1976, 1980; Harriss et al., 1982;

Svensson and Rosswall, 1984; Sebacher et al., 1986; Harriss et al., 1988b; Whalen and Reeburgh, 1988; Moore and Knowles, 1989; Morrissey and Livingston, 1992; Roulet et al., 1992a; Moore and Roulet, 1993; Klinger et al., 1994; Moore et al., 1994; Liblick et al., 1997), nutrient input and organic accumulation (Harriss and Sebacher, 1981; Svensson and Rosswall, 1984; Wilson et al., 1989; Morrissey and Livingston, 1992), substrate characteristics (Sebacher et al., 1986; Valentine et al., 1994), successional status (Moore et al., 1994; Klinger et al., 1994), vegetation characteristics and phenology (Dacey and Klug, 1979; Cicerone and Shetter, 1981; Sebacher et al., 1985; Wilson et al., 1989; Whiting and Chanton, 1992; Schimel, 1995), redox potential (Svensson and Rosswall, 1984), salinity (Bartlett et al., 1985), net primary productivity (Aselmann and Crutzen, 1989; Whiting et al., 1991; Whiting and Chanton, 1993; Cao et al., 1996; Potter, 1997), and methane oxidation (King, 1992, 1994). Most researchers concur that factors influencing fluxes are not entirely independent and that variables that serve as environmental integrators of methane production and consumption processes may be more successful predictors of fluxes (Moore et al., 1990; Whalen and Reeburgh, 1992; Bubier and Moore, 1994; Christensen et al., 1995; Reeburgh et al., 1998).

A series of large-scale interdisciplinary field campaigns designed to characterize the chemistry and dynamics of the tropical regional troposphere have provided fundamental information on the role of tropical wetland environments in the global methane budget. The ABLE 2 and CAMREX missions were carried out in Amazonian Brazil in the 1985 dry season (ABLE 2A) and the 1987 wet season (ABLE 2B) (Bartlett et al., 1988, 1990; Crill et al., 1988b; Devol et al., 1988, 1990; Harriss et al., 1988a, 1990). The TROPOZ (Tropospheric Ozone) and DECAFE (Dynamique et Chimie de l'Atmosphere en Foret Equatoriale) experiments were carried out in wet and dry seasons in Central Africa during several field seasons during 1987-1989 (Delmas et al., 1992; Fontan et al., 1992; Tathy et al., 1992). Until these campaigns, low latitude wetlands had received little attention. Furthermore, early studies in low latitudes concentrated on subtropical environments of the southeastern US characterized by very low emissions (Harriss and Sebacher, 1981; Harriss et al., 1982; Bartlett et al., 1985; Barber et al., 1988). Results from the campaigns and other studies confirmed tropical wetlands as the dominant natural source of methane. This larger role is due primarily to substantially higher emission rates measured at tropical wetland sites than the rates of more subtropical/temperate environments used as tropical proxies in early estimates.

Results from the Arctic Boundary Layer Expeditions (ABLE 3A, Alaska, summer 1988 and ABLE 3B, Canada, summer 1990) (Harriss et al., 1988a, 1994) and the Northern Wetlands Study (NOWES) in Canada in summer 1990 (Glooschenko et al., 1994), as well as additional measurements in Siberia (Christensen et al., 1995) showed a pattern of fluxes from northern wetlands lower than those initially reported by Sebacher et al. (1986) and Crill et al. (1988a) (Whalen and Reeburgh, 1988, 1990, 1992; Moore et al., 1990; Morrissey and Livingston, 1992; Ritter et al., 1992; Roulet et al., 1992a, 1993; Christensen, 1993; Edwards et al., 1994; Moore et al., 1994; Ritter et al., 1994; Roulet et al., 1994). These lower values are consistent with some of those measured in the early 1980s (Svensson, 1980; Svensson and Rosswall, 1984). This shift partially reflects full-season field measurements that captured variations around the peak summer fluxes measured earlier, as well as the inclusion of less productive high-latitude sites that may occupy substantial areas in boreal and arctic regions (Ritter et al., 1992; Dise, 1993; Christensen et al., 1995; Reeburgh et al., 1998).

Using the measurement compilation of Bartlett and Harriss (1992) as a base, Matthews (1993) summarized continental/regional wetland areas and available emission measurements in wetlands of these regions. At that time, northern high-latitude systems, which account for about half the world's wetland area, were more comprehensively covered than were tropical and subtropical habitats. In North America, this coverage was heavily weighted toward Alaskan measurements and included a suite of wetland habitats measured over complete seasonal cycles and over several years. In contrast, South American measurements, although representing both wet and dry seasons, were exclusively in environments closely associated with the Amazon River. Recently, Sippel et al. (1994) and Hamilton et al. (1995, 1996) conducted studies in the large wetland complexes of the Pantanal. Russian wetlands (e.g., the West Siberian Lowlands), accounting for ~25% of the global area, were not represented at all in CH_4 flux measurements until those reported by Panikov et al. (1993) and Christensen (1993, 1995). Observational data from African wetlands remains scanty aside from the measurements of Delmas et al. (1992), Fontan et al. (1992) and Tathy et al. (1992) for forested wetlands and may remain so because of obstacles to field campaigns in these African environments (i.e., political turmoil). Methane flux measurements in Asian wetlands are still absent from the published literature.

3.1.1 Tropical Measurements. Most tropical studies have been carried out in Amazonian riverine habitats - flooded forests, floating grass mats, and lakes and channels. Bartlett et al. (1990) concluded that wet- and dry-season methane fluxes from Amazon floodplain environments may be relatively constant, but added the cautionary note that emissions during transition periods are not yet measured and may be higher. In general, methane fluxes from open water are lower and less variable than those from flooded forests and floating mats. The work of Wassmann et al. (1992) suggests differences in the seasonal pattern of peak fluxes from open water and several vegetated environments. Fluxes are similar during wet (Bartlett et al., 1990; Devol et al., 1990) and dry seasons (Bartlett et al., 1988; Devol et al., 1988) partially due to very high variability. Although there are substantial seasonal changes in inundated areas of tropical riverine systems associated with the Amazon (Melack et al., 1994; Hess et al., 1995), these habitats are not characterized by the large seasonal pulses of organic inputs from litterfall or temperature-regulated pulses of microbial activity found in higher latitudes. However, episodic events can play a significant role in seasonal emissions in the tropics and elsewhere (Table 2). Field techniques designed to measure separately the direct and ebullitive contributions to methane fluxes in tropical riverine and lake environments, as well as calculations of the role of bubbling in overall fluxes, confirm that episodic ebullition may commonly account for 20-75% or more of the total seasonal emission of methane in these environments further complicating measurement and modeling of fluxes from these dynamic environments (Bartlett et al., 1988, 1990; Moore et al., 1990; Wilson et al., 1989; Devol et al., 1988, 1990; Crill et al., 1988a,b; Wilson et al., 1989; Keller, 1990; Wassmann et al., 1992; Keller and Stallard, 1994). The relative role of bubbling apparently varies with ecosystem; ebullient fractions increase from open water to grass mats and flooded forest. Furthermore, bubbling events appear to be more pronounced during periods of falling and low water and in shallow lake waters.

Table 2. Contribution of episodic events to seasonal methane emission from a series of wetland habitats.

Wetland Habitat	Mechanism/Magnitude
[1] Subarctic boreal fens	degassing pulse with lowered water table after 3 weeks of low rainfall; accounted for 18-65% of seasonal emission, depending upon habitat
[2] Temperate freshwater swamp	(direct measurement of ebullient flux) ebullition: observed in 19% of measurements, accounted for 34% of the seasonal emission
[3] Amazon floodplain	ebullition loss: 73% of emission in rising-water season, 59% of emission in low-water season
[4] Amazon lake (Lago Calado)	(direct measurement of diffusive and ebullitive flux) ebullition loss: 70% of flux from open-water lake areas
[5] Amazon floodplain	(direct measurement of diffusive and ebullient flux)ebullition loss: 49% of open-water flux, 54% of flooded-forest flux, 64% of grass-mat flux
[6] Amazon floodplain	ebullition loss: 80% of open-water flux, 91% of flooded-forest flux, 67% of floating grass-mats flux, 80% of total flux from Varzea
[7] Tropical lake (Gatun Lake)	(two techniques for direct measurement of ebullient flux) ebullition loss: 98% of total flux

[1] Moore et al. (1990); [2] Wilson et al. (1989); [3] Devol et al. (1988, 1990); [4] Crill et al. (1988a); [5] Bartlett et al., 1988); [6] Wassmann et al. (1992); [7] Keller and Stallard (1994).

3.1.2 Subtropical Measurements. Few measurements are available for subtropical wetlands, and are confined to the southeastern US (Virginia, South Carolina, Georgia and Florida). Wilson et al. (1989) demonstrated that seasonal trends in methane flux from the Newport News Swamp in Virginia were strongly correlated with temperature best represented by a step function. They suggest pulses of organic substrate inputs as the likely driver for the series of emission peaks observed during the year. The spring peak reflects mineralization of labile organic matter accumulated during the winter followed by temperature-triggered decomposition to substrates for methanogenesis while summer and autumn peaks are related to root exudates and litter input. The remaining subtropical measurements show considerably smaller methane fluxes from low latitude swamps. Until ~1990, the studies of Harriss and Sebacher (1981), Harriss et al. (1982, 1988b) were the only published low latitude measurements to serve as proxies for tropical wetlands.

3.1.3 Temperate Measurements. Early methane flux measurements in temperate and low boreal regions suggested these wetlands as extremely productive environments. For example, Harriss et al. (1985) and Baker-Blocker et al. (1977) showed methane fluxes in Minnesota and Michigan in the range of 200 to ~600 mg CH_4 m^{-2} d^{-1} in summer and fall. High fluxes measured in diverse wetlands in Minnesota (increasing from forested fens and bogs, to nonforested bogs and sedge meadows) were the ecosystem fluxes used to represent boreal wetlands by Matthews and Fung (1987). Later studies at Minnesota sites (Crill et al., 1988a) and a New Hampshire site (Frolking and Crill, 1994) confirm high fluxes from open bogs, circumneutral fens, and a poor fen similar to results obtained in Alaskan alpine fens (Sebacher et al., 1986). For example, Frolking and Crill (1994) reported monthly mean fluxes ranging from 21 mg CH_4 m^{-2} d^{-1} in February, 1992 to 649 mg m^{-2} d^{-1} in July, 1991. Annual totals were ~69 g CH_4 m^{-2} in both years although timing and rapidity of onset varied between the years . Finally, Dise (1992) reported that winter methane fluxes from Minnesota peatlands may account for up to 20% of the annual emission from these environments, highlighting the importance of acquiring measurements that represent all seasons.

3.1.4 Boreal and Arctic Measurements. Whalen and Reeburgh (1988) measured year-round methane fluxes at a series of permanent sites in a subarctic muskeg and along a pond margin in Alaska. These tussock and carex sites, chosen as representative of arctic tussock tundra and wet meadow tundra in the region, showed complex seasonal emission patterns. For example, although situated in similar climatic regimes, tussock sites showed positive fluxes from July through October with highest values in July-August while the carex sites showed positive fluxes from June through December peaking in August.

The first reports from the Hudson Bay Lowlands (HBL) revealed overall lower flux rates for these low boreal habitats than previously measured (Hamilton et al., 1994; Roulet et al., 1992a). Ponds (fen and beaver) exhibited fluxes in the upper range of HBL wetlands while bogs and fens averaged fluxes of 13 and 3 mg CH_4 m^{-2} d^{-1}, respectively, over the season from May to October. Individual sites within these wetland types periodically exhibited larger emission rates, e.g., a thicket swamp with emissions of 40-60 g CH_4 m^{-2} d^{-1} in May and June, rising to 120-160 mg m^{-2} d^{-1} in July and remaining between 60-120 mg m^{-2} d^{-1} through August (Roulet et al., 1992a). Furthermore, Moore et al. (1990) found that episodic degassing pulses associated with lowered water tables following several weeks of reduced precipitation could account for ~20-65% of the seasonal emission of methane from subarctic boreal fens in Canada (Table 2).

The dry upland tundra and large lakes measured by Bartlett et al. (1992) during the July-early August period of ABLE 3A showed low summer fluxes (mean 2-4 mg $CH_4 m^{-2} d^{-1}$), while the smaller lake, lake vegetation and wet meadow tundra exhibited mean values of 77, 89, and 144 mg CH_4 m^{-2} d^{-1}, respectively. These chamber measurements were in reasonable agreement with eddy correlation and aircraft measurements during the same summer period. For example, Ritter et al. (1992) reported a mean of 51 mg m^{-2} d^{-1} for the tundra environments from aircraft measurements while the eddy-correlation measurement means of Fan et al. (1992) were as follows: 11 ± 3 mg m^{-2} d^{-1} for dry tundra, 29 ± 3 mg m^{-2} d^{-1} for wet tundra, and 57 ± 6 mg m^{-2} d^{-1} for lakes, giving an area-weighted regional mean of 25 ± 1 mg m^{-2} d^{-1}. Closer agreement among these measurement techniques was obtained in a similar suite of measurements taken during ABLE 3B/NOWES expeditions in 1990 although overall lower fluxes were reported. For example, Moore et al. (1994) reported chamber flux measurements for a series of wetland environments including recently emerged coastal marsh, coastal fen, tamarack fen, and interior fen (<2 g $CH_4 m^{-2}$ for the season), 2-5 g $CH_4 m^{-2}$ over the season from shallow ponds and pools, and 2-17 g $CH_4 m^{-2}$ over the season for degrading peats. Hamilton et al. (1994) reported large fluxes from ponds (110-118 mg $CH_4 m^{-2} d^{-1}$) which covered only 8-12% of the area but contributed 30% to the methane flux in the study area. Eddy correlation measurements of Edwards et al. (1994) show a mean for all environments of 16 mg $CH_4 m^{-2} d^{-1}$, equal to enclosure measurements extrapolated for the region by Roulet et al. (1994).

High boreal methane fluxes taken in the Northwest Territories during the warm and dry summer of 1995 ranged from -1.3 to 1144 mg $CH_4 m^{-2} d^{-1}$, with a mean of 77 mg CH_4 m^{-2} d^{-1} (Liblick et al. (1997). A poor fen and collapsed bog exhibited the highest seasonal means (99-210 mg $CH_4 m^{-2} d^{-1}$), followed by open rich graminoid fen (47-81 $CH_4 m^{-2} d^{-1}$); lowest positive fluxes were observed at shrub fens (0.5-23 CH_4 $m^{-2} d^{-1}$) while woody bogs consumed methane. The overall range of fluxes reported by Liblick et al. (1997) is similar to those observed in similar environments elsewhere in Canada (Bubier et al., 1995a), although open fen emissions are substantially lower, perhaps because of the lower water table in the dry summer of 1995 in the Northwest Territories. Panikov et al. (1993) undertook several spot measurements of Siberian wetlands in the summer of 1990 which exhibited extreme variability perhaps due, in part, to the measurement technique. Christensen et al. (1995) reported the first comprehensive suite of methane flux measurements from the Siberian Lowlands, taken on a transect extending from the European to the Siberian arctic in summer of 1994. Mesic tundra sites averaged fluxes of 2.3 ± 0.7 mg $CH_4 m^{-2} d^{-1}$, and wetland sites averaged 46.8 ± 5.9 mg CH_4 m^{-2} d^{-1}; the mesic fluxes were lower than similar environments in Canada while fluxes from the high-latitude sites between $67°$-$77°N$ were higher than those from similar Canadian sites at lower latitudes.

Inclusion of the Siberian wetlands closed the last major gap in methane flux measurements in high-latitude environments, confirming them as generally similar to those measured in Canada and Alaska.

3.2 Regional and Global Emission Estimates

Although there is general agreement concerning the global area and distribution of wetlands, uncertainties remain as to the dynamics of wetland areas and methane production. Aselmann and Crutzen (1989) estimated total methane emission from natural wetlands to be 80 Tg with a range of 40-160 Tg, similar to the 110 Tg estimate of Matthews and Fung (1987) (Table 3). However, the relative contribution of high- and low-latitude ecosystems to the total emission was reversed in these two early studies. The high daily fluxes and emission periods of 100 or 150 days applied to boreal wetlands by Matthews and Fung (1987) resulted in about 60% of the total emission confined to the region from 50°-70°N; these large high-latitude emissions were later assessed to be inconsistent with the atmospheric measurements of methane concentrations and their seasonal cycles (Fung et al., 1991). Tropical/subtropical emission periods of 180 days together with lower flux rates for these low-latitude wetlands (area-weighted mean of swamps = ~100 mg m^{-2} d^{-1}) resulted in about 30% of the total wetland area in the low latitudes contributing about 25% of the total annual emission in the Matthews and Fung (1987) study. Aselmann and Crutzen (1989) assumed ~20% higher daily flux rates for the swamps and marshes predominating in the low latitudes along with emission periods of more than 250 days, resulting in a total emission largely concentrated in the tropics. Boreal and polar wetlands, with production periods averaging almost six months and area-weighted daily flux rates about one-quarter those assumed by Matthews and Fung (1987), played a smaller role in the emission of methane, contributing about one-third of the annual emission. The large sensitivity of emission estimates to the assumed length of methane-production seasons highlights the crucial importance of improving information on seasonal as well as interannual variations in areas and methane-production conditions for wetland ecosystems.

Bartlett et al. (1990) estimated the global methane emission from natural wetlands using areas, ecosystem classes and inundation periods of Matthews and Fung (1987) combined with updated fluxes for major ecosystems based on measurements from the Amazon Boundary Layer Experiment campaigns in 1985 (dry season) and 1987 (wet season) and new measurements from northern wetlands. The re-evaluated fluxes were higher than those of Matthews and Fung (1987) for tropical/subtropical swamps; alluvial formations were also recognized to be higher methane emitters with a flux of 160 mg m^{-2} d^{-1} in contrast to 30 mg m^{-2} d^{-1} assumed by Matthews and Fung (1987). Fluxes for boreal habitats from more comprehensive measurements were also lower, equal to about 50-75% of those used by Matthews and Fung (1987). In the estimate of Bartlett et al. (1990), tropical/subtropical ecosystems contributed about two-thirds to the total emission of methane from wetlands similar to the pattern indicated by Khalil and Rasmussen (1983) using scarce information, and Aselmann and Crutzen (1989), later confirmed in Fung et al.'s (1991) model evaluation of seasonal and spatial emissions using atmospheric CH$_4$ measurements as validation (Table 3). However, dominance of the low-latitude source in the work of Bartlett et al. (1990) and Fung et al. (1991) is a function of higher daily flux rates for tropical wetland ecosystems based on tropical measurements whereas the relative dominance of the tropics/subtropics in the analysis of Aselmann and Crutzen (1989) resulted from the combined effects of larger tropical wetland areas, moderately higher daily flux rates and substantially longer production seasons for these low-latitude wetlands.

Wetland areas and their geographic distribution contribute little to variations among the global estimates shown in Table 3 since the wetland data bases commonly used in these estimates are very similar. Differences among the emission studies reflect differing assumptions about the duration of methane productive seasons as well as incorporation of methane fluxes from the expanding suite of field measurements. While estimates of annual methane emission from wetlands are converging around 100 Tg, with about two-thirds emanating from the low latitudes, the very close agreement apparent in table 3 stems in part from reliance on the same inundation periods and areas in several of the studies.

Whalen and Reeburgh (1992) estimated annual tundra emissions for 1987-1990 by extrapolating mean fluxes measured in a series of habitats for each of the four years. This range of tundra estimates encompasses most of the boreal/arctic values in Table 3, which represent the full complement of northern wetland ecosystems. The tundra area of 7.3 x $10^{12}m^2$ used in this study is ~3 times the boreal/arctic wetland area suggested by most others. However, the measurements of Whalen and Reeburgh (1992) and others suggest that substantial portions of drier tundra may emit methane at low rates. These drier areas are not typically included in estimates although Ritter et al. (1992) suggested that regional emission estimates from ABLE 2A were an upper limit partially because the sampled areas were likely biased toward particularly productive habitats. Several recent estimates of tundra methane fluxes suggest emissions substantially lower than the 25-35 Tg commonly estimated (Table 3) while the inverse modeling of Hein et al. (1997) estimates extremely large overall emissions totaling 232 ± 27 Tg, which is inconsistent with modeling studies (Fung et al., 1991).

4. Modeling Wetlands and Their Methane Emissions

As noted above, the length of productive seasons as well as seasonal and interannual variations in extent of inundation remain major uncertainties in understanding the role of wetlands in the global methane cycle. Furthermore, data-based emission estimates have assumed constant daily fluxes that vary only with ecosystem type. Early efforts to introduce seasonally varying flux rates observed in the field incorporated temperature-flux relationships for northern ecosystems (Fung et al., 1991). However, while this approach provided reasonable emission patterns on a seasonal basis, other processes such as seasonal water-table variations or substrate input were not represented. While uncertainties in global methane estimates resulting from spatial and ecosystem variability have been reduced by expanding the measurement base, the most promising approach to reducing uncertainties in the seasonality of wetland areas and methane-producing inundated conditions is through integration of remote-sensing and modeling techniques.

This section is divided into three areas dealing chronologically with modeling methane emissions and wetlands: (1) early hybrid techniques of extrapolation and modeling of methane emissions from wetlands, (2) ecological process modeling of methane emissions from wetlands, and (3) modeling the distribution of wetlands.

4.1 Hybrid Extrapolation/Modeling of Methane Emissions

Early efforts to estimate emissions employed the simple approach of multiplying one or a few flux measurements of methane to wetland areas, assuming some hydroperiod. For example, Matthews and Fung (1987) used global vegetation, soils, and inundation data sets to target and characterize the distribution and environmental/ecological features of wetlands. They initially assumed methane production periods as a function of latitude, and applied daily methane emission rates for various wetland ecosystems to the appropriate ecosystems identified in the data set to estimate spatial and temporal distributions of methane emissions. Later, Fung et al. (1991) used a simple model for hydroperiod based on temperature in northern wetlands and precipitation-evaporation relationships in low-latitude wetlands; daily flux rates for wetland ecosystems were estimated using Q_{10} relationships and temperature. This technique, which provided a more reasonable seasonal cycle for wetlands, was applied to the wetland data set of Matthews and Fung (1987). Aselmann and Crutzen (1989) estimated a range for methane emissions from wetlands by assuming that 2-7% of primary productivity was emitted to the atmosphere as methane; the mean total was 80 Tg with a range of 40-160 yr CH_4y^{-1}, boreal environments contributed ~30% to the total. Later studies focused incorporating new measurements to improve ecosystem flux estimates, but followed similar techniques of extrapolation or simple modeling to globalize the measurements.

4.2 Process Modeling of Methane Production, Transport, and Emission

Very recently, researchers have initiated efforts to model methane emissions over large areas using climatic, edaphic, and biological parameters. These models are designed to simulate more accurately the seasonal cycles of methane efflux, as well as flux variations within wetland ecosystems. For example, Cao et al. (1996) modeled supply of substrate, relationships between flux and climatic/environmental controls such as temperature and soil moisture, consumption of methane, and surface methane flux. Applying the model to the data set of Matthews and Fung (1987) gave a global total emission of 92 Tg $CH_4 y^{-1}$, with ~22 Tg from >50°N, ~14 Tg from temperate regions (30-50°N/S) and ~55 Tg from the tropics ±30° (Table 3) These values indicate a mean of ~3.5% for NPP ultimately released as methane, in the mid-range of such estimates (Christensen et al., 1996). Walter et al.'s (1996) methane model focused on soil physics and was validated with methane flux measurements from several northern and tropical wetland sites. It has been applied globally to the wetland data set of Matthews and Fung (1987). Total emission appears to be anomalously high at >300 Tg y^{-1}. Christensen et al. (1996) modeled methane emissions from northern wetlands using steady-state seasonal NPP and heterotrophic respiration (HR) from the BIOME ecosystem model (Prentice et al., 1993), and accounting for peatland carbon storage. Methane emission is a proportion of HR with a constant of proportionality estimated from observations and varying with ecosystem. Applying the model to the same wetland data set used by Cao et al. (1996) and Walter et al. (1996), they estimate that northern wetlands emit 20 Tg CH_4

y^{-1}, generally consistent with other studies (Table 3). Recently, Potter (1997) published a 1-D methane model that includes water table and thaw depth, substrate production and decomposition, methane production and transport, and surface methane flux parameterized for a suite of tundra plant communities; the model has been validated with the seasonal tundra measurements of Whalen and Reeburgh (1992).

All the models summarized above perform reasonably well in simulating the seasonality and magnitude of methane emissions from wetlands. However, all have been applied either at single sites or to the externally-prescribed distribution of the same wetland data set so that the modeled features of methane emission encompass no impact from seasonal, interannual, or longer-term trends in the distribution of wetlands themselves. Currently, the principal barrier to development of comprehensive wetland-methane models is the difficulty of modeling the distribution of wetlands themselves, as well as their seasonal and interannual variations with climate although such capability is required in order to assess responses and feedbacks between climate change and wetlands for past and future climates.

4.3 Modeling Wetland Distribution

Parameters known to influence the distribution of wetlands include climate, topography, slope, vegetation, and soils etc. However, these parameters have not yet been used successfully to model wetland location, extent or behavior due to the complex dynamics among the variables, as well as to the very local nature of their variation and influence. Global estimates of some of these parameters are now becoming available, and efforts have recently begun to incorporate them into simulations of terrestrial hydrologic processes.

Coe (1997, 1998) reports on a global model designed to simulate mean annual terrestrial hydrologic processes including rivers, lakes, and wetlands as a linked dynamic system. The model has two main components currently operating at 5' latitude by 5' longitude resolution. The land surface component, derived from digital elevation data, determines potential surface water areas, maximum water volume within potential lakes and wetlands, and the direction in which excess water is transported across the land surface as rivers. The water volume component uses a linear reservoir model, and estimates of runoff, precipitation, and evaporation to determine the water volume available to fill potential water areas and to form rivers. These linked components dynamically simulate the transport of water across the land surface and to the oceans in the form of rivers, as well as the storage of surface water in lakes and wetlands. However, wetland simulation is poor in many regions where wetlands are common, and wetlands are arbitrarily defined as water bodies with a depth of 1 meter or less. Coe (1998) suggests that accurate wetland simulation is hampered primarily by limits imposed by the 5' resolution of the digital elevation model (DEM) used to date because many wetland complexes are composed of features too small to resolve at this resolution.

For methane studies, a linked hydrological model such as that of Coe (1998) is the most promising approach to simulating variations in wetland extent and depth on seasonal and interannual timescales. With such a hydrological model available, the

Table 3. Regional wetland areas and associated methane emissions from studies published between 1983 and 1997.

Tropical		Temperate		Boreal/Arctic		Global		References and Comments
Area 10^{12} m^2	Emission Tg	Area 10^{12} m^2	Emission Tg	Area 10^{12} m^2	Emission Tg	Area 10^{12} m^2	Emission Tg	
-	90	(Included in boreal)		-	66	-	156	Khalil & Rasmussen, 1983
-	-	-	-	4.5-9.0	45-106	-	-	Sebacher et al., 1996: peatlands
-	38±17	-	-	-	-	-	47±22	Seiler & Conrad, 1987
2.0	34	0.6	1.2	2.7	65	5.3	111	Matthews & Fund, 1987
0.1-0.5	3-17	-	-	-	-	-	-	Barlett et al., 1988: Amazon floodplain
-	-	-	-	-	72	-	-	Crill et al, 1988a
-	8-13	-	-	-	-	-	-	Devol et al., 1988: Amazon floodplain
2.1	45	1.1	11	2.4	25	5.7	80	Aselmann & Crutzen, 1989
2.0	55	0.6	17	2.7	39	5.3	111	Bartlett et al, 1990
0.1	5	-	-	-	-	-	-	Devol et al., 1990: Amazon floodplain
-	-	-	-	1.5	14-19	-	-	Moore et al., 1990: fens
2.0	71	0.6	12	2.7	32	5.3	115	Fung et al., 1991
0.1	2-3	-	-	-	-	-	-	Tathy et al., 1991: Congo Basin
-	-	-	-	7.3	44	-	-	Ritter et al., 1992: tundra
-	-	-	-	7.3	14-42 (1987)	-	-	Whalen & Reeburgh, 1992: tundra

Tropical		Temperate		Boreal/Arctic		Global		References and Comments
Area $10^{12} m^2$	Emission Tg	Area $10^{12} m^2$	Emission Tg	Area $10^{12} m^2$	Emission Tg	Area $10^{12} m^2$	Emission Tg	
-	-	-	-	7.3	26-78 (1988)	-	-	
-	-	-	-	7.3	24-67 (1989)	-	-	
-	-	-	-	7.3	69-135 (1990)	-	-	
2.0	66	0.6	5	2.7	34	5.3	105	Bartlett & Harriss, 1992
				0.3	0.5			Roulet et al., 1994: Hudson Bay Lowland
-	-	-	-	-	20±13	-	-	Christensen et al., 1996: tundra
2.0	55.2	0.6	13.8	2.7	21.8	5.3	92	Cao et al., 1996: process model
-	-	-	-	7.3	5.5-5.8	-	-	Reeburgh et al., 1998: dry tundra
-	100	-	87	--	45	-	232±27	Hein et al., 1997: inverse model

importance of specific plant communities to methane emissions makes it likely that the hydrological model will be integrated with vegetation models for methane studies (Bubier et al., 1995b; Christensen et al., 1995; Potter, 1997).

5. Wetlands and Climate Change

About half of the global wetland area, and the majority of peatlands, occur in latitudes between 50° and 70°N, regions expected to undergo temperature increases on the order of several degrees C during the next 100 years (Hansen et al., 1988, 1997; Gorham et al., 1991; Houghton et al., 1996). These changes may lead to (1) a lengthened thaw season and associated increase in biological activity, (2) larger areas subjected to thaw and anaerobic conditions, (3) increased net primary productivity due either to direct fertilization from increases in CO_2 concentration or to indirect temperature effects, (4) changes in above- and below-ground carbon allocation, and/or (5) changes in plant distributions and successions. Based on temperature increases alone, methane emissions would probably increase in high latitudes providing a positive feedback on the climate system. However, other factors may moderate this response. For example, nutrient limitation may restrict productivity increases and microbial adaptation to the current thermal regime may be inelastic. Available water supply in high latitudes may decline in response to increased evaporation under warmer conditions which could produce lower water tables and dry soil conditions in previously waterlogged or inundated environments, thereby increasing methane oxidation, reducing methane emissions and perhaps causing former wetlands to act as methane sinks. Seasonal precipitation is the major controller of tropical and subtropical methane emissions, affecting both area and length of inundation periods. While General Circulation Models generally predict greater precipitation for high-latitude wetland areas in the future, low-latitude regions may be subject to reduced precipitation. However, predictions of hydrologic perturbations over the next 50-100 years are highly uncertain, leaving open the question of whether current wetlands will become larger or smaller methane sources or perhaps sinks in the future. As noted above, difficulties in integrating the climatic, edaphic, and topographic underpinnings of wetland behavior indicate that very local conditions determine whether current wetland sites may expand or contract, and whether per-unit-area fluxes increase or decline.

A suite of investigators have evaluated potential changes in methane emissions from high latitude wetland ecosystems in response to predicted climate alterations using current relationships between methane flux and climate variables (Table 4). Frolking (1993) and Harriss and Frolking (1992) evaluated possible inter-annual oscillations in high latitude summer methane emission from wetlands as a function of 20th century temperature variations. Compared to baseline long-term summer averages, temperature anomalies range from about -2°C for the coolest years to +2°C for the warmest. Based on reconstructed summer temperature anomalies in high latitude wetland regions, and correlations between temperature and methane flux measurements in Alaska and Minnesota, they estimate that summer methane emissions from boreal wetlands during the last century have varied approximately 5

Tg around a mean of 32 Tg. This result suggests that wetland emissions are moderately sensitive to the magnitude of temperature variations that might be expected in the early stages of warming predicted for the next century. Frolking (1993) further suggests that a 3-5°C temperature rise, with no change in water status, might increase boreal emissions to more than twice their current estimated value. Livingston and Morrissey (1991) evaluated the sensitivity of regional North Slope (Alaska) methane emissions to potential changes in both temperature and water status. *In situ* flux measurements show a three-fold increase in response to the ~5°C mid-summer soil temperature elevation observed in the study area between the 1987 and 1989 field seasons. Extrapolation to the North Slope region suggests a potential four-to-five-fold increase in emission under conditions wetter than those of 1987 and a doubling of methane emission with a 4°C elevation even under conditions moderately drier than 1987. Based on methane measurements at a broad suite of tundra sites in Alaska, Whalen and Reeburgh (1992) estimated annual tundra emissions for 1987-1990 based on mean emission values for tundra habitats measured over the period. Interannual fluxes varied substantially over the period which encompassed consistent summer temperature anomalies of +2°C and precipitation variations ranging from ~60% (1987-1989) to 190% (1990) of long-term averages. Estimates of total tundra emissions vary by a factor of four, from 14-42 Tg CH_4 in the driest year (1987) to 69-135 Tg in 1990, the anomalously wet year. These latter two analyses based on field measurements indicate potentially large sensitivity of high latitude methane emissions to combined effects of temperature and precipitation changes. The analyses of Roulet et al. (1992b) on the sensitivity of methane emissions to temperature and to water-table variations in Canadian fens indicate a moderate dependence of the emissions on temperature but a very strong dependence of emission on water table depth. Their modeling study suggests an emission increase of ~15% with a 2°C elevation in temperature at 10 cm soil depth but ~75-80% declines in emissions from floating and non-floating fens following a 14 cm drop in water table.

In the Northwest Territories (Liblick et al. (1997), as well as in NOWES sites (Moore et al., 1994), largest methane emissions were associated with degrading and collapsing peats. Based on this evidence, Liblick et al. (1997) hypothesize that increased permafrost melting will initially increase boreal methane emissions due to formation of collapse scars, but that such increases will eventually reverse with lowered water tables and associated increases in methane oxidation. Similarly, Hamilton et al. (1994) suggest that high-latitude warming may increase landscape ratios of ponds (with high fluxes) to vegetated areas (lower fluxes), increasing overall fluxes.

Christensen and Cox (1995) developed a model of permafrost thermodynamics and methane emission and drove it with current and $2xCO_2$ climates from the UK Meteorological Office GCM, as well as with North Slope field measurements. Simulated methane emissions in the $2xCO_2$ experiment increased due to slightly warmer temperatures. However, the dominant cause of increased emissions was from a 42% increase in mean thaw depth as a result of deeper maximum thaws and a longer thaw season. This result, along with the measurements of high-latitude peatlands, illustrate the sensitivity of emissions to the combined effect of temperature and moisture impacts of changing climates.

Table 4. Methane emission and climate change (adapted from Matthews, 1993, and Öquist et al., 1996).

Description	Methane-Emission Response
1 Relationship between regional North Slope (Alaska) tundra emissions and temperature; simulated changes in water status via changes in vegetation distributions; extrapolation using 1987-1989 measurements	local: 4-fold increase with 4°C summer soil temperature rise; regional: 4-5-fold increase under 4°C warmer and wetter conditions; 2-fold increase under 4°C warmer and drier conditions: large sensitivity to warming
2 Sensitivity of high latitude emissions to historical summer temperature variations; no evaluation of water status; temperature-flux correlations and regional 20th century temperature anomalies used to model historical flux anomalies	± 2°C anomalies gave ±5 Tg variation around mean annual emission of ~30 Tg; suggests that 3-5°C rise could double emissions: moderate sensitivity to early warming
3 Four-year time series of flux measurements at fixed Alaskan tundra sites extrapolated to estimate interannual variations in high-latitude tundra emissions	4-fold emission variation over four years: 14-42 Tg (1987, driest), 26-78 Tg (1988), 24-67 Tg (1989), 69-135 Tg (1990, wettest): large sensitivity to combined temperature and water effects
4 Measurements of methane emission and related environmental factors in low boreal Canadian fens; water-table position predicted with hydrologic/thermal model for peatlands	15% emission increase with 2°C elevation in soil temperature; 75-80% emission decline with 14 cm drop in water table: moderate sensitivity to warming, large sensitivity to water status
5 Measurements of methane emission and temperature from drained boreal wetlands in Finland	4 cm decline in water table reduced methane flux by 80%; 20 cm drop in water table reduced flux to zero: large sensitivity to water status
6 Measurements of methane emission and water table in drained boreal wetlands in Canada	100% reduction of emission with 10 cm decline in water table; >10 cm water table drop changed wetland to small methane sink: large sensitivity to water status
7 Sensitivity study of modeled methane emission to temperature increase	2° T increase increased emission by 36%, 29%, and 12% in northern, temperate and tropical wetlands; large sensitivity to temperature in northern environments

[1] Livingston and Morrissey (1991); [2] Frolking (1993), Harriss and Frolking (1992); [3] Whalen and Reeburgh (1992); [4] Roulet et al. (1992c); [5] Martikainen et al. (1992); [6] Roulet et al. (1993); [7] Cao et al. (1996).

A primary determinant of high latitude wetland responses to climate change lies in the extent to which anaerobic conditions are maintained under circumstances of reduced moisture availability. Seasonal, areal and vertical components of methane production and consumption activities will likely adjust differently to moisture changes.

6. Summary

6.1 Area, Distribution, and Seasonality of Wetlands and Emissions

Presently-available information on the distribution of wetlands exceeds large-scale information on methane production, oxidation and emission characteristics. Independent data sources generally agree on latitudinal and environmental profiles of wetlands. A comprehensive suite of methane flux measurements conducted in a broad array of geographically and ecologically diverse ecosystems has characterized large-scale features of the role of natural wetlands in the global methane cycle. Model studies including photochemical and transport processes, as well as regional emissions estimates derived from tower- and aircraft-base eddy-correlation techniques, suggest patterns consistent with the ground-based measurements.

Tropical and subtropical wetlands, which account for about one-third of the global wetland area, likely contribute ~50-75% to the annual emission of methane from natural wetlands. Precipitation-driven inundation periods may last from a few to 12 months a year in these environments. More than half the world's wetlands occur in boreal and arctic habitats north of 50°N. As in the tropical riverine wetlands, many of these environments are characterized by complex landscapes composed of herbaceous and open-water features with distinctive methane emissions. At present boreal methane emissions, telescoped into a few months of the year during thaw seasons, probably contribute about one-third of the world's total wetland emissions.

The largest uncertainties in the role of wetlands and their methane emission in the global methane cycle are variations in inundation periods, seasonal changes in wetland habitat areas, interannual variations in inundation extent. these uncertainties in wetlands themselves translate into uncertainties in emissions. The dependence on climate of wetland variations and emissions remains the major uncertainty. Remote sensing and model development are contributing to reducing these uncertainties in seasonal and interannual variations. Remotely-sensed data can provide information on seasonal and interannual expansions and contractions of wetlands, and may allow simple parameterizations of wetlands in GCMs and/or biogeochemical models.

6.2 Flux Measurements

Methane fluxes have been measured in all major wetlands. the recent measurements in the Siberian lowlands and the South American Pantanal closed the last gaps in ecosystem coverage of methane emissions although seasonal variations in the former should be better quantified . The seasonality of methane emissions for most wetland habitats is well measured as are interannual variations in high-latitude fluxes. Many Amazonian environments have been measured in both wet and dry seasons although African wetlands are poorly represented by field measurements with respect to seasonal and interannual variations, and no measurements exist for herbaceous wetlands in Africa. Low-latitude Asian wetlands are not measured at all but occupy a small area globally. While sites in Manitoba and Quebec initially dominated measurements of methane emissions in Canada, field studies from the Northwest territories, which account for 25% of Canada's wetlands, have recently become available. Continued measurements in Canada confirm general patterns and levels of fluxes measured earlier in the 1990s.

Measurements of methane fluxes in natural wetlands continue to confirm that methane fluxes can vary by orders of magnitude among ecosystems at local scales, and by factors of three to four interannually in response to temperature and water-status variations. Moreover, episodic events have been shown to contribute substantially to total seasonal emissions in a broad array of wetlands accounting for 20-75% of seasonal methane emissions in tropical environments; similar fractions due to degassing pulses associated with lowered water tables have been observed in Canadian wetlands. Given the relatively comprehensive coverage of wetland habitats with respect to flux measurements, researchers are now focusing more on quantifying complex interactions among dynamic environmental variables and their influence on fluxes. Variables that serve as environmental integrators of production and consumption processes remain the most promising predictors of methane fluxes.

6.3 Regional and Global Emission Estimates

Tropical and subtropical wetlands, which account for about one-third of the global wetland area, likely contribute ~50-75% to the annual emission of methane from natural wetlands. Precipitation-driven inundation periods are highly variable and may last from a few to 12 months a year in these environments. More than half the world's wetlands occur in boreal and arctic habitats north of 50°N. As in the tropical riverine wetlands, many of these environments are characterized by complex landscapes composed of herbaceous and open-water features with distinctive methane emissions. At present boreal methane emissions, telescoped into a few months of the year during thaw seasons, probably contribute about one-third of the world's total wetland emissions. However, their response to climate changes predicted for the next century is highly uncertain. Depending on local interactions among temperature, water status, nutrients, microbial populations etc., boreal/arctic ecosystems may become larger or smaller methane sources, or methane sinks.

6.4 Modeling Emissions, Wetlands and Climate Change

Uncertainties in the response of methane emission from wetlands to climate changes predicted for the next century remain large. Depending on very local interactions among temperature, water status, nutrients, microbial populations etc., current wetlands may become larger or smaller methane sources, or methane sinks.

Researchers have used interannual variations in climate during field studies as proxies to predict likely scenarios of climate change impacts on emissions. However, modeling and field studies do not indicate any clear trends in emissions under changed climates primarily because few studies include the combined effects and interactions of hydrological and thermal parameters. Direct effects of increasing temperature, as well as indirect effects of increased thaw season and depth, would increase CH_4 emissions if local hydrological regimes remained constant. However, reduction in precipitation or lowering of water tables in present wetlands would likely reduce emissions under constant or increasing temperatures.

Recently-developed methane-emission models perform reasonably well when applied to independently-defined wetland distributions. However, simulating the response of wetlands and their methane emissions to climate change requires linked hydrological and ecosystem models preferably integrated into GCMs, i.e., modeling the distribution and variations of wetlands as well as their methane emissions. Modeling wetlands directly is a more difficult problem because of extremely local influences (e.g., topography) on wetland distribution and area. However, simulation of interactions among terrestrial methane sources, atmospheric composition, and climate hinges upon such integrated modeling.

References

Achutuni, R., R. A. Scofield, N. C. Grody and C. Tsai, Global monitoring of large flooding using the DMSP SSM/I soil wetness index, paper presented at Annual Meeting, American Meteorol. Soc., Atlanta, Ga., 1996.

Aselmann, I., and P. Crutzen, Global distribution of natural freshwater wetlands and rice paddies: Their net primary productivity, seasonality and possible methane emissions, *J. Atmos. Chem., 8*, 307-358, 1989.

Auerbach , N. A., D. A. Walker, and J. G. Bockheim, *Land cover map of the Kuparuk River Basin, Alaska*, Institute of Arctic and Alpine Research, University of Colorado, 1997.

Baker-Blocker A, T. M. Donohue, and K. H. Mancy, Methane flux from wetland areas, *Tellus, 29*, 245-250, 1977.

Barber, T. R., R. A. Burke, Jr., and W. M. Sackett, Diffusive flux of methane from warm wetlands, *Global Biogeochem. Cycles, 2*, 411-414, 25, 1988.

Bartlett, D. S., K. B. Bartlett, J. M. Hartman, R. C. Harriss, D. I. Sebacher, R. Pelletier-Travis, D. D. Dow, and D. P. Brannon, Methane emissions from the Florida Everglades: Patterns of variability in a regional wetland ecosystem, *Global Biogeochem. Cycles, 3*, 363-374, 1989.

Bartlett, K. B., and R. C. Harriss, Review and assessment of methane emissions from wetlands, *Chemosphere, 26*, 261-320, 1993.

Bartlett, K. B., P. M. Crill, D. I. Sebacher, R. C. Harriss, J. O. Wilson, and J. M. Melack, Methane flux from the central Amazonian floodplain, *J. Geophys. Res., 93*, 1571-1582, 1988.

Bartlett, K. B., P. M. Crill, J. A. Bonassi, J. E. Richey, R. C. Harriss, Methane flux from the Amazon River floodplain: Emissions during rising water, *J. Geophys. Res., 95*, 16,773-16,788, 1990.

Bartlett, K. B., P. M. Crill, R. L. Sass, R. C. Harriss, N. B. Dise, Methane emissions from tundra environments in the Yukon-Kuskowim Delta, Alaska, *J. Geophys. Res., 97*, 16,645-16,660, 1992.

Bartlett, K. B., R. C. Harriss, D. I. Sebacher, Methane flux from coastal salt marshes, *J. Geophys. Res., 90*, 5710-5720, 1985.

Blake, D. R., and F. S. Rowland, Continuing worldwide increase in tropospheric methane, 1978 to 1987, *Science, 239*, 1129-1131, 1988.

Blake, D. R., and F. S. Rowland, Worldwide increase in tropospheric methane, 1978 to 1983, *J. Atmos. Chem., 4*, 43-62, 1986.

Blake, D. R., E. W. Mayer, S. C. Tyler, Y. Makide, D. C. Montague, and F. S. Rowland, Global increase in atmospheric methane concentrations between 1978 and 1980, *Geophys. Res. Lett., 9*, 477-480, 1988.

Blake, D. R., Increasing concentrations of atmospheric methane. PhD thesis, 213p, University of California at Irvine, 1984.

Bridgham, S. D., C. A. Johnson, J. pastor, and K. Updegraff, Potential feedbacks of northern wetlands on climate change, *BioScience, 45*, 262-274, 1995.

Bubier, J. L., and T. R. Moore, an ecological perspective on methane emissions from emissions from northern wetlands, *Trends Ecol. Evol., 9*, 460-464, 1994.

Bubier, J. L., T. Moore, and S. Juggins, Predicting methane emission from bryophyte distribution in northern Canadian peatlands, *Ecology, 76*, 677-693, 1995b.

Bubier, J. L., T. R. Moore, L. Bellisario, N. Comer, and P. M. Crill, Ecological controls on methane emissions from a northern peatland complex in the zone of discontinuous permafrost, Manitoba, Canada, *Global Biogeochem. Cycles, 9*, 455-470, 1995a.

Burke, R. A., Jr., T. R. Barber, and W. M. Sackett, Methane flux and stable hydrogen and carbon isotope composition of sedimentary methane from the Florida Everglades, *Global Biogeochem. Cycles, 2*, 329-340, 1988.

Cao, M., S. Marshall, and K. Gregson, Global carbon exchange and methane emissions from natural wetlands: Application of a process-based model, *J. Geophys. Res., 101*, 14,399-14,414, 1996.

Chappellaz, J. A., I. Y. Fung, and A. M. Thompson, The atmospheric CH_4 increase since the Last Glacial Maximum, 1. Source estimates, *Tellus, 45B*, 228-241, 1993.

Chappellaz, J. A., J. M. Barnola, D. Raynaud, Y. S. Korotkevich, and C. Lorius, Ice-core record of atmospheric methane over the past 160,000 years, *Nature, 345*, 127-131, 1990.

Choudhury, B. J., Microwave vegetation index: A new long-term global data set for biospheric studies, *Int. J. Remote Sens., 9*, 185-186, 1988.

Choudhury, B. J., Monitoring global land surface using Nimbus-7 37 GHz data. Theory and examples, *Int. J. Remote Sens., 10*, 1579-1605, 1989.

Christensen, T. R., Methane emission from Arctic tundra, *Biogeochemistry, 21*, 117-139, 1993.

Christensen, T. R., and P. Cox, Response of methane emission from arctic tundra to climatic change: Results from a model simulation, *Tellus, 47B*, 301-309, 1995.

Christensen, T. R., S. Jonasson, T. V. Callaghan, and M. Havström, Spatial variation in high-latitude methane flux along a transect across Siberian and European tundra environments, *J. Geophys. Res., 100*, 21,035-21,045, 1995.

Christensen, T. R., I. C. Prentice, J. Kaplan, A. Haxeltine, and S. Stitch, Methane flux from northern wetlands and tundra, an ecosystem source modeling approach, *Tellus, 48B*, 652-661, 1996.

Cicerone, R. J., and J. D. Shetter, Sources of atmospheric methane: Measurements in rice paddies and a discussion, *J. Geophys. Res., 86*, 7203-7209, 1981.

Cicerone, R. J., and R. S. Oremland, Biogeochemical aspects of atmospheric methane, *Global Biogeochem. Cycles, 2*, 299-327, 1988.

Coe, M. T., A linked global model of terrestrial hydrologic processes: Simulation of modern rivers, lakes, and wetlands, *J. Geophys. Res., 103*, 8885-8899, 1998.

Coe, M.T., Simulating continental surface waters: An application to Holocene northern Africa, *J. Clim., 10*, 1680-1689, 1997.

Craig H., and Chou, C. C., Methane: The record in polar ice cores, *Geophys. Res. Lett., 9*, 1221-1224, 1982.

Crill, P. M., K. B. Bartlett, HR. C. Harriss, E. Gorham, E. S. Verry, D. I. Sebacher, L. Madzar, and W. Sanner, Methane flux from Minnesota peatlands, *Global Biogeochem. Cycles, 2*, 371-384, 1988a.

Crill, P. M., K. B. Bartlett, J. O. Wilson, D. I. Sebacher, and R. C. Harriss, Tropospheric methane from an Amazonian floodplain lake, *J. Geophys. Res., 93*, 1564-1570, 1988b.

Crutzen, P. J., Methane sources and sinks, *Nature, 350*, 380-381, 1991.

Dacey, J. W. H., and M. Klug, Methane flux from lake sediments through water lilies, *Science, 203*, 1253-1255, 1979.

DeFries, R., and J. R. G. Townshend, NDVI-derived land cover classifications at the global scale, *Int. J. Rem. Sens., 15*, 3567-3586, 1994.

Delmas, R. A., J. Servant, J.-P. Tathy, B. Cros, M. and Labat, Sources and sinks of methane and carbon dioxide exchanges in mountain forest in Equatorial Africa, *J. Geophys. Res., 97*, 6169-6179, 1992.

Devol, A. H., J. E. Richey, B. R. Forsberg, and L. A. Martinelli, Seasonal dynamics of methane emissions from the Amazon River floodplain to the troposphere, *J. Geophys. Res., 95*, 16,417-16,426, 1990.

Devol, A. H., J. E. Richey, W. A. Clark, and S. L. King, Methane emissions to the troposphere from the Amazon Floodplain. J. Geophys. Res., 93, 1583-1592, 1988.

Dise, N. B., E. Gorham, and E. S. Verry, Environmental factors controlling methane emissions from peatlands in northern Minnesota, *J. Geophys. Res., 98*, 10,583-10,594, 1993.

Dise, N. B., Winter fluxes if methane in Minnesota peatlands, *Biogeochemistry, 17*, 71-83, 1992.

Dlugokencky, E. J., L. P. Steele, P. M. Lang, and K. A. Masarie, The growth rate and distribution of atmospheric methane, J. Geophys. Res., 99, 17,021-17,043, 1994a.

Dlugokencky, E., K. A. Masarie, P. M. Lang, P. P. Tans, Continuing decline in the growth rate of atmospheric methane, *Nature, 393* (6684), 447-450, 1998.

Dlugokencky, E., K. A. Masarie, P. M. Lang, P. P. Tans, L. P. Steele, and E. G. Nisbet, A dramatic decrease in the growth rate of atmospheric methane in the northern hemisphere during 1992, *Geophys. Res. Lett., 21*, 45-48, 1994b.

Edwards, G. C., H. H. Neumann, G. Den Hartog, G. W. Thurtell, and G. Kidd, Eddy correlation measurements of methane fluxes using a tunable diode laser at the Kinosheo lake tower site during the northern wetlands study (NOWES), *J. Geophys. Res., 99*, 1511-1517, 1994.

Ehhalt, D. H. and U. Schmidt, Sources and sinks of atmospheric methane, *Pageoph., 116*, 452-464, 1978.

Ehhalt, D. H., R. J. Zander, and R. A. Lamontagne, On the temporal increase of tropospheric CH_4, *J. Geophys. Res., 88*, 8442-8446, 1983.

Ehhalt, D. H., The atmospheric cycle of methane, *Tellus, 26*, 58-70, 1974.

Etheridge, D. M., G. I. Pearman, and P. J. Fraser, Changes in tropospheric methane between 1841 and 1978 from a high accumulation-rate Antarctic ice core, *Tellus, 44*, 282-294, 1992.

Fan, S.-M., S. C. Wofsy, P. S. Bakwin, D. J. Jacob, S. M. Anderson, P. L. Kebabian, J. B. McManus, C. E. Kolb, D. R. Fitzjarrald, Micrometeorological measurements of CH_4 and CO_2 exchange between the atmosphere and the Subarctic tundra, *J. Geophys. Res., 97*, 16,627-16,643, 1992.

Fontan, J., A. Druilhet, B. Benech, R. Lyra, B. Cros, The DECAFE experiments: Overview and meteorology, *J. Geophys. Res., 97*, 6123-6136, 1992.

Frolking S., Methane from northern peatlands and climate change, in. S. Vinson and T. P. Kolchugina (eds.),*Carbon Cycling in Boreal Forest and Sub-Arctic Ecosystems*, Conference Proceedings, EPA/600/R-93/084, p. 109-124, Corvallis Oregon, 1993.

Frolking, S., and P. Crill, Climate controls on temporal variability of methane flux from a poor fen in southeastern New Hampshire: Measurement and modeling, *Global Biogeochem. Cycles, 8*, 385-397, 1994.

Fung, I., J. John, J. Lerner, E. Matthews, M. Prather, L. P. Steele, and P. J. Fraser, Three-dimensional model synthesis of the global methane cycle, *J. Geophys. Res., 96*, 13,033-13,065, 1991.

Funk, D. E., E. Pullman, K. Peterson, P. Crill, and W. D. Billings, Influence of water table on carbon dioxide, carbon monoxide and methane flux from taiga bog microcosms, *Global Biogeochem. Cycles, 8*, 271-278, 1994.

Giddings, L. and B. J. Choudhury, Observation of hydrological features with Nimbus-7 37 GHz data applied to South America, *Int. J. Remote Sens., 10*, 1673-1686, 1989.

Glaser, P. H., *The Ecology of Patterned Boreal Peatlands of Northern Minnesota: A community Profile*, US Fish and Wildlife Service Biological Report 85 (7.14). US Department of the Interior, Washington DC, 98p, 1987.

Glooschenko, W. A., N. T. Roulet, L. A. Barrie, H. I. Schiff, and H. G. McAdie, The Northern Wetlands Study (NOWES): An overview, *J. Geophys. Res., 99*, 1423-1428, 1994.

Glooschenko, W. A., N. T. Roulet, L. A. Barrie, H. I. Schiff, and H. McAdie, The Northern Wetlands Study (NOWES): An overview, *J. Geophys. Res., 99*, 1423-1428, 1994.

Gore, A. J. P. (ed.), *Ecosystems of the World, Mires: Swamp, Bog, Fen and Moor, Case Studies, 4B*, Elsevier, New York, 479p, 1983b.

Gore, A. J. P. (ed.), *Ecosystems of the World, Mires: Swamp, Bog, Fen and Moor, General Studies, 4A*, Elsevier, New York, 440p, 1983a.

Gorham, The role of northern peatlands in the carbon cycle, and probable responses to climatic warming, *Ecol. Appl., 1*,182-195, 1991.

Hamilton, J. D., C. A. Kelly, J. W. M. Rudd, R. H. Hesslein, and N. T. Roulet, Flux to the atmosphere of CH_4 and CO_2 from wetland ponds on the Hudson Bay Lowlands (HBLs), *J. Geophys. Res., 99*, 1495-1510, 1994.

Hamilton, S., S. Sippel, and J. Melack, Oxygen depletion and carbon dioxide and methane production in water of the Pantanal wetland of Brazil, *Biogeochemistry, 30*, 115-141, 1995.

Hamilton, S., S. Sippel, and J. Melack, Inundation patterns in the Pantanal wetland of South America determined from passive microwave remote sensing, *Arch. Hydrobiol., 137*, 1-23, 1996.

Hansen J., I. Fung, A. Lacis, D. Rind, S. Lebedeff, R. Ruedy, G. Russell, and P. Stone, Global climate changes as forecast by Goddard Institute for Space Studies three-dimensional model, *J. Geophys. Res., 93*, 9341-9364, 1988.

Hansen, J., Mki. Sato, R. Ruedy, A. Lacis, K. Asamoah, K. Beckford, S. Borenstein, E. Brown, B. Cairns, B. Carlson, B. Curran, S. de Castro, L. Druyan, P. Etwarrow, T. Ferede, M. Fox, D. Gaffen, J. Glascoe, H. Gordon, S. Hollandsworth, X. Jiang, C. Johnson, N. Lawrence, J. Lean, J. Lerner, K. Lo, J. Logan, A. Luckett, M.P. McCormick, R. McPeters, R. Miller, P. Minnis, I. Ramberran, G. Russell, P. Russell, P. Stone, I. Tegen, S. Thomas, L. Thomason, A. Thompson, J. Wilder, R. Willson, and J. Zawodny, Forcings and chaos in interannual to decadal climate change, *J. Geophys. Res., 102*, 25,679-25,720, 1997.

Harriss, R. C., and D. I. Sebacher, Methane flux in forested freshwater swamps of the southeastern United States, *Geophys. Res. Lett., 8*, 1002-1004, 1981.

Harriss, R. C., and S. Frolking, The sensitivity of methane emissions from northern freshwater wetlands to global warming. In P. Firth and S. G. Fisher (eds.), *Climate Change and Freshwater Ecosystems*, p. 48-67, Springer-Verlag, New York, 1992.

Harriss, R. C., D. I. Sebacher, and F. P. Day, Jr., Methane flux in the Great Dismal Swamp, *Nature, 297*, 673-674, 1982.

Harriss, R. C., E. Gorham, D. I. Sebacher, K. B. Bartlett, P. A. Flebbe, Methane flux from northern peatlands, *Nature, 315*, 652-653, 1985.

Harriss, R. C., S. C. Wofsy, M. Garstang, E. V. Browell, L. C. B. Molion, R. J. McNeal, J. M. Hoell, Jr, R. J. Bendura, S. M. Beck, R. L. Navarro, J. T. Riley, and R. L. Snell, The Amazon Boundary Layer Experiment (ABLE 2A): Dry season 1985, *J. Geophys. Res., 93*, 1351-1360, 1988a.

Harriss, R. C., D. I. Sebacher, K. B. Bartlett, D. S. Bartlett, P. M. Crill, Sources of atmospheric methane in the south Florida environment, *Global Biogeochem. Cycles, 2*, 231-243, 1988b.

Harriss, R. C., M. Garstang, S. C. Wofsy, S. M. Beck, R. J. Bendura, J. R. B. Coelho, J. W. Drewry, J. M. Hoell, Jr, P. A. Matson, R. J. McNeal, L. C. B. Molion, R. L. Navarro, V. Rabine, and R. L. Snell, The Amazon Boundary Layer Experiment (ABLE 2B): Wet season 1987, *J. Geophys. Res., 95*, 16,721-16,736, 1990.

Harriss, R. C., K. B. Bartlett, S. Frolking, and P. M. Crill, Methane emissions from northern high-latitude wetlands, in R. S. Oremland (ed.), *Biogeochemistry of Global Change: Radiatively Active Trace Gases*, Chapman and Hall, New York, p. 449-486, 1993.

Harriss, R. C., S. C. Wofsy, J. M. Hoell, Jr., R. J. Bendura, J. W. Drewry, R. J. McNeal, D. Pierce, V. Rabine, And R. L. Snell, The Arctic Boundary Layer Expedition (ABLE-3B): July-August, 1990, *J. Geophys. Res., 99*, 1635-1643, 1994.

Hein, R., P. J. Crutzen, and M. Heimann, An inverse modeling appoach to investigate the global atmospheric methane cycle, *Global Geochem. Cycles, 11* (1), 43-76, 1997.

Hess, L, J. Melack, S. Filoso, and Y. Wang, Delineation of inundated area and vegetation along the Amazon floodplain with the SIR-C synthetic aperture radar, *IEEE Trans. Geosci. Rem. Sens., 33*, 896-904, 1995.

Hess, L., J. Melack, and D. Simonett, Radar detection of flooding beneath the forest canopy: A review, *Int. J. Rem. Sens., 11*, 1313-1325, 1990.

Hogan, K. B., and R. C. Harriss, Comment on 'A dramatic decrease in the growth rate of atmospheric methane in the northern hemisphere during 1992' by E. J. Dlugokencky, et al., *Geophys. Res. Lett., 23*, 2761-2764, 1996.

Holzapfel-Pschorn, A. and W. Seiler, Methane emission during a cultivation period from an Italian rice paddy, *J. Geophys. Res., 91*, 11,803-11,814, 1986,

Houghton, J. T., L. G. Meira, B. A. Callander, N. Harris, A. Kattenberg, and K. Maskell (eds.), *Climate Change 1995: The Science of Climate Change*, Cambridge University Press, Cambridge, 1996.

Justice, C. O., J. R. Townshend and B. J. Choudhury, Comparison of AVHRR and SMMR data for monitoring vegetation phenology on a continental scale, *Int. J. Remote Sens., 10*, 1607-1632, 1989.

Keller, M. M., and R. F. Stallard, Methane emission from bubbling in Gatun Lake, Panama, *J. Geophys. Res., 99*, 8307-8319, 1994.

Keller, M. M., Biological sources and sinks of methane in tropical habitats and tropical atmospheric chemistry, PhD thesis, Princeton Univ and Natl. Center Atmos. Res., 216p., 1990.

Kerr, Y. H. and E. G. Njoku, On the use of passive microwave at 37 GHz in remote sensing of vegetation, *Int. J. Remote Sens., 14*, 1931-1943, 1993.

Khalil, M. A. K, and R. A. Rasmussen, Atmospheric methane: Trends over the last 10,000 years, *Atmos. Environ., 21*, 2445-2452, 1987.

Khalil, M. A. K., and R. A. Rasmussen, Causes of increasing methane: Depletion of hydroxyl radicals and the rise of emissions, *Atmos. Environ., 19*, 397-407, 1985.

Khalil, M. A. K., and R. A. Rasmussen, Secular trends of atmospheric methane (CH_4). *Chemosphere, 11*, 877-883, 1982.

Khalil, M. A. K., and R. A. Rasmussen, Sources, sinks and seasonal cycles of atmospheric methane, *J. Geophys. Res., 88*, 5131-5144, 1983.

Khalil, M. A. K., R. A. Rasmussen, and M. J. Shearer, Trends of atmospheric methane in the 1960s and 1970s, *J. Geophys. Res., 94*, 18,279-18,288, 1989.

King, G. M., and W. J. Wiebe, Methane release from soils of a Georgia salt marsh, *Geochim. Cosmochim. Acta, 42*, 343-348, 1978.

Klinger, L. F., Introduction to special section on the Northern Wetlands Study and the Arctic Boundary Layer Expedition 3B: An international and interdisciplinary field campaign, *J. Geophys. Res., 99*, 1421-1422, 1994.

Klinger, L., P. R. Zimmerman, J. P. Greenberg, L. E. Heidt, and A. B. Guenther, Carbon trace gas fluxes along a successional gradient in the Hudson Bay Lowland, *J. Geophys. Res., 99*, 1469-1494, 1994.

Koyama, T., Biogeochemical studies on lake sediments and paddy soils in the production of atmospheric methane and hydrogen, in Y. Miyake and T. Koyama (eds.), *Recent Researches in the Fields of Hydrosphere, Atmosphere and Nuclear Geochemistry*, p. 143-177, Muruzen Co. Ltd., Tokyo, 1964.

Koyama, T., Gaseous metabolism in lake sediments and paddy soils and the production of atmospheric methane and hydrogen, J. *Geophys. Res., 68*, 3971-3973, 1963.

Lang, P. M., L. P. Steele, and R. C. Martin, Atmospheric methane data for the period 1986-1988 from the NOAA/CMDL global cooperative flask sampling network, *Tech. Mem. ERL CMDL-2*, Natl. Oceanic and Atmos. Admin., Boulder, Colorado, 1990b.

Lang, P. M., L. P. Steele, R. C. Martin, and K. A. Masarie, Atmospheric methane data for the period 1983-1985 from the NOAA/CMDL global cooperative flask sampling network, *Tech. Mem. ERL CMDL-1*, Natl. Oceanic and Atmos. Admin., Boulder, Colorado, 1990a.

Liblick, L. K., T. R. Moore, J. L. Bubier, and S. D. Robinson, Methane emissions from wetlands in the zone of discontinuous permafrost: Fort Simpson, Northwest Territories, Canada, *Global Biogeochem. Cycles, 11*, 485-494, 1997.

Livingston, G. P., and L. A. Morrissey, Methane emissions from Alaskan arctic tundra in response to climatic change. In G. Weller, C. L. Wilson, and B. A. B. Severin (eds.), *Role of Polar Regions in Global Change: Proceedings of a Conference*, p. 372-377, Geophysical Institute and Center for Global Change and Arctic System Research, University of Alaska Fairbanks, Fairbanks, Alaska, 1991.

Matthews E., and I. Fung, Methane emission from natural wetlands: Global distribution, area, and environmental characteristics of sources, *Global Biogeochem. Cycles, 1*, 61-86, 1987.

Matthews, E., Wetlands. In M. A. K. Khalil (ed.) *Atmospheric Methane: Sources, Sinks, and Role in Global Change*, Berlin, Springer-Verlag, NATO ASI Series, I 13, p. 14-61, 1993.

Mayer, E. W., D. R. Blake, S. C. Tyler, Y. Makide, D. S. Montague, and F. S. Rowland, Methane: Interhemispheric concentration gradient and atmospheric residence time, *Proc. Natl. Acad. Sci. USA, 79*, 1366-1370, 1982.

Melack, M. M., L. L. Hess, and S. Sippel, Remote sensing of lakes and floodplains in the Amazon basin, *Rem. Sens. Rev., 10*, 127-142, 1994.

Moore T., N. Roulet, and R. Knowles, Spatial and temporal variations of methane flux from subarctic/northern boreal fens, *Global Biogeochem. Cycles, 4*, 29-46, 1990.

Moore, T. R., A. Heyes, and N. T. Roulet, Methane emissions from wetlands, southern Hudson Bay lowland, J. Geophys. Res., 99, 1455-1467, 1994.

Moore, T. R., and N. T. Roulet, Methane flux: Water table relations in northern wetlands, *Geophys. Res. Lett., 20*, 587-590, 1993.

Moore, T. R., and R. Knowles, Methane and carbon dioxide evolution from subarctic fens, *Can. J. Soil Sci., 67*, 77-81, 1987.

Moore, T. R., and R. Knowles, Methane emissions from fen, bog and swamp peatlands in Quebec, *Biogeochemistry, 11*, 45-61, 1990.

Moore, T. R., and R. Knowles, The influence of water table levels on methane and carbon dioxide emissions from peatland soils, *Can. J. Soil Sci., 69*, 33-38, 1989.

Morrissey, L. A., and G. P. Livingston, Methane emissions from Alaska arctic tundra: An assessment of local spatial variability, *J. Geophys. Res., 97*, 16,661-16,670, 1992.

Morrissey, L. A., and R. A. Ennis, Vegetation mapping of the National Petroleum Reserve in Alaska using Landsat digital data, *US Geol. Surv. Open File Report 81-315*, US Geol. Surv., Reston, VA, 25p, 1981.

Morrissey, L., G. Livingston, and S. Durden, Use of SAR in regional methane exchange studies, *Int. J. Rem. Sens., 15*, 1337-1342, 1994.

Morrissey, l., S. Durden, G. Livingston, J. Stearn, and L. Guild, Differentiating methane source areas in Arctic environments with multispectral ERS-1 SAR data, *IEEE Trans. Geosci. Rem. Sens., 34*, 667-673, 1996.

National Wetlands Working Group, *Wetlands of Canada*, Ecological Land Classification Series No. 24. Sustainable Development Branch, Environment Canada, Ottawa and

Polyscience Publications, Montreal, 452p, 1988.

Neale, C. M., M. J. McFarland and K. Chang, Land surface-type classification using microwave temperatures from the Special Sensor Microwave/Imager, *IEEE Trans. Geosci. Remote Sens., 28,* 829-838, 1990.

Öquist, M. G., B. H. Svensson, P. Groffman, M. Taylor, K. B. Bartlett, M. Boko, J. Brouwer, O. F. Canzini, C. B. Craft, J. Laine, D. Larsen, P. J. Martikainen, E. Matthews, W. Mullié, S. Page, C. J. Richardson, J. Rieley, N. Roulet, J. Silviola, and Y. Zhang, Non-Tidal Wetlands, in R. Watson, M. C. Zinyowera, and R. H. Moss (eds.), *Climate Change 1995: Impacts, Adaptations, and Mitigation of Climate Change: Scientific-Technical Analyses, IPCC Second Assessment Report,* Cambridge, Cambridge Univ. Press, p. 215-239, 1996.

Ormsby, J. P., A. J. Blanchard, Detection of lowland flooding using active microwave systems, *Photogram. Eng. Rem. Sens., 51,* 317-328, 1985.

Pearman, G. I., D. Etheridge, F. De Silva, and P. J. Fraser, Evidence of changing concentrations of atmospheric CO_2, N_2O and CH_4 from air bubbles in Antarctic ice, *Nature, 320,* 248-250, 1986.

Potter, C. S., An ecosystem simulation model for methane production, *Global Biogeochem. Cycles, 11,* 495-506, 1997.

Prentice, I. C., S. T. Sykes, and W. Cramer, A simulation model for the transient effects of climate change on forest landscapes, *Ecol. Modeling, 65,* 51-70, 1993.

Prigent, C., W. B. Rossow, and E. Matthews, Microwave land surface emissivities estimated from SSM/I observations, *J. Geophys. Res.,* 102, 21,867-21,890, 1997.

Rasmussen, R. A., and M. A. K. Khalil, Atmospheric methane (CH_4): Trends and seasonal cycles, *J. Geophys. Res., 86,* 9826-9832, 1981.

Rasmussen, R. A., and M. A. K. Khalil, Atmospheric methane in the recent and ancient atmospheres: Concentrations, trends and interhemispheric gradient, *J. Geophys. Res. 89,* 11,599-11,605, 1984.

Reeburgh, W. S., J. Y. Kling, S. K. Regli, G. W. King, N. A. Auerbach, and D. A. Walker, A CH_4 emission estimate for the Kuparuk River Basin, Alaska, *J. Geophys. Res. 103*(D22), 29,005-29,013 1998.

Ritter, J. A., J. D. W. Barrick, C. E. Watson, G. W. Sachse, G. L. Gregory, B. E. Anderson, M. A. Woerner, And J. E. Collins, Jr., AIRBORNE BOUNDARY LAYER Flux measurements of trace species over the Canadian boreal forests and northern wetland regions, *J. Geophys. Res., 99,* 1671-1685, 1994.

Ritter, J. A., J. D. W. Barrick, G. W. Sachse, G. L. Gregory, M. A. Woerner, C. E. Watson, G. F. Hill and J. E. Collins, Airborne flux measurements of trace species in an arctic boundary layer, *J. Geophys. Res.,* 97, 16,601-16,625, 1992.

Rose, P. W., and P. C. Rosendahl, Classification of Landsat data for hydrologic application, Everglades National Park, *Photogram. Eng. Rem. Sens., 49,* 505-511, 1983.

Roulet, N. T., A. Jano, C. A. Kelly, L. F. Klinger, T. R. Moore, R. Protz, J. A. Ritter, and W. R. Rouse, Role of the Hudson Bay lowland as a source of atmospheric methane, *J. Geophys. Res., 99,* 1439-1454, 1994.

Roulet, N. T., A. Jano, C. Kelly, L. Klinger, T. R. Moore, R. Protz R, J. Ritter, and W. R. Rouse, The Hudson Bay Lowland as a source of atmospheric methane, *J. Geophys. Res., 99,* 1439-1454, 1993.

Roulet, N. T., R. Ash, and T. R. Moore, Low boreal wetlands as a source of atmospheric methane, *J. Geophys. Res., 97,* 3739-3749, 1992a.

Roulet, N. T., T. Moore, J. Bubier, and P. LaFleur, Northern fens: Methane flux and climatic change, *Tellus, 44B,* 100-105, 1992b.

Running, S. W., T. R. Loveland, and L. L. Pierce, A vegetation classification logic based in remote sensing for use in global biogeochemical models, *Ambio, 23,*77-81, 1994.

Sahagian, D., and J. Melack (Eds.), *Global Wetland Distribution and Functional Characterization: Trace Gases and the Hydrologic Cycle,* Wetlands Workshop Report, IGBP GAIM-DIS-BAHC-IGAC-LUCC Workshop, Santa Barbara, CA, May 1996.

232   Matthews

Schimel, J. P., Plant transport and methane production as controls on methane flux from arctic wet meadow tundra, *Biogeochemistry, 28*, 183-200, 1995.

Sebacher, D. I., R. C. Harriss, K. B. Bartlett, Methane emissions to the atmosphere through aquatic plants, *J. Environ. Qual., 14*, 40-46, 1985.

Sebacher, D. I., R. C. Harriss, K. B. Bartlett, S. M. Sebacher, and S. S. Grice, Atmospheric methane sources: Alaskan tundra bogs, an alpine fen, and a subarctic boreal marsh, *Tellus, 38B*, 1-10, 1986.

Seiler, W., and R. Conrad, Contribution of tropical ecosystems to the global budgets of trace gases, especially CH_4, H_2, CO and N_2O, in R. E. Dickinson (ed.), The Geophysiology of Amazonia: Vegetation and Climate Interactions, p 133, John Wiley, New York, 1987

Seiler, W., Contribution of biological processes to the global budget of CH_4 in the atmosphere, in M. Klug and C. Reddy (eds.), *Current Perspectives in Microbial Ecology*, p. 468, Amer. Soc. Microbiol., Washington, D. C., 1984.

Shannon, R. D., and J. R. White, A three-year study of controls on methane emissions from two Michigan peatlands, *Biogeochemistry, 27*, 35-60, 1994.

Sippel, S., S. Hamilton, and J. Melack, Determination of inundation area in the Amazon River floodplain using SMMR 27 GHz polarization difference, *Rem. Sens. Env., 48*, 70-76, 1994.

Stauffer B, E. Lochbronner, H. Oeschger, and J. Schwander, Methane concentration in the glacial atmosphere was only half that of the pre-industrial Holocene, *Nature, 332*, 812-814, 1988.

Stauffer, B., F. Fischer, A. Neftel, and H. Oeschger, Increase of atmospheric methane recorded in Antarctic ice core, *Science, 229*, 1386-1388, 1985.

Steele, L. P., P. J. Fraser, R. A. Rasmussen, M. A. K. Khalil, T. J. Conway, A. J. Crawford, R. H. Gammon, K. A. Masarie, K. W. Thoning, The global distribution of methane in the troposphere, *J. Atmos. Chem., 5*, 125-171, 1987.

Svensson, B. H., Carbon dioxide and methane fluxes from ombrotrophic parts of a subarctic mire, *Ecol. Bull. (Stockholm), 30*, 235-250, 1980.

Svensson, B. H., Methane production in tundra peat. In H. G. Schlegel, G. Gottschalk, and N. Pfennig (eds.), Microbial Production and Utilization of Gases (H_2, CH_4, CO), p 135, Gottingen, 1976.

Svensson, B. H., T. Rosswall, In situ methane production from acid peat in plant communities with different moisture regimes in a subarctic mire, *Oikos, 43*, 341-350, 1984.

Tathy J.-P., B. Cros, R. A. Delmas, A. Marenco, J. Servant, M/ Labat, Methane emission from flooded forest in Central Africa, *J. Geophys. Res., 97*, 6159-6168, 1992.

Tucker, C. J., Comparing SMMR and AVHRR data for drought monitoring, *Int. J. Remote Sens., 10*, 1663-1672, 1989.

Twenhofel, W. H., *Principles of Sedimentation*, McGraw-Hill, New York, 1926, 1951.

UNESCO, *International Classification and Mapping of Vegetation*, UNESCO, Paris, 93p, 1973.

Valentine, D., E. Holland, and D. Schimel, Ecosystem and physiological controls over methane production in northern wetlands, *J. Geophys. Res., 99*, 1563-1571, 1994.

Walker, D. A., W. Acevedo, K. R. Everett, L. Gaydos, J. Brown, P. J. Webber, *Landsat-assisted environmental mapping in the Arctic National Wildlife Refuge, Alaska*, US Cold Regions Res. Eng. Lab., Hanover, NH, 1982.

Walter, B., M. Heimann, R. Shannon, and J. White, A process-based model to derive methane emissions from natural wetlands, *Geophys. Res. Lett., 23*, 3731-3734, 1996.

Wassmann, R., U. G. Thein, M. J. Whiticar, H. Rennenberg, W. Seiler, and W. J. Junk, Methane emissions from the Amazon floodplain: Characterization of production and transport, *Global Biogeochem. Cycles, 6*, 3-13, 1992.

Whalen, S. C., and W. S. Reeburgh, A methane flux transect along the trans-Alaska pipeline haul road, *Tellus, 42B*, 237-249, 1990.

Whalen, S. C., and W. S. Reeburgh, Interannual variations in tundra methane emission: A 4-year time-series at fixed sites, *Global Biogeochem. Cycles, 6*, 139-159, 1992.

Whalen, S. C., W. S. Reeburgh, A methane flux time series for tundra environments, *Global Biogeochem. Cycles, 2*, 399-409, 1988.

Whiting, G. J., and J. P. Chanton, Plant-dependent CH$_4$ emission in a subarctic Canadian fen, *Global Biogeochem. Cycles, 6*, 225-231, 1992.

Whiting, G. J., and J. P. Chanton, Primary production control of methane emission from wetlands, *Nature, 364*, 794-795, 1993.

Whiting, G. J., J. P. Chanton, D. S. Bartlett, J. D. Happell, Relationships between CH$_4$ emission, biomass and CO$_2$ exchange in a subtropical grassland, *J. Geophys. Res., 96*, 13,067-13,071, 1991.

Wilson, J. O., P. M. Crill, K. B. Bartlett, D. I. Sebacher, R. C. Harriss, and R. L. Sass, Seasonal variation of methane emissions from a temperate swamp, *Biogeochemistry, 8*, 55-71, 1989.

Zoltai, S. C., and F. C. Pollett, Wetlands in Canada: Their classification, distribution and use, in A. J. P. Gore (ed.), *Ecosystems of the World, Mires: Swamp, Bog, Fen and Moor, Case Studies, 4B*, p 245-268, Elsevier, New York, 1983.

13. Waste Management

Susan A. Thorneloe[1], Morton A. Barlaz[2], Rebecca Peer[3], L.C. Huff[4], Lee Davis[3,] Joe Mangino[3]
[1]United States Environmental Protection Agency, Office of Research and Development
Air and Energy Engineering Research Laboratory, Research Triangle Park, North Carolina, U.S.A.
[2]North Carolina State University, Civil Engineering Department, Raleigh, North Carolina, U.S.A.
[3]Radian Corporation, Research Triangle Park, North Carolina, U.S.A.
[4]ERG, Research Triangle Park, North Carolina, U.S.A.

I. Introduction

Landfills, wastewater treatment lagoons, and livestock waste management are operations representing sources of methane. Estimates of CH_4 emissions from these sources suggest approximately 70 (54-95) Tg/yr globally or 14% of total global CH_4 emissions of 500 Tg/yr (IPCC, 1992). This chapter begins with a brief overview of how CH_4 is generated from the anaerobic decomposition of waste and then discusses generation of CH_4 in detail in landfills, wastewater treatment lagoons, and livestock waste management. Current techniques for estimating CH_4 emissions from waste are summarized, and sources of uncertainty are identified.

The potential control of CH_4 emissions from waste management has been targeted by the United States (U.S.) and other countries as part of greenhouse gas reduction programs designed to meet the goals of treaties signed at the United Nations Conference on Environment and Development (UNCED) held in 1992. Consequently, reducing the uncertainty associated with CH_4 emission estimates is a high priority.

2. Methane Production During the Anaerobic Decomposition of Waste

The anaerobic decomposition of organic matter is a complex process that requires that several groups of microorganisms act synergistically under favorable environmental conditions (see Boone, this volume). The pathway described below has been demonstrated to apply to anaerobic decomposition in sludge digesters and in livestock waste management systems. This anaerobic pathway is also expected to occur in landfills and anaerobic wastewater lagoons (Barlaz et al., 1989a).

Mohammad Aslam Khan Khalil (Ed.)
Atmospheric Methane
© *Springer-Verlag Berlin Heidelberg 2000*

Three trophic groups of anaerobic bacteria must be present to produce CH_4 from biological polymers such as, cellulose, hemicellulose, and protein: (1) hydrolytic and fermentative microorganisms, (2) obligate proton-reducing acetogens, and (3) methanogens (Wolfe, 1979; Zehnder et al., 1982). The hydrolytic and fermentative group is responsible for the hydrolysis of biological polymers. The initial products of polymer hydrolysis are soluble sugars, amino acids, long-chain carboxylic acids, and glycerol. Following polymer hydrolysis, the hydrolytic and fermentative microorganisms ferment the initial products of decomposition into short-chain carboxylic acids, alcohols, carbon dioxide (CO_2), and hydrogen. Acetate, a direct precursor of CH_4, is also formed.

The second group of bacteria -- obligate proton-reducing acetogens -- convert the fermentation products of the hydrolytic and fermentative microorganisms to CO_2, hydrogen, and acetic acid. The conversion of fermentation intermediates, such as butyrate, propionate, and ethanol, is thermodynamically favorable only at very low hydrogen concentrations. Thus, these substrates are utilized only when the obligate proton-reducing acetogenic bacteria can function in syntrophic association with hydrogen scavengers, such as CH_4-producing or sulfate-reducing organisms.

The third group of bacteria necessary for the production of CH_4 are methanogens. Major substrates utilized by methanogens for the production of CH_4 are acetate, formate, methanol, methylamines, and hydrogen plus CO_2 (Wolin and Miller, 1985).

While CH_4 and CO_2 are the terminal products of anaerobic decomposition, CO_2 and water are the terminal products of aerobic decomposition. Aerobic decomposition occurs in management facilities where waste is exposed to air, such as when compost is turned for aerating, and in uncontrolled dumps, such as when refuse is spread in thin layers or otherwise exposed to oxygen (as by scavenging). When refuse is buried in large piles, whether at an open dump or in a sanitary landfill, the oxygen entrained at burial is consumed rapidly, and the majority of decomposition will occur under anaerobic conditions (Bhide et al., 1990).

3. Methane Production from Waste Burial

The proportion of waste generated in developing countries has been projected to increase over the next several decades, while the proportion of waste generated by developed countries is expected to decline. This trend can be attributed to projections of higher population increases in developing countries and not to increased per capita waste generation. Despite efforts toward source reduction and recycling programs, per capita waste generation is expected to increase in the U.S. (Kaldjian, 1990) and in other industrialized countries. The much lower rates of population growth in industrialized countries are expected to result in slower growth in municipal solid waste (MSW) production, as compared to developing countries. However, this scenario will not be realized if per capita income decreases in developing countries. Recently, declining economic conditions have resulted in reduced MSW generation in Caracas, Venezuela, Mexico City, Mexico, and Buenos Aires, Argentina (Bartone et al., 1991).

Methods of managing MSW vary widely, ranging from open dumps and open burning to sanitary landfills with leachate collection systems and landfill gas control. The majority of the world's MSW is managed using either sanitary landfills or open dumps. In the U.S., recent estimates indicate that 72% of MSW is buried in landfills (Kaldjian, 1990). Anaerobic decomposition prevails in landfills. Both anaerobic and aerobic processes occur at open dumps. The CH_4 potential of other waste management processes such as incineration, recycling, and composting is considered insignificant in comparison to landfills and open dumps.

Landfilled waste contains numerous constituents that have the potential to biodegrade under anaerobic conditions. However, optimal conditions for anaerobic decomposition within a landfill may not exist and may thus result in overestimated emissions. Many methodologies for estimating emission assume that optimal conditions exist. In a recent study field data were gathered to develop an empirical model that is intended to reflect actual emissions to the atmosphere (Peer et al., 1993). This model adjusts for gas recovery efficiency and CH_4 oxidation. Estimates are presented later in this chapter, along with updated estimates using the approach developed by Bingemer and Crutzen (1987).

3.1 Factors Affecting CH_4 Potential of Buried Waste

The traditional method of classifying MSW according to sortable categories (such as paper, plastic, food waste, yard waste, glass, metals, rubber, wood, textiles, dirt, and miscellaneous (Kaldjian, 1990)) is appropriate for recycling studies and overall solid waste management planning. However, data specific to the chemical composition of refuse are more applicable to analyses of refuse decomposition. Studies of refuse in Madison, Wisconsin, showed cellulose plus hemicellulose to be about 60% of landfill waste and to account for 91% of the CH_4 potential of refuse (Barlaz, 1985, 1988; Barlaz et al., 1989b). The components of MSW that contain significant biodegradable fractions are food waste, yard waste, and paper, which have a combined cellulose and hemicellulose content of 50 to 100%. Lignin is the other major organic component of refuse; however, lignin does not undergo significant decomposition under anaerobic conditions (Young and Frazer, 1987).

Methane formation does not occur immediately after refuse is placed in a landfill or dump. It can take months or years for the proper environmental conditions and the required microbiological populations to become established. Numerous factors control decomposition, including moisture content, nutrient concentrations, presence and distribution of microorganisms, particle size, water flux, pH level, and temperature. For a review of the factors affecting CH_4 production see Halvadakis et al. (1983), Pohland and Harper (1987), and Barlaz et al. (1990).

The two factors that appear to have the greatest effect on CH_4 production are moisture content and pH. The effect of refuse moisture content has been summarized by Halvadakis et al. (1983), although some of their data relate to manure and not to municipal waste. The broadest data sets are those of Emberton (1986) and Jenkins and Pettus (1985). Emberton measured CH_4 production rates in excavated landfill samples under laboratory conditions. Jenkins and Pettus sampled refuse from landfills and tested how CH_4 production was affected by the moisture content of refuse. In both studies, the CH_4 production rate increased with increasing moisture content, despite differences in refuse density, age, and composition. It is difficult to translate the results of these laboratory studies to actual landfills. An attempt by the U.S. EPA's Air and Energy Engineering Research Laboratory (AEERL) to identify a statistically significant correlation between landfill gas recovery and precipitation (which affects refuse moisture content) found no such correlation (Peer et al., 1992).

A second key factor influencing the rate and onset of CH_4 production is pH. The optimum pH level for activity by methanogenic bacteria is between 6.8 and 7.4. CH_4 production rates decrease sharply with pH values below about 6.5 (Zehnder et al., 1982). When refuse is buried in landfills, there is often a rapid accumulation of carboxylic acids; this results in a pH decrease and a long time lapse between refuse burial and the onset of CH_4 production.

Neutralizing leachate and recycling it back through refuse has been shown to enhance the onset and rate of CH_4 production in laboratory studies (Pohland, 1975; Buivid et al., 1981; Barlaz et al., 1987, International Energy Agency, 1992). Given that moisture and pH have been reported as the two most significant factors limiting CH_4 production, the stimulatory effect of leachate neutralization and recycling is expected. Neutralization of leachate provides a means of externally raising the pH of the refuse ecosystem. Recycling neutralized leachate back through a landfill increases and stabilizes refuse moisture content and substrate availability and provides mixing in what would otherwise be an immobilized batch reactor.

Notably, field experience with leachate recycling systems is limited and more information is needed to fully document their value. In addition, the lapsed time preceding the onset of CH_4 production in landfills is an important aspect when considering the management of individual landfills for biogas recovery or emissions mitigation. The age at which landfills and uncontrolled dumps begin to produce CH_4 is of lesser importance when evaluating global CH_4 emissions from MSW management systems. In this case, the total CH_4 production potential is more critical.

3.2 Determination of the CH_4 Potential of Landfills and Dumps

Knowledge of the chemical composition of refuse buried in a landfill makes it possible to estimate the volume of CH_4 that may be produced. The mass of CH_4 that would be produced if all of a given constituent were converted to CH_4, CO_2, and ammonia may be calculated from Equation 1 (Parkin and Owen, 1986):

$$C_nH_aO_bN_c + [n - 1/4a \ 1/2b + 3/4c]H_2O \rightarrow \qquad (1)$$

$$[1/2n - 1/8a + 1/4b + 3/8c]CO_2 + [1/2n + 1/8a - 1/4b - 3/8c]CH_4 + cNH_3$$

Using this stoichiometry, the CH_4 potential of cellulose ($C_6H_{10}O_5$) and hemicellulose ($C_5H_8O_4$) is 415 and 424 liters (l) CH_4 at standard temperature and pressure (0°C, 1 atmosphere) per dry kilogram (kg), respectively (18.5 and 18.9 grams [g] CH_4/dry kg).

These methane potentials represent maximum CH_4 production if 100% of the cellulose and hemicellulose were converted to CH_4. However, decomposition of these constituents in landfills is well below 100%, mainly because (1) some cellulose and hemicellulose is surrounded by lignin or other recalcitrant materials (such as plastic) and, therefore, is not biologically available; and (2) without active intervention, buried refuse is not evenly exposed to moisture, microorganisms, and nutrients. Barlaz et al. (1989b) applied mass balances to shredded refuse incubated in laboratory-scale lysimeters with leachate recycle. Carbon recoveries of 87 to 111% were obtained, where a perfect mass balance would give a carbon recovery of 100%. Mineralization of 71% of the cellulose and 77% of the hemicellulose was measured in a container sampled after 111 days. Mass balances were useful for documenting the decomposition of specific chemical constituents and demonstrating the relationship between cellulose and hemicellulose decomposition and CH_4 production.

Mass balances may be used to estimate the CH_4 potential remaining in a landfill by sampling the refuse, performing the appropriate chemical analyses, and calculating the CH_4 potential. Ideally, the initial chemical composition and CH_4 potential of the refuse would be known, in which case comparing that initial CH_4 potential with the potential at the time of sampling would provide information on the fraction of the refuse that has been degraded. Indisputably, representative sampling of a full-scale sanitary landfill is not realistic. However, it is possible to obtain multiple samples at presumably representative locations within a landfill to get an estimate of the range and extent of decomposition. Samples should be as large as can reasonably be handled and reduced.

Another technique for assessing the CH_4 potential of refuse is the biochemical methane potential (BMP) test (Shelton and Tiedje, 1984; Bogner, 1990). In the BMP test, the anaerobic biodegradability of a small sample of refuse (5 to 10 g) is measured in a small batch reactor (100 to 200 m l). While the BMP represents an upper bound of CH_4 potential from refuse, it will be lower than the stoichiometric estimate described above. BMPs also require representative sampling in landfills.

Comparison of CH_4 production data between field-scale landfills and laboratory experiments is difficult because there are essentially no data in the open literature on CH_4 production rates in field-scale facilities. Data from field-scale landfills are complicated by questions regarding the mass of refuse responsible for production of a measured volume of gas and the efficiency of gas collection. There are more CH_4 production data collected under laboratory conditions than field conditions. However, the laboratory data are not always comparable to experimental conditions. For instance, moisture, particle size, and temperature are not uniform between studies. In addition, most laboratory experiments were conducted to explore techniques for enhancing CH_4 production. The enhanced CH_4 production rates would not normally be expected at field-scale landfills.

CH_4 yields of 1.9 to 5.4 g CH_4/dry kg refuse have been reported in laboratory tests conducted with leachate recycle and neutralization (Buivid et al., 1981; Barlaz et al., 1987; Kinman et al., 1987; Barlaz, 1988). These studies show significant variation in CH_4 production rate and CH_4 yield. Some of the differences can be explained by differences in experimental design. For example, the data reported by Barlaz et al. (1987) and Barlaz (1988) differ in reactor volume (100 vs. 2 ℓ), temperature (25 vs. 41°C), and the rate of leachate recycle. Also, Buivid et al. (1981) used refuse with an abnormally high paper content.

CH_4 yields were measured in field-scale test cells as part of the Controlled Landfill

Project in Mountain View, California (Pacey, 1989). Yields of 38.6 to 92.2 ℓ (1.7 to 4.1 g) CH_4 /dry kg of refuse were measured after 1597 days. However, mass balance data on cellulose losses from individual cells suggested that more CH_4 was produced than was measured in certain test cells.

A number often used in engineering practice as an estimate of CH_4 production in field scale landfills is 0.1 ft^3 CH_4/wet lb-yr, over a 10 to 20 year production period. This value has not been documented in open literature. Assuming refuse is buried at 20% moisture content, this converts to 7.8 ℓ (0.3 g) CH_4/dry kg-yr, a number comparable to some of the lower values reported in the literature.

Even in landfills with venting systems, some of the CH_4 is likely to escape from the landfill through the final cover. The fraction released through the final cover will be a function of the type of gas venting system in place and the type of cover. Probably not all the CH_4 that escapes from landfills is released to the atmosphere as CH_4. CH_4 that passes through the cover soil may be converted to CO_2 in the presence of oxygen by aerobic methanotrophic bacteria. CH_4 oxidation has been documented in landfill cover soil studied under laboratory conditions (Whalen et al., 1990). However, there are no data on the quantitative significance of CH_4 oxidation above landfills. CH_4 escaping through cracks in a landfill cover likely will not reside in the cover for a period sufficient to undergo significant biodegradation.

3.3 Emissions Estimate Methodology for Landfills and Open Dumps

Two techniques for estimating emissions from landfills are reviewed here: the Organization of Economic Cooperation and Development (OECD) and EPA/AEERL methods. These methods were developed to estimate global CH_4 emissions. Models that estimate CH_4 production from individual landfills are reviewed by Augenstein and Pacey (1990) and Peer et al., (1993). While global estimates focus on ultimate CH_4 release, models of individual landfills emphasize the rate and duration of CH_4 production as these factors affect the economics of landfill gas recovery projects.

3.4 OECD Methodology

OECD (1991) used the mass balance approach developed by Bingemer and Crutzen (1987), where an instantaneous release of CH_4 is assumed to enter the atmosphere during the same year that refuse is placed in a landfill. This method also assumes that (1) all of the CH_4 that is produced escapes to the atmosphere (none is oxidized on its way to the atmosphere) and (2) all developing nations generate and dispose of MSW at the same per capita rate.

To calculate the annual emission from MSW, OECD used the following equation from Bingemer and Crutzen: CH_4 Emission = Total MSW Generated (kg/yr) x MSW Landfilled (%) x DOC in MSW (%) x Fraction Dissimilated DOC (%) x 0.5 g CH_4/g Biogas x Conversion Factor (16 g CH_4/12 g C) – Recovered CH_4 (kg/yr) where DOC is degradable organic carbon; Fraction Dissimilated DOC is the portion of carbon in substrates that is converted to landfill gas; and "Recovered CH_4" is the amount of CH_4 that is recovered through gas recovery systems and never emitted to the atmosphere.

The uncertainties of this approach are attributed to assumptions regarding anaerobic decomposition. Many factors inhibit this process, and this approach tends to overestimate potential emissions. Moreover, this methodology does not adjust for CH_4 oxidation, which is known to occur.

3.5 EPA/AEERL's Regression Model Methodology

The EPA/AEERL methodology uses an empirical model derived using landfill gas recovery data. The quantity of CH_4 estimated by this model is much less than that predicted by stoichiometric analyses or by laboratory studies   (EMCON, 1982; Barlaz et al., 1989b, 1990; Peer et al., 1992). The data gathered from U.S. landfills that were used to develop this model represent a broad range of climate zones and waste composition as described below (Campbell et al., 1991; Peer et al., 1992). This model is intended to reflect the amount of gas that is ultimately released to the atmosphere by adjusting for gas recovery efficiency and CH_4 oxidation. For the estimates presented in this chapter, it was assumed that the recovery efficiency is 80% and that 10% of non-recovered CH_4 is oxidized. Refinements of this methodology include adjustments for recovery efficiency and CH_4 oxidation based on factors derived through an uncertainty analysis. Comparison of the refined estimates with the earlier estimates indicates a slight increase.

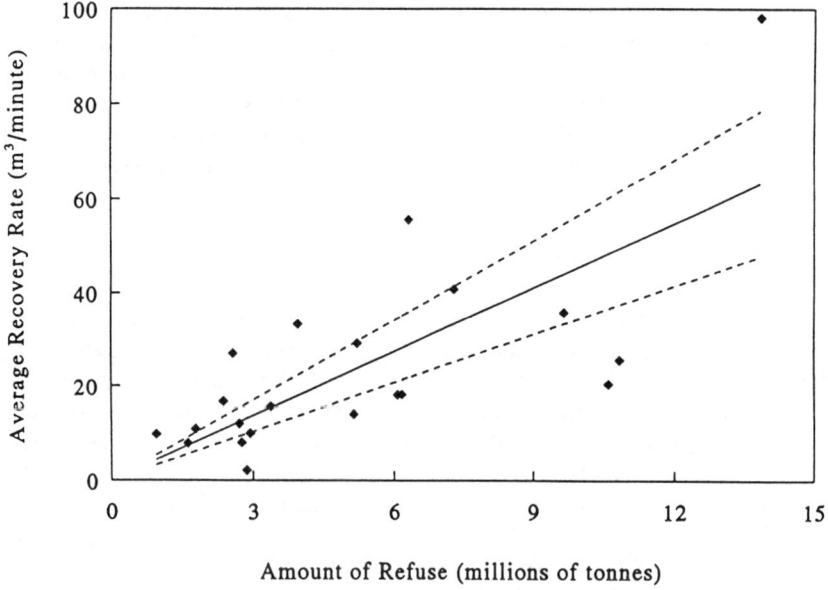

Fig. 1. Methane recovery data as a function of buried refuse (Peer et al., 1992).

Data from 21 landfills were used to determine if there is a correlation between CH_4 recovery and landfill characteristics such as waste quantity, age, depth, and climate. Selection of variables for the regression models was based on the results of the correlation and scatter plots of selected variables (such as climate, landfill depth, refuse mass, and refuse age). The main conclusion of the study was that the annual CH_4 recovery rate was linearly correlated with the mass of refuse in the landfill, and with landfill depth (Peer et al., 1992). The data, regression line, and the 95% confidence limits of the regression coefficient are shown in Figure 1. The regression was significant ($P < 0.01$), but much of the variability in the data is unexplained (adjusted $R^2 = 0.50$). The intercept was not significant, so the final model was forced through the zero. Peer et al. (1992) provide detailed information on the characteristics of each site that may contribute to data variability.

No statistically significant relationships were identified between annual CH_4 recovery and climate variables such as precipitation, temperature, and dew point. The effect of refuse age on gas production was also analyzed. Gas recovery correlated most strongly with refuse between 10 and 20 years old. Although these results were not conclusive, they suggest that the generation time for gas production is 20 to 30 years (Peer et al., 1992). This generation time is within the range of generation times assumed in many landfill gas recovery models (EMCON, 1982; Augenstein and Pacey, 1990).

One advantage of using the EPA/AEERL model is that refuse mass is the only variable required to estimate CH_4 emissions. Furthermore, it will be relatively easy to update the relationship between CH_4 recovery and refuse mass as more data become available. The confidence limits of the regression coefficient can be used to bound emission estimates:

(1) The upper and lower 95% confidence limits are 6.5 and 2.5 m^3 CH_4/min/10^6 Mg wet refuse (4.6 to 1.8 g CH_4/min/kg of wet refuse), respectively.
(2) Assuming an average generation time of 25 years gives an average CH_4 recovery of 59.4 m^3 CH_4/Mg wet refuse (42 g CH_4/kg of wet refuse).
(3) A range of 33 to 86 m^3 CH_4/10^6 Mg wet refuse (24 to 61 g CH_4/kg wet refuse) results.

These values were derived using data collected at U.S. MSW landfills. The use of this factor assumes that other countries have a waste composition similar to U.S. landfills. However, U.S. MSW generally has a higher organic content than most other countries (Bingemer and Crutzen, 1987). Therefore, the use of this factor may overestimate landfill emissions for other countries. Future refinements of the model will adjust for waste composition using gas potential data for different biodegradable waste streams.

3.6 Assumptions and Data Used to Estimate Waste Generation Rates

To estimate global CH_4 emissions it was necessary to make several assumptions regarding waste generation and disposal. Emission factors in all industrialized countries were assumed to be equal to those in the U.S. The CH_4 emission factors for the less-developed countries (LDCs), where adequate data were not available, were assumed to be 25 to 75% of the average U.S. estimate. Several factors were taken into account to develop this range. The composition of waste is different in third world countries. For example, much of the garbage is scavenged before it is placed in the landfill, especially paper, textiles, and metal products. More putrescibles end up in dumps and probably do not generate as much CH_4 because more oxidation (aerobic process) takes place. In addition, garbage is often burned, which decreases the amount of material available for anaerobic decomposition, but may increase CH_4 emissions from inefficient combustion. Finally, most landfills in the LDCs are open dumps which are scavenged by humans and wild and domestic animals. While anaerobic conditions may form in some of these dumps, the potential for aerobic decomposition is much greater than in sanitary landfills. Based on these assumptions, the 25 to 75% range was chosen as the default value for LDCs, since no data on actual CH_4 emissions are available.

Estimates of waste generation and burial are presented in Table 1. This table was developed using the data presented in the references shown. A recent review of global waste management trends found that information on MSW generation in developing countries is difficult to obtain; in many cases, it is anecdotal (Davis et al., 1992). As shown in the reference list for Table 1, much country-specific data were available to determine MSW generation and land disposal values, especially for developed countries. Where no data were available, data from similar countries were used.

Most of the available data for developing countries are provided on a per capita basis for only the larger cities; this information was combined with population (United Nations, 1990) and percent of the country that is considered urban (Population Reference Bureau, Inc., 1989), to determine the amount of waste generated in urban centers of these countries. An estimate for rural per capita refuse generation rate (Kessler, 1990) was then combined with the rural population value to determine rural waste generation. Once the amount of waste generated for the entire country was estimated, a default value of 50% disposed on land (whether in landfills or open dumps) was used to estimate the amount of waste that may degrade anaerobically. The 50% default was chosen to represent waste that is actually collected in some manner and disposed of in landfills or large enough open dumps for anaerobic conditions to occur. The remaining 50% of waste generated is assumed to be (1) incinerated or combusted, (2) dumped in rivers or other bodies of water, or (3) scattered or buried in small piles that degrade anaerobically. These disposal methods do not produce large amounts of CH_4.

The amount of MSW landfilled in the U.S. is approximately 189 million tonnes (U.S. EPA, 1988). Paper was the largest single component of the DOC fraction in both the U.S. and Canada. Per capita MSW generation was in the range of 1.7 to 1.8 kg/person/day for both the U.S. and Canada (Kaldjian, 1990; El Rayes and Edwards, 1991), and the average DOC content of paper or MSW is 20%. The average MSW generation rate in other OECD countries is 1.1 kg/person/day. MSW in these countries has a DOC content of approximately 15.3%. The value used for the U.S. is for MSW only; an additional 15 Tg/yr of biodegradable industrial solid waste is also landfilled (U.S. EPA, 1987), which is unaccounted for in the present estimates of landfill CH_4. In most cases, country-specific information does not state whether industrial waste is included with MSW.

Table 1. Waste totals (Tg) by geographic region used to develop emission estimates

Geographic Region	Waste	Waste Landfilled	References*
Africa	80	38	1-14
Asia and the Middle East	363	175	15-28
Europe	224	165	29-46
North America	301	216	47-49
Oceania	14	12	50-51
South and Central America	64	47	52-57

*Reference Key:

1. El Halwagi et al., 1988.
2. El Halwagi et al., 1986.
3. Kaltwasser, 1986.
4. UNDP et al., 1987.
5. Holmes, 1984.
6. Monney, 1986.
7. Cointreau, 1984.
8. Cointreau, 1987.
9. The World Bank, 1985.
10. Mwiraria et al., 1991.
11. United Republic of Tanzania, 1989.
12. Verrier, 1990.
13. World Resources Institute, 1990.
14. Rettenberger and Weiner, 1986.
15. Bhide and Sundaresan, 1983.
16. United Nations, 1989.
17. Maniatis et al., 1987.
18. Lohani and Thanh, 1980.
19. Ahmed, 1986.
20. Pairoj-Boriboon, 1986.
21. Gadi, 1986.
22. Mei-Chan, 1986.
23. U.S. EPA, 1990.
24. Diaz and Goulueke, 1987.
25. Xianwen and Yanhua, 1991.
26. Cossu, 1990a.
27. Hayakawa, 1990.
28. Swartz, 1989.
29. Lechner, 1990.
30. World Resources Institute, 1990.
31. Carra and Cossu, 1990.
32. Ettala, 1990.
33. Stegmann, 1990.
34. Ernst, 1990.
35. Cossu and Urbini, 1990.
36. Beker, 1990.
37. Carra and Cossu, 1990.
38. Gandolla, 1990.
39. Cossu, 1990b.
40. Swartz, 1989.
41. Richards, 1989.
42. U.S. EPA, 1990.
43. Scheepera, 1990.
44. Bartone and Haley, 1990.
45. Bartone, 1990a.
46. Bingemer and Crutzen, 1987
47. U.S. EPA, 1988.
48. U.S. EPA, 1990.
49. El Rayes and Edwards, 1990.
50. Bateman, 1988.
51. Richards, 1989
52. Kessler, 1990.
53. U.S. EPA, 1990.
54. World Resources Institute, 1990.
55. Diaz and Golueke, 1987
56. Bartone et al., 1991.
57. Yepes and Campbell, 1990.

Information on the amount of MSW generated and landfilled in the European countries that are not OECD members and in the former Soviet Union is limited. Nozhevnikova et al. (1993) used 0.8 kg/person/day for MSW generation in the former USSR. Average MSW generation for Greece, the former Soviet Union, and Eastern Europe is approximately 0.6 kg/person/day (Bingemer and Crutzen, 1987; Frantzis, 1988; Papachristou, 1988; Peterson and Perlmutter, 1989), and the available data indicate that putrescibles make up a large portion of the MSW (estimates range from

32 to 60%). This MSW contains approximately 15% DOC (Bingemer and Crutzen, 1987; Frantzis, 1988; Papachristou, 1988; Peterson and Perlmutter, 1989; Zsuzsa, 1990).

For most Asian countries, estimates of MSW generation were identified for one or two major cities, but not for the entire country. National per capita MSW generation estimates were identified for Indonesia, Sri Lanka, the Philippines, Singapore, Taiwan, and Pakistan. These estimates range from 0.4 kg/person/day for the Philippines to 1.0 kg/person/day for Singapore. The average per capita MSW generation for these countries is estimated to be 0.6 kg/person/day (Davis et al., 1992).

Few data are available on MSW production and management in Central America, South America, the Caribbean Islands, and Mexico. Most of the available information is only for the larger cities. The average per capita MSW generation rate in seven South and Central American countries (Brazil, Colombia, Chile, Paraguay, Peru, Venezuela, Costa Rica), and Mexico is estimated to be 0.8 kg/person/day. The components are mainly vegetable and putrescible waste paper and cardboard. The average DOC for the seven South and Central American countries and Mexico is 17% (Davis et al., 1992).

Information on MSW generation and disposal for African and Middle Eastern countries is also very limited. In Africa, it appears that toxic and hazardous industrial and commercial wastes are purposely or inadvertently disposed of with the MSW stream. Some information pertaining to generation rates for African countries was located; but information for only two Middle Eastern countries, Israel and Yemen, was obtained. Based on the very limited information for these two continents, it is estimated that per capita generation rates range from 0.3 to 1.1 kg/person/day, and the DOC content ranges from 3 to 20%.

3.7 Global and Country-specific Estimates of CH_4 Emissions from Landfills and Dumps

3.7.1. OECD Proposed Methodology.   The MSW generation and landfill disposal data in Table 1 were used to calculate CH_4 emissions for each country using the OECD methodology discussed earlier.

This methodology is identical to that of Bingemer and Crutzen (1987), but makes use of more recent waste generation data and the results of an exhaustive study by the EPA/AEERL to gather country-specific waste generation and disposal data. As shown in Table 2, the global estimate of landfill CH_4 emissions using this methodology is 57 Tg/yr. The estimate for the U.S. (i.e., 21 Tg/yr) has been adjusted for the amount of CH_4 that is recovered for energy utilization (1.2 Tg/yr, Thorneloe, 1992).

3.7.2 EPA's Regression Model Methodology.   Based on the EPA's regression model methodology, global landfill CH_4 emissions are estimated to range from 11 to 32 Tg/yr, with a midpoint of 21 Tg/yr. Additional refinements in the methodology may lend to adjustment of these estimates. Table 2 presents country-specific global estimates using the methodology described in Peer et al. (1992) and provides the lower, most probable, and upper-bound estimates of CH_4 emissions by continent. This estimate for the U.S. (i.e., 4 to 12 Tg/yr) has also been adjusted for the amount of CH_4 that is recovered for utilization based on a recently completed survey (i.e., 1.2 Tg/yr, Thorneloe, 1992). Emission estimates for countries other than the U.S. have not been adjusted for the amount of CH_4 that is recovered for utilization. Richards (1989) and Thorneloe (1992) estimate that worldwide there are 269 sites in 20 countries where landfill gas is recovered, including 114 sites in the U.S. No estimates are available, however, on the amount of CH_4 recovered in other countries.

Table 2. Methane emission estimates from buried refuse using the OECD and regression models

Country	OECD Method	Regression Model Method		
		Lower Bound	Midpoint of CH_4 Emissions	Upper Bound
Africa				
Congo	0.01	0.01	<0.01	<0.01
Egypt	0.32	0.05	0.10	0.15
Gambia	0.01	0.00	0.00	0.00
Ghana	0.05	0.01	0.02	0.02
Kenya	0.09	0.01	0.03	0.04
Liberia	0.01	0.00	0.00	0.01
Morocco	0.12	0.02	0.04	0.06
Nigeria	0.48	0.07	0.15	0.22
South Africa (Customs Union)	0.43	0.07	0.13	0.20
Sudan	0.09	0.01	0.03	0.04
Tanzania, United Republic	0.09	0.01	0.03	0.04
Uganda	0.06	0.01	0.02	0.03
Zimbabwe	0.08	0.01	0.02	0.03
Other Africa	1.27	0.19	0.39	0.58
TOTAL AFRICA	3.11	0.46	0.96	1.42

Table 2. Methane emission estimates from buried refuse using the OECD and regression models

Country	OECD Method	Regression Model Method		
		Lower Bound	Midpoint of CH_4 Emissions	Upper Bound
Asia				
Bangladesh	0.42	0.06	0.13	0.19
China (Mainland, NMP)	3.87	0.59	1.18	1.77
India	0.80	0.12	0.24	0.37
Iran, Islamic Republic of	0.31	0.05	0.09	0.14
Iraq	0.12	0.02	0.04	0.05
Israel	0.05	0.01	0.01	0.02
Japan	1.04	0.28	0.50	0.72
Korea, People's Demo. Rep.	0.14	0.02	0.04	0.06
Korea, Republic of	0.64	0.10	0.20	0.29
Kuwait	0.02	0.00	0.00	0.01
Malaysia	0.08	0.01	0.02	0.04
Mongolia	0.01	0.00	0.00	0.00
Myanmar	0.16	0.02	0.05	0.07
Pakistan	0.75	0.11	0.23	0.34
Philippines	0.42	0.06	0.13	0.19
Saudi Arabia	0.10	0.01	0.03	0.04
Sri Lanka	0.10	0.01	0.03	0.04
Thailand	0.28	0.04	0.09	0.13
Turkey	0.20	0.07	0.12	0.18
United Arab Emirates	0.01	0.00	0.00	0.01
Vietnam	0.22	0.03	0.07	0.10
Other Asia	2.04	0.32	0.62	0.94
TOTAL ASIA	11.8	1.93	3.82	5.70
Latin America				
Argentina	0.28	0.04	0.09	0.13
Brazil	2.23	0.34	0.68	1.02
Colombia	0.44	0.07	0.13	0.20
Costa Rica	0.02	0.00	0.01	0.01
Mexico	0.65	0.10	0.20	0.30
Venezuela	0.05	0.01	0.02	0.02
Other South America	0.45	0.07	0.13	0.21
TOTAL LATIN AMERICA	4.12	0.63	1.26	1.89

Table 2. Methane emission estimates from buried refuse using the OECD and regression models

Country	OECD Method	Regression Model Method		
		Lower Bound	Midpoint of CH_4 Emissions	Upper Bound
North America				
Canada	2.02	0.47	0.84	1.21
U. S. A.	21.04	3.86	8.00	12.14
Other North America	0.36	0.06	0.11	0.17
TOTAL NORTH AMERICA	23.4	4.39	8.95	13.5
Europe				
Albania	0.01	0.00	0.00	0.00
Austria	0.17	0.04	0.08	0.12
Belgium	0.14	0.04	0.07	0.10
Bulgaria	0.06	0.01	0.02	0.02
Czechoslovakia	0.15	0.02	0.04	0.06
Denmark	0.07	0.02	0.04	0.05
Finland	0.24	0.06	0.12	0.17
France	0.88	0.23	0.42	0.61
German Democratic Republic	0.17	0.02	0.04	0.06
Germany, Federal Republic of	1.80	0.48	0.87	1.25
Greece				
Hungary	0.48	0.13	0.23	0.33
Ireland	0.24	0.03	0.06	0.09
Italy	0.11	0.03	0.05	0.08
Netherlands	1.45	0.39	0.70	1.01
Norway	0.44	0.12	0.21	0.31
Poland	0.11	0.03	0.05	0.07
Romania	0.37	0.05	0.10	0.14
Spain	0.13	0.02	0.03	0.05
Sweden	0.85	0.23	0.41	0.59
Switzerland & Liechtenstein	0.08	0.02	0.04	0.06
U. S. S. R.	0.12	0.03	0.06	0.08
United Kingdom	2.49	0.32	0.65	0.97
Yugoslavia	2.85	0.76	1.37	1.98
Other Europe	0.17	0.02	0.04	0.07
	0.08	0.02	0.04	0.05

Table 2. Methane emission estimates from buried refuse using the OECD and regression models

Country	OECD Method	Regression Model Method		
		Lower Bound	Midpoint of CH_4 Emissions	Upper Bound
TOTAL EUROPE	13.7	3.12	5.74	8.32
Oceania				
Australia	1.16	0.27	0.48	0.70
New Zealand*	0.15	0.04	0.07	0.11
Other Oceania	0.03	0.00	0.01	0.01
TOTAL OCEANIA	1.34	0.31	0.56	0.82
GLOBAL TOTAL	57	11	21	32

* This source has been estimated to emit 0.2 Tg/yr by Lassey et al. (1992).

3.8 Uncertainty Associated with Estimating Landfill CH_4 Emissions

There are several sources of uncertainty in estimating emissions of CH_4 from landfills, including:

(1) The quantity of CH_4 that is actually produced from the waste in the landfill;
(2) The quantity of CH_4 that is actually emitted to the atmosphere (the question as to how much is emitted and how much is oxidized as it passes through the landfill); and
(3) The quantity and composition of landfilled waste.

Two issues contribute to the difficulty of estimating CH_4 potential from open dumps: (1) the physical characteristics (size, configuration, temperature, moisture, compaction) of open dumps are unknown, and (2) the quantity and composition of open-dumped waste are also unknown. However, CH_4 is generated from open dumps. Bhide ct al. (1990) reported biogas recovery from two uncontrolled landfills in Nagpur, India. Each of these sites was about 8 hectares in surface area and about 3 to 5 m deep. Neither site contained any cover material, and the older of the two landfills accepted waste from 1971 to 1984. Most of the organic matter had decomposed by the time the tests were performed, but biogas was obtained from wells 50 mm in diameter at a rate of 0.240 m³/hr (CH_4 content not identified). Waste had been deposited in the second site "only recently" and the rate of biogas recovery was from 5 to 9 m³/hr. The CH_4 content of the biogas from the second site was 30 to 40%. The work of Bhide et al. (1990) suggests that open dumps are a source of CH_4. Therefore, they have been included in the emission estimates.

The CH_4 potential of other types of landfills, such as those containing industrial and hazardous wastes, is not well understood. Industrial waste contains waste streams that will decompose under anaerobic conditions. Certain industrial waste streams, such as those of the food industry, may have high organic content and are, therefore, potentially significant sources of CH_4. However, landfills containing hazardous waste will have a low CH_4 potential because of the low moisture content and the requirement that only solid materials are accepted. In addition, the chemicals in the waste stream may be toxic to the microbes. Therefore, we believe the global emissions are negligible compared to those from MSW landfills or for industrial landfills.

The disposal of industrial and hazardous waste with MSW was common in the U.S. until 1975. Many closed landfill sites in the U.S. and worldwide contain biodegradable mixtures of these waste streams. Some industrialized countries such as the United Kingdom consider landfills as an acceptable treatment option for hazardous and industrial wastes. However, regulations being considered by the European Commission may prevent this practice. Waste streams in developing countries are less controlled and mingling of MSW, industrial wastes, and raw sewage in landfills is common (Cointreau, 1982). Co-disposal sites will generate CH_4 and may have emission potentials similar to MSW landfills having no history of co-disposal. Currently, published estimates of CH_4 emissions specific to open dumps and industrial and hazardous waste landfills are not available. Estimates presented in Table 2 include open dump emissions, but do not specifically consider the CH_4 potential of landfilled industrial and hazardous waste.

3.9 Trends in Waste Management and Their Impact on CH_4 Emissions

As methods of MSW disposal change, there will be changes in CH_4 emissions. Trends in global waste management and their impact on CH_4 release are discussed here.

3.9.1 U.S. and Canada. Landfilling is the predominant MSW management method in both the U.S. and Canada. However, there is a trend in both countries toward more recycling, more incineration (especially in the U.S.), and less landfilling (U.S. EPA, 1988; Swartz, 1989; Kaljian, 1990; El Rayes and Edwards, 1991; Alter, 1991). Both countries also have a growing number of landfill gas recovery sites. The U.S. generates 12 times as much MSW as Canada.

Although the percentage of MSW to be placed in landfills is predicted to decline, increases in the amount of MSW generated and in the percent DOC will cause the fraction of degradable MSW in landfills to remain close to current levels. For example, if reliance on landfill disposal in the U.S. were to decline to 50% in the year 2010, 114 of the predicted 227 million Mg of MSW generated would be placed in landfills. Assuming an increase in carbon content to 22.2%, it is reasonable to assume that the U.S. will place 25 million Mg of DOC in landfills in 2010. Therefore, it can also be assumed that the current rate of landfill gas recovery will continue for several more decades (Boyner et al., 1988; Willumsen, 1990).

Although landfill gas production in the U.S. and Canada is expected to remain relatively steady over the next two decades, the startup of new landfill gas recovery systems is expected to shift the balance of landfill gas emissions. The amounts of CH_4 emitted to the atmosphere will decrease as more is controlled through flaring or utilization. Concurrently, the amount of CO_2 released will increase (Bonomo and Higginson, 1988; Boyner et al., 1988; El Rayes and Edwards, 1991; Willumsen, 1990; Rathje, 1991; Thorneloe, 1992). U.S. landfills are currently recovering about 1.2 Tg/yr of CH_4 and producing 344 MW_e of power (Thorneloe, 1992). Clean Air Act regulations proposed in May 1991 are expected to have a major impact on reducing landfill CH_4 emissions from both new and existing MSW landfills in the U.S. The proposed air emission regulations are expected to result in an additional emission reduction ranging from 5 to 7 Tg/yr of CH_4 (Federal Register, 1991; U.S. EPA, 1991).

3.9.2 OECD Countries. In the future, most OECD countries are considering policies that would increase the amount of MSW handled by recycling and incineration and to decrease the amount placed in landfills. Landfills will be large, regional sites that also will be used to dispose of incinerator ash. The decreases in the amount of MSW landfilled, coupled with increases in landfill gas recovery and incineration, are anticipated to lead to reduced CH_4 emissions and increased emissions of CO_2 and other combustion gases.

3.9.3 European countries and the former Soviet Union. Sanitary landfills or open dumps (e.g., placing refuse in a scattered fashion near residences or along roadsides) are used almost exclusively for MSW management in Greece, Hungary, Portugal, Poland, Romania, Bulgaria, Yugoslavia, and the former Soviet Union (Curi, 1988; Bartone, 1990a; Bartone and Haley, 1990; Mnatsaknian, 1991). In the future, Poland plans to close open dumps and dispose of MSW in larger, regional sanitary landfills. The former Soviet Union hopes to establish an effective recycling program. No landfill gas recovery sites were identified in these countries.

3.9.4 Asian Countries. Some Asian countries are upgrading their collection methods by introducing compactor trucks and covered containers. These changes could serve to decrease the amount of scavenging and increase the amount of MSW that is dumped in an uncontrolled fashion. Economic constraints, in tandem with a history of slow MSW management development, indicate that the use of sanitary landfills will not increase markedly in the near future for most of Asia. However, increases in population and total MSW will likely lead to increased CH_4 emissions.

3.9.5 South America, Central America, Mexico, and the Caribbean Islands. In the future, Brazil hopes to build recycling and composting plants, and sanitary landfills. Mexico also hopes to increase its number of sanitary landfills. Few landfill gas recovery sites are currently operating in South America and Mexico; there seems to some interest in increasing the number of landfill gas recovery sites in Brazil (Richards, 1988; Kessler, 1990).

3.9.6 Africa and the Middle East. In Africa the only MSW management methods reported to hold promise for expanded and successful application are recycling, composting, or possibly biogas recovery if markets and appropriate technologies can be developed to support these systems (Conner, 1978; Cointreau, 1982; Betts, 1984; Oluwande, 1984; Vogler, 1984; El-Halwagi et al., 1988). Much of the recyclable material in the MSW stream is currently being recovered (Cointreau, 1982; Wright et al., 1988), at least when comparing the amount of material recycled in low-income countries with that of middle-income and developed nations. However, recyclable materials are still available in the waste streams and, because wages are low in the developing countries, further recycling may be a viable waste management option (Cointreau, 1982).

4. Wastewater Treatment and Septic Sewage Systems

Wastewater from domestic, commercial, and industrial facilities is also a source of CH_4. Wastewater treatment lagoons, particularly those that treat wastewater with high biochemical oxygen demand (BOD), are suspected of emitting significant amounts of CH_4. Global CH_4 emissions from wastewater treatment lagoons are estimated to be about 25 Tg/yr (Orlich, 1990). This estimate is based on data from one wastewater lagoon study in Thailand and the assumption that 300 ℓ (13.4 g at standard temperature and pressure) of CH_4 is produced per kg of BOD. This estimate is considered uncertain due to a lack of field data on the CH_4 production potential for different types of lagoons. In addition, uncertainty results from limitations in available data on quantities of wastewater being treated and the characteristics of lagoons worldwide that affect CH_4 production.

Lagoons are commonly used for wastewater treatment and disposal in developing countries, primarily because land is available, operations are relatively simple, minimal energy is needed, and capital and operating expenses are low. Anaerobic conditions are favored because of the limited maintenance and control of these lagoons and limited use of expensive aeration devices. Moreover, the average temperature in many tropical and subtropical developing countries is close to the optimum biological temperature of methanogenic bacteria (i.e., 35°C). This results in greater biological activity and a higher CH_4 production potential, as compared to lagoons found at cooler latitudes (Gloyna, 1971).

Mean daily gas production rates were estimated by Toprak (1993) to be 51,000 m^3/day for wastewater treatment lagoons for a treatment plant being constructed in Izmir to be completed in the year 2000. This city is one of the fastest growing metropolitan areas in Turkey. Plans are being considered to utilize the CH_4 as an alternative source of energy.

The World Bank predicts that, because lagoons are relatively inexpensive and easy to operate, they will continue to be a preferred wastewater treatment method for developing countries (Bartone, 1990c). Furthermore, with increasing population growth and urbanization in developing countries, the number and variety of sources discharging into lagoons will increase, resulting in higher BOD loading rates. Increased CH_4 emissions from lagoons will result as these changes occur.

The U.S. EPA estimates that 130 billion ℓ (34 billion gal.) of domestic, commercial, and industrial wastewater is treated in the U.S. each day (U.S. EPA,

1987). Approximately 433 domestic and industrial lagoon systems receive various types and quantities of industrial wastewater, in addition to domestic wastewater. Furthermore, another estimated 5,000 municipal lagoons contain domestic wastes from residential, commercial, and institutional sources (U.S. EPA, 1987). Insufficient data are available to characterize the CH_4 emission potential of wastewater treatment lagoons both in the U.S. and globally. An empirical model for wastewater treatment lagoons using field data would be of value in relating BOD and any other significant factors to CH_4 emissions.

The CH_4 potential is expected to be minimal for the lagoons that receive pretreated wastewater. However, when no pretreatment occurs or pretreatment is minimal, lagoons may be a significant source of CH_4. CH_4 formation varies depending on temperature, retention time, BOD loading, lagoon depth, oxygen content, and the frequency at which the lagoon is dredged. The majority of the U.S. lagoons are facultative; that is, aerobic decomposition occurs in the upper strata and anaerobic degradation occurs in the lower strata. It is likely that facultative lagoons proceed to a more anaerobic state as BOD loading increases and surface aeration diminishes (e.g., due to low wind speed). In particular, information on lagoons used for wastewater from food-processing industries and other industries that have high BOD wastewater streams would be of value.

Most of the world's population, including about 25% of the U.S. population, relies on individual septic systems. This is the most common treatment and disposal method for domestic wastewater. Septic tanks are anaerobic digesters of the simplest form and release CH_4 to the atmosphere at a rate that is dependent on temperature, retention time, and system configuration. However, a certain portion of the CH_4 will be oxidized as the gas diffuses through the soil. These findings were confirmed in work by Khalil et al. (1990), who showed that very little CH_4 escapes to the atmosphere from underground biogas pits in China.

The process of CH_4 production in septic tanks is similar in principle to small biogas pits that are used extensively in China and India to recover gas for energy. The primary difference between the septic tanks used in the U.S. and the biogas pits of China and India is that septic tanks are not designed to produce gas for energy recovery. Rather, septic tanks typically treat only domestic sewage generated by household residences and commercial establishments; CH_4 emitted from the tanks is not normally collected.

A layer of sludge accumulates on the bottom of septic tanks where anaerobic processes occur. CH_4 may be released from septic systems through manhole openings, cracked lids, and vent pipes. Because these tanks are underground and typically receive sewage directly from a source in close proximity, it is not likely that the system will be upset by external factors. For the anaerobic process to produce significant amounts of CH_4, however, temperatures inside the tank would need to be above 15°C. Below 15°C, the anaerobic process would slow to a point where the septic tank would merely act as a sludge storage area.

5. Livestock Waste

A review of CH_4 emissions from livestock waste is provided by Johnson et al. (this volume). They estimate that less than 5% of animal CH_4 is accounted for by the anaerobic treatment of animal excreta. The fermentation of sewage sludge may be extensive, such as in anaerobic digesters designed for CH_4 recovery and utilization (Thorneloe, 1992). CH_4 from the fermentation of excreta from free-range livestock is thought to be inconsequential (Johnson et al., this volume). The majority of the world's livestock are free ranging. Recent research by Lodman et al., (1993) indicates that CH_4 emissions from livestock waste represent less than 1% of the potential CH_4 resulting from anaerobic lagoons. Another recent study (Williams, 1993) found that CH_4 emissions from dairy cow patties contribute less than 1% of the potential. In the U.S., livestock accounts for about 6 Tg/yr of CH_4, most of which originates in the rumen. CH_4 emissions from livestock manure are estimated to be 2.5 Tg/yr in the U.S. (Johnson et al., this volume).

In another study, Safley et al., (1992) estimated that CH_4 emissions from livestock manure contribute 21 to 35 Tg/yr, with an average of 28 Tg/yr. The estimate for the U.S. is 4 Tg/yr, as compared to Johnson et al.'s estimate of 0.6 Tg/yr. The major difference between these estimates is the use of an emission factor representing free-range livestock waste. The factor used in the Safley study is at least 10 times higher than that derived by Lodman et al., (1993) and Williams (1993). Using Safley et al.'s methodology and reducing the emission factors for pasture/range, drylot, and daily spread animal waste disposal by a factor of 10, leads to a global estimate of approximately 15 Tg/yr from animal waste.

The principal factors controlling potential CH_4 production from manure include the quantity and characteristics of the animal waste, the type of waste management system, and the temperature and moisture content of the waste (Safley et al., 1992). Cattle in the U.S. produce larger quantities of organic waste than any other type of livestock. The average head of cattle produces 24 kg of wet feces per day (Overcash et al., 1983), including 2.8 kg of organic matter. A portion of this organic matter can be decomposed by methanogenic bacteria. Other animals such as sheep, goats, horses, and fowl, have a larger fraction of volatile solids in their feces, but because cattle produce a larger quantity of manure per individual and are more populous than other livestock, they contribute the largest portion of CH_4 from livestock manure. Volatile solids are that portion of organic matter that can be decomposed by microorganisms (Safley et al., 1992). Safley et al. estimate that global cattle populations contribute 53% of the CH_4 emissions from livestock waste.   In addition to the quantity and characteristics of livestock manure, animal waste management systems largely determine the potential for CH_4 generation. Livestock are managed in conditions ranging from open pasture and range to complete confinement. Concurrently, manure may simply be left on the pasture or range where it is deposited, or it may be prepared using dry storage methods or liquid treatments. The CH_4 production potential of all waste management systems is highly dependent on temperature and moisture (Safley and Westerman, 1987; Johnson et al., this volume). Generally, warm temperatures and high moisture content provide for maximum CH_4 production.

Deep pit stacking and daily spreading of solid and semi-solid manure have the least CH_4-producing potential of all livestock waste management systems (Safley et al., 1992). Under these two management systems, manure generally has a very low moisture content. By contrast, anaerobic lagoons have the greatest CH_4-producing potential of all livestock waste management systems. Due to the high moisture content of the waste, almost all of the CH_4-producing potential of waste can be realized with proper design and operation of the anaerobic lagoons. Safley et al., (1992) estimate that the CH_4-producing potential of anaerobic lagoons is 70% higher than any other form of livestock waste management system. In warm and tropical latitudes, anaerobic lagoons have their greatest CH_4-producing potential. Although liquid/slurry management systems have 70% less CH_4-producing potential than anaerobic lagoons (Safley et al., 1992), they rank second among livestock waste management systems for CH_4-producing potential, because the moisture content of the waste is high. Unfavorable temperatures and short residence time of the waste in storage are probably the factors that limit the CH_4 production potential of liquid/slurry systems. By virtue of their relatively widespread use in western and eastern Europe, Asia, and North America, liquid/slurry management systems are estimated to contribute 26% of the CH_4 from livestock waste management systems. Together, liquid/slurry systems and anaerobic lagoons have been estimated to account for 10 Tg/year, or almost 36% of the total CH_4 emissions from livestock waste management (Safley et al., 1992).

Because more livestock waste is deposited in pasture and range systems than in any other management system globally, assumptions regarding the CH_4-producing potential of this waste have significant effects on the total emission estimate. Research being conducted at Colorado State University indicates that manure CH_4 production varies under feedlot conditions and under simulated grazing conditions. Temperature, moisture, and animal diet were the variables that had the greatest influence on CH_4 production. More accurate estimates of how manure-handling systems affect fermentation would help obtain a better estimate of CH_4 production.

Uncertainty in the estimates of global CH_4 emissions from livestock waste results primarily from limitations in available data. Data are particularly limited for developing countries and for free-range livestock waste management. Even in developed countries uncertainty is associated with animal population estimates, animal sizes and diet, and the types and numbers of animal waste management systems. In addition, refinements to the current estimates using field test data on the CH_4-production potential of livestock waste both under free range conditions and in livestock management facilities would be of value. Currently, the only published global estimate for this source is by Safley et al. Initial revised estimates by EPA/AEERL suggest that this source contributes 2 to 5 Tg/yr of CH_4 globally. Data from field and laboratory studies could help reduce the current uncertainty of this estimate.

6. Summary

Global and U.S. estimates of CH_4 emissions from landfills, wastewater treatment, and livestock waste are presented in Table 3. The recent estimate of 11-33 Tg/yr by EPA/AEERL for landfills is thought to more accurately reflect CH_4 emissions from landfilled waste than previous estimates. The global estimates for wastewater treatment and livestock waste are very uncertain due to lack of data, particularly on the emission potential of wastewater treatment lagoons and free-range livestock waste. There is also a lack of country-specific data for this source category. Uncertainties in these estimates result from optimistic assumptions regarding the extent of anaerobic decomposition and limitations in available data characterizing (1) waste quantities and composition, and (2) treatment or disposal practices.

Using the ranges presented in Table 3, two alternative estimates of the contribution of waste sources to global CH_4 emissions can be derived by calculating the joint probability distributions of the estimates as described by Khalil (1992). Assuming that any value within the range shown for landfills, wastewater and sewage treatment, and livestock waste are equally probable, a Monte Carlo model was used to generate random values for each source. Repeating this process for 200 iterations gives a distribution of estimates. Using (1) Bingemer and Crutzen's (1987) estimate for landfills, (2) Orlich's (1990) estimate for wastewater treatment, and (3) Safley et al.'s (1992) estimate for livestock waste, a global estimate of 103 Tg/yr with a 95% confidence interval of 75 to 103 Tg/yr results. Using the EPA/AEERL estimate for landfills (i.e., 11 to 33 Tg/yr) and Orlich's and Safley's estimates, results in a global estimate for waste management of 72 Tg/yr with a 95% confidence interval of 54 to 95 Tg/yr. Using Lodman et al.'s (1993) and Williams' (1993) emission factors with Safley et al.'s methodology, plus EPA/AEERL's landfill estimate and Orlich's wastewater estimate, gives a global estimate of 60 Tg/yr.

Table 3. Global and U.S. estimates of CH_4 emissions from waste management

	Avg.	Range	
Global (Tg/yr)			
Landfills	22 [a,c]	11-33	Thorneloe, this work
Wastewater Treatment	25 [a]	12-38 [d]	Orlich, 1990
Livestock Waste	25	20-35	Safley et al., 1992
	15 [e]		
U.S. (Tg/yr)			
Landfills	6 [a]	3-8 [b]	Augenstein & Pacey, 1990
	23 [c,a]	--	Bingemer & Crutzen, 1987 [c]
	9 [a]	4-14	Thorneloe, this work
Livestock Waste	4	--	Safley et al., 1992
	0.7	--	Lodman et al., 1993

a   Potential emissions, not corrected for amount that is flared or utilized. Approximately 1.2 million tonnes of CH_4 is being recovered from U.S. landfills (Thorneloe, 1992).
b   Uses estimated annual placement rates from 1950 to 1990.
c   Previous global estimate are 50 Tg/yr (30-70 Tg/yr range) by Bingemer and Crutzen (1987), which comes to 60 Tg/yr with country-specific data on MSW generation. The estimate of Richards (1989) is 15 Tg/yr (10-20 Tg/yr range).
d   Assumes 50% uncertainty of Orlich's (1990) estimates of 25 Tg/yr.
e   Uses Safley et al.'s (1992) methodology and Lodman et al.'s (1993) pasture/range emission factor.

References

Ahmed, M.F. 1986. Recycling of solid wastes in Dhaka. In: *Waste Management in Developing Countries, 1* (K.J. Thome-Kozmiensky, ed.), EF-Verlag fur Energie und Umvelttechnik GmbH, Berlin. pp. 169-173.

Alter, H. 1991. The future course of solid waste management in the U.S. *Waste Mgmt. & Res., 9*:3-20.

Augenstein, D., J. Pacey. 1990. Modeling landfill CH_4 generation. In: *International Conference on Landfill Gas: Energy and Environment*, 10/17/90, Bournemouth, England.

Barlaz, M.A. 1985. Factors affecting refuse decomposition in sanitary landfills. M.S. Thesis, Dept. of Civil and Environmental Engineering, Univ. of Wisconsin - Madison.

Barlaz, M.A. 1988. Microbiological and chemical dynamics during refuse decomposition in a simulated sanitary landfill. Ph.D. Dissertation, Department of Civil and Environmental Engineering, University of Wisconsin - Madison.

Barlaz, M.A., M.W. Milke, R.K. Ham. 1987. Gas production parameters in sanitary landfill simulators. *Waste Mgmt. & Res., 5*:27.

Barlaz, M.A., D.M. Schaefer, R.K. Ham. 1989a. Bacterial population development and chemical characteristics of refuse decomposition in a simulated sanitary landfill. *Appl. Env. Microbiol., 55* (1):55-65.

Barlaz, M.A., R.K. Ham, D.M. Schaefer. 1989b. Mass balance analysis of decomposed refuse in laboratory scale lysimeters. *ASCE J. of Environ. Engineering, 115* (6):1,088-1,102.

Barlaz, M.A, R.K. Ham, D.M. Schaefer. 1990. CH_4 production from municipal refuse: A review of enhancement techniques and microbial dynamics. *CRC Critical Reviews in Environmental Control, 19*, Issue 6.

Bartone, C.R. 1990a. Economic and Policy Issues in Resource Recovery from Municipal Solid Wastes. *Resour. Conserva. Recycl, 4*:7-23.

Bartone, C.R. 1990b. Investing in Environmental Improvements Through Municipal Solid Waste Management. Paper presented at the WHO/PEPAS Regional Workshop on National Solid Waste Action Planning, Kuala Lumpur, 02/26/90-03/02/90.

Bartone, C.R. 1990c. Urban wastewater disposal and pollution control: Emerging issues for sub-Saharan Africa. In: *Proceedings of the African Infrastructure Symposium*, The World Bank, Baltimore, Maryland, 01/08-09/90, p. 6.

Bartone, C.R., C. Haley. 1990. The Bled Symposium: Introduction. *Resour. Conserva. Recycl., 4*:1-6.

Bartone, C.R., L. Leite, T. Triche, R. Schertenleib. 1991. Private sector participation in municipal solid waste service: Experiences in Latin America. *Waste Mgmt. & Res., 9*:495-509.

Bateman, C.S. 1988. Landfill gas development in Australia. In: *Proceedings of the International Conference on Landfill Gas and Anaerobic Digestion of Solid Waste* (Y.R. Alston and G.E. Richards, eds.), October 4-7, Harwell Laboratory, Oxfordshire, UK.

pp. 156-161.

Beker, D. 1990. Sanitary landfilling in the Netherlands. In: *International Perspectives on Municipal Solid Wastes and Sanitary Landfilling* (J.S. Carra and R. Cossu, eds.), Academic Press, New York, NY. pp. 139-155.

Betts, M.P. 1984. Trend in solid waste management in developing countries. In: *Managing Solid Wastes in Developing Countries* (J.R. Holmes, ed.), John Wiley & Sons, Ltd. Chichester, England, pp. 291-302.

Bhide, A.D., B.B. Sundaresan. 1983. Solid waste management in developing countries. In: *Indian National Scientific Documentation Centre*, New Delhi, India.

Bhide, A.D., S.A. Gaikwad, B.Z. Alone. 1990. CH_4 from land disposal sites in India. In: *Proceedings of the International Workshop on CH_4 Emissions from Natural Gas Systems, Coal Mining and Waste Management Systems*. Environment Agency of Japan, the U.S. Agency for International Development, and the U.S. Environmental Protection Agency, Washington, D.C., 04/09-13/90.

Bingemer, H.G., P.J. Crutzen. 1987. The production of CH_4 from solid wastes. *J. Geophys. Res., 92* (D2):2,181-2,187.

Bogner, J.E. 1990. Controlled study of landfill biodegradation rates using modified BMP assays. *Waste Mgmt. & Research, 8*:329-352.

Bonomo, L., A.E. Higginson. 1988. *International Overview on Solid Waste Management*. Harcourt Brace Jovanovich, New York, NY, p. 268.

Boyner, J., M. Vogt, R. Piorkowski, C. Rose, M. Hou. 1988. U.S. Landfill Gas Research. In: *Proceedings of the International Conference on Landfill Gas and Anaerobic Digestion of Solid Waste* (Y.R. Alston, and G.E. Richards, eds.), 10/04-07/88, Harwell Laboratory, Oxfordshire, UK, pp. 313-338.

Buivid, M.G., et al. 1981. Fuel gas enhancement by controlled landfilling of municipal solid waste, *Resource Recovery and Conservation, 6*:3.

Campbell, D., D. Epperson, L. Davis, R. Peer, W. Gray. 1991. *Analysis of Factors Affecting Methane Gas Recovery from Six Landfills*. Prepared for Air and Energy Engineering Research Laboratory, U.S. Environmental Protection Agency, Research Triangle Park, NC. EPA-600/2-91-055 (NTIS PB92-101351).

Carra, J.S., R. Cossu (editors). 1990. *International Perspectives on Municipal Solid Wastes and Sanitary Landfilling*, Academic Press, New York, NY, pp. 1-14.

Cointreau, S.J. 1982. *Environmental Management of Urban Solid Wastes in Developing Countries*. A Project Guide, Urban Development Technical Paper Number 5, World Bank, Washington, D.C., pp. 1-17.

Cointreau, S.J. 1984. Solid waste collection practice and planning in developing countries. In: *Managing Solid Wastes in Developing Countries* (J.R. Holmes, ed.), John Wiley & Sons, Ltd. Chichester, England, pp. 151-182.

Cointreau, S.J. 1987. *Solid Waste Management Study for the Greater Banjul Area, The Gambia*. Ministry of Economic Planning and Industrial Development, Banjul, The Gambia. 09/87.

Conner, M.A. 1978. Modern Technology for Recovering Energy and Materials from Urban Wastes - Its Applicability in Developing Countries. *Resour. Conserv. Recycl., 2*:85-92.

Cossu, R. 1990a. Sanitary landfilling in Japan. In: *International Perspectives on Municipal Solid Wastes and Sanitary Landfilling* (J.S. Carra and R. Cossu, eds.), Academic Press, New York, NY, pp. 110-138.

Cossu, R. 1990b. Sanitary landfilling in the United Kingdom. In: *International Perspectives on Municipal Solid Wastes and Sanitary Landfilling* (J.S. Carra and R. Cossu, eds.), Academic Press, New York, NY, pp. 199-220.

Cossu, R., G. Urbini. 1990. Sanitary landfilling in Italy. In: *International Perspectives on Municipal Solid Wastes and Sanitary Landfilling* (J.S. Carra and R. Cossu, eds.), Academic Press, New York, NY, pp. 94-109.

Curi, K. 1988. Comparison of solid waste management in touristic areas of developed and developing areas. In: *Proceedings of the 5th International Solid Waste Conference, Internal Solid Waste and Public Cleansing Association*, Copenhagen, Denmark, 09/88.

Diaz, L.F., C.G. Golueke. 1987. Solid waste management in developing countries. *BioCycle, 28*:50-55.

El-Halwagi, M.M., S.R. Tewfik, M.H. Sorour, A.G. Abulnour. 1986. Municipal solid waste management in Egypt: Practices and trends. In: *Waste Management in Developing Countries, 1* (K.J. Thome-Kozmiensky, ed.), EF-Verlag fur Energie und Emvelttechnik GmbH, Berlin, pp. 283-288.

El-Halwagi, M.M., S.R. Tewfik, M.H. Sorour, A.G. Abulnour. 1988. Municipal solid waste management in Egypt. In: *Proceedings of the 5th International Solid Waste Conference, International Solid Waste and Public Cleansing Association*, Copenhagen, Denmark, September, pp. 415-424.

El Rayes, H., W.C. Edwards (B.H. Levelton & Associates, Ltd.) . 1991. Inventory of CH_4 Emissions from Landfills in Canada. Prepared for Environment Canada, Hull, Quebec, pp. 25-69.

Emberton, J.R. 1986. The biological and chemical characterization of landfills. In: *Proceedings of Energy from Landfill Gas*, Solihull, West Midlands, UK, 10/30-31/86.

EMCON Associates. 1982. CH_4 *Generation and Recovery from Landfills*, Ann Arbor Science Publishers, Inc., Ann Arbor, MI.

Ernst, A. 1990. A review of solid waste management by composting in Europe. *Resour. Conserva. Recycl., 4*:135-149.

Ettala, M.O. 1990. Sanitary landfilling in Finland. In: *International Perspectives on Municipal Solid Wastes and Sanitary Landfilling* (J.S. Carra and R. Cossu, eds.), Academic Press, New York, NY, pp. 67-77.

Federal Register, Vol. 56, No. 104. May 30, 1991, pp. 24,468-24,528.

Frantzis, I. 1988. Recycling in Greece. *BioCycle*:30-31.

Gadi, M.T. 1986. In: *Waste Management in Developing Countries, 1* (K.J. Thome-Kozmiensky, ed.), EF-Verlag fur Energie and Umvelttechnik GmbH, Berlin. pp. 188-194.

Gandolla, M. 1990. Sanitary landfilling in Switzerland. In: *International Perspectives on Municipal Solid Wastes and Sanitary Landfilling* (J.S. Carra and R. Cossu, eds.), Academic Press, New York, NY, pp. 190-198.

Gloyna, E.F. 1971. Waste stabilization ponds. World Health Organization, Geneva, p. 76.

Halvadakis, C.P., et al. 1983. Landfill Methanogenesis: Literature Review and Critique, Technical Report No. 271, Department of Civil Engineering, Stanford University.

Hayakawa, T. 1990. The status report on waste management in Japan - special focus on methane emission prevention. In: *Proceedings of the International Workshop on Methane Emissions from Natural Gas Systems, Coal Mining, and Waste Management Systems*, Washington, D.C., April 9-13, pp. 509-523.

Holmes, J.R. 1984. Solid waste management decisions in developing countries. In: *Managing Solid Wastes in Developing Countries* (J.R. Holmes, ed.), John Wiley & Sons, Ltd. Chichester, England, pp. 1-17.

IPCC. 1992. Climate Change 1992. The Supplementary Report to the IPCC Scientific Assessment. Published for The Intergovernmental Panel on Climate Change (IPCC), World Meteorological Organization/United Nations Environment Programme. Cambridge University Press. Edited by J.T. Houghton, G.J. Jenkins, and J.J. Ephraums.

International Energy Agency. 1992. Landfill Gas Enhancement Test Cell Data Exchange-- Final Report of the Landfill Gas Expert Working Group. Editor: Pat Lawson, AEA-EE-0286.

Jenkins, R.L., J.A. Pettus. 1985. In: *Biotechnological Advances in Processing Municipal Wastes for Fuels and Chemicals* (A.A. Antonopoulos, ed.), Argonne Natl. Lab. Report ANL/CNSV - TM - 167, p. 419.

Kaldjian, P. 1990. *Characterization of Municipal Solid Waste in the United States: 1990 Update*, EPA-530/SW-90-042, PB90-215112. Office of Solid Waste, Washington, DC.

Kaltwasser, B.J. 1986. Solid waste management in medium sized towns in the Sahel area. In: *Waste Management in Developing Countries, 1* (K.J. Thome-Kozmiensky, ed.), EF-Verlag für Energie und Umvelttechnik GmbH, Berlin, pp. 299-307.

Kessler, T. 1990. Brazilian trends in landfill gas exploitation. *ETATEC Consultores s/c Ltda*, Sào Paulo, Brazil.

Khalil, M.A.K. 1992. A statistical method for estimating uncertainties in the total global budgets of atmospheric trace gases. *J. Environ. Sci. Health, A27*:777-770.

Khalil, M.A.K., R.A. Rasmussen, M.-X. Wang. 1990. Emissions of trace gases from Chinese rice fields and biogas generators: CH_4, N_2O, CO, CO_2, chlorocarbons, and hydrocarbons. *Chemosphere, 20*:207-226.

Kinman, R.N., et al. 1987. *Waste Management and Research, 5*:13.

Lassey, K.R., D.C. Lowe, M.R. Manning. 1992. A Source Inventory for Atmospheric CH_4 in New Zealand and Its Global Perspective. *J. Geophys. Res., 97*:3,751-3,765.

Lawson, P. 1992. Landfill gas expert working group summary report, 1989-1991. *International Energy Agency*, AEA-EE-0305.

Lodman, D.W., M.E. Branine, B.R. Carmean, P. Zimmerman, G.M. Ward, D.E. Johnson. 1993. Estimates of CH_4 Emissions from Manure of U.S. Cattle. *Chemosphere, 26* (1-4):189-200.

Lohani, B.N., N.C. Thanh. 1980. Problems and practices of solid waste management in Asia. *J. Envion. Sci.*, 06/80, pp. 29-33.

Maniatis, K., S. Vanhille, A. Hartawijaya, A. Buekens, W. Verstraete. 1987. Solid waste management in Indonesia: Status and Potential. *Resour. Conserva. Recycl., 15* (87):277-290.

Mei-Chan, L. 1986. Waste management in the Taiwan area. In: *Waste Management in Developing Countries, 1* (K.J. Thome-Kozmiensky, ed.), EF-Verlag fur Energie und Umvelttechnik GmbH, Berlin, pp. 247-250.

Mnatsaknian, R.A. 1991. Legislation and public control of waste sites in the U.S.S.R. In: *Proceedings of the Third International Landfill Symposium*, Sardinia, Italy, 10/91, pp. 1,747-1,753.

Monney, J.G. 1986. Municipal solid waste management--Ghana's experience. In: *Waste Management in Developing Countries, 1* (K.J. Thome-Kozmiensky, ed.), EF-Verlag fur Energie und Umvelttechnik GmbH, Berlin, pp. 32-1327.

Mwiraria, M., J. Broome, R. Semb, W.P. Meyer. 1991. Municipal Solid Waste Management in Uganda and Zimbabwe. Draft report of the United Nations Development Program and The World Bank. 05/18/91.

Nozhevnikova, A.N., A.B. Lifshitz, V.S. Lebedev, G.A. Zavarzin. 1993. Emission of methane into the atmosphere from landfills in the former USSR. *Chemosphere, 26* (1-4):401-418.

Oluwande, P.A. 1984. Assessment of solid waste management problems in China and Africa. In: *Managing Solid Wastes in Developing Countries* (J.R. Holmes, ed.), John Wiley & Sons, Ltd., Chichester, England, pp. 71-89.

OECD. 1991. Estimation of Greenhouse Gas Emissions and Sinks. Final Report from the OECD Experts Meeting, 02/18-21/91. Prepared for the IPCC. Revised 08/91.

Orlich, J. 1990. CH_4 Emissions from Landfill Sites and Waste Water Lagoons. Proceedings of the International Workshop on CH_4 Emissions from Natural Gas Systems, Coal Mining, and Waste Management Systems, 04/09-13/90. Environment Agency of Japan, U.S. Agency for International Development, and the U.S. Environmental Protection Agency, Washington, DC.

Overcash, M.R., F.J. Humenik, J.R. Miner. 1983. *Livestock Waste Management*. Vol. II. CRC Press, Boca Raton, FL.

Pacey, J. 1989. Enhancement of degradation: Large-scale experiments. In: *Sanitary Landfilling: Process Technology and Environmental Impact* (T. Christensen, R. Cossu, and R. Stegmann, eds.), Academic Press, London, pp. 103-119.

Pairoj-Boriboon, S. 1986. State-of-the-art of waste management in Thailand. In: *Waste*

Management in Developing Countries, 1 (K.J. Thome-Kozmiensky, ed.), EF-Verlag für Energie und Umvelttechnik GmbH, Berlin, pp. 208-219.

Papachristou, E. 1988. Solid wastes management in Rhodos. In: *Proceedings of the 5th International Solid Wastes Conference, International Solid Waste and Public Cleansing Association*, Copenhagen, Denmark, September.

Parkin, G.F., W.F. Owen. 1986. Fundamentals of anaerobic digestion of wastewater sludges. *J. Environmental Engineering Division, ASCE, 112* (5), p. 867.

Peer, R.L., D.L. Epperson, D.L. Campbell, P. Von Brook. 1992. Development of an empirical model of methane emissions from landfills, EPA-600/R-92-037, PB92-152875. Prepared for the U.S. Environmental Protection Agency, Air and Energy Engineering Research Laboratory, Research Triangle Park, NC.

Peer, R.L., S.A. Thorneloe, D.L. Epperson. 1993. A comparison of methods for estimating global methane emissions from landfills. *Chemosphere, 26* (1-4):387-400.

Peterson, C., A. Perlmutter. 1989. Composting in the Soviet Union. *BioCycle*, July, 74-75.

Pohland, F.G. 1975. Sanitary landfill stabilization with leachate recycle and residual treatment. EPA-600/2-75-043, PB248524.

Pohland, F.G., S.R. Harper. 1987. Critical review and summary of leachate and gas production from landfills. EPA/600/2-86/073, PB86-240181.

Population Reference Bureau, Inc. 1989. World Population Data Sheet of the Population Reference Bureau, Inc. Demographic data and estimates for the countries and regions of the world. Washington, D.C.

Rathje, W.L. 1991. Once and future landfills. *National Geographic, May*, pp. 117-134.

Richards, K.M. 1988. Landfill gas - A global review. In: *Biodeterioration 7* (D.R. Houghton, R.N. Smith, and H.O.W. Eggins, eds.), Elsevier Applied Science, London, pp. 774-790.

Richards, K.M. 1989. Landfill gas: Working with Gaia. *Biodeterioration Abstracts, 3*:317-331.

Safley, L.M., P.W. Westerman. 1987. Biogas production from anaerobic lagoons. *Biological Wastes, 23*:181.

Safley, L.M., M.E. Casada, J.W. Woodbury, K.F. Roos. 1992. Global CH_4 emissions for livestock and poultry manure. EPA/400/1-91/048, U.S. Environmental Protection Agency, Air and Radiation, Washington, D.C, 02/92.

Scheepera, M.J.J. 1990. Landfill gas in the Dutch perspective. *International Conference on Landfill Gas: Energy and Environment '90*, Session 3.2.

Shelton, D.R., J.M. Tiedje. 1984. General method for determining anaerobic biodegradation potential. *Appl. Env. Microbiol., 47*:850-857.

Stegmann, R. 1990. Sanitary landfilling in the Federal Republic of Germany. In: *International Perspectives on Municipal Solid Wastes and Sanitary Landfilling* (J.S. Carra and R. Cossu, eds.), Academic Press, New York, NY, pp. 51-66.

Swartz, A. 1989. Overview of International Solid Waste Management Methods. State Government Technical Brief 98-89-MI-2. The American Society of Mechanical Engineers, Washington, D.C.

Thorneloe, S. June 1992. Landfill Gas Recovery/Utilization - Options and Economics, Published in Proceedings for the 16th Annual Conference by the Institute of Gas Technology on Energy from Biomass and Waste, 03/92, Orlando, FL.

Thorneloe, S.A. August 1992. Emissions and Mitigation at Landfills and Other Waste Management Facilities. Presented at the EPA Symposium on Greenhouse Gas Emissions and Mitigation Research, Washington, DC. To be published in conference proceedings.

Toprak, H. 1993. CH_4 emissions originating from the anaerobic waste stabilization ponds. *Chemosphere, 26* (1-4):633-640.

United Nations. 1989. City Profiles. Prepared by the United Nations Centre for Regional Development and the Kitakyushu City Government. 64 pp.

United Nations. 1990. 1988 Demographic Yearbook. Fortieth Issue. Department of

International Economic and Social Affairs, Statistical Office, New York, NY

United Nations Development Programme (UNDP), The World Bank, and the Canadian International Development Agency. 1987. Master Plan for Resource Recovery and Waste Disposal, City of Abidjan. Final Report. Prepared by Roche Ltd. Consulting Group, Sainte-Foy, Quebec, Canada. February.

United Republic of Tanzania. 1989. Masterplan of Solid Waste Management for Dar es Salaam. Volume II: Annexes. Ministry of Water, Department of Sewerage and Sanitation. Prepared by HASKONING, Royal Dutch Consulting Engineers and Architects, Nijmegen, The Netherlands, and M-Konsult Ltd., Consulting Engineers, Dar es Salaam, Tanzania. pp. 1-39, 47-49, 74-94, 109-124, 142-145.

U.S. Environmental Protection Agency. 1987. *Report to Congress, Municipal Wastewater Lagoon Study*, Office of Municipal Pollution Control.

U.S. Environmental Protection Agency. 1988. *Report to Congress, Solid Waste Disposal in the United States*, Volume 1, EPA/530-SW-88-011, PB89-110381, Office of Solid Waste, Washington, DC.

U.S. Environmental Protection Agency. 1991. *Air Emissions from Municipal Solid Waste Landfills - Background Information for Proposed Standards and Guidelines*, EPA-450/3-90-011a, PB91-197061. Office of Air Quality Planning and Standards, Research Triangle Park, NC.

Verrier, S.J. 1990. Urban Waste Generation, Composition and Disposal in South Africa. In: *International Perspectives on Municipal Solid Wastes and Sanitary Landfilling* J.S. Carra and R. Roccu, eds.), Academic Press, Harcourt Brace Jovanovich, London, England, pp. 161-176.

Vogler, J.A. 1984. Waste recycling in developing countries: A review of the social, technological, and market forces. In: *Managing Solid Wastes in Developing Countries* (J.R. Holmes, ed.), John Wiley and Sons, Ltd., Chichester, England, pp. 241-266.

Whalen, S.C., W.S. Reeburgh, K.S. Sanbeck. 1990. Rapid methane oxidation in a landfill cover. *Applied and Environmental Microbiology, 56*:3,405-3,411.

Williams, D.J. 1993. Methane emissions from the manure of free range dairy cows. *Chemosphere, 26* (1-4):179-188.

Willumsen, H. 1990. Landfill gas. *Resour. Conserva. Recycl., 4*:121-133.

Wolfe, R.S. 1979. Methanogenesis. In: *Microbial Biochemistry, International Review of Biochemistry*, Vol. 21 (J.R. Quayle, ed.), University Park Press, Baltimore, MD.

Wolin, M.J. and T.L. Miller. 1985. In: *Biology of Industrial Microorganisms* (A.L. Dernsin and N.A. Solomon, eds.), Benjamin/Cumings Publishing Company, Inc., Menlo Park, CA, pp. 189-221.

World Bank, The. 1985. Metropolitan Area of Douala. Study of Waste Management and Resource Recovery. Part A. Phase I. Prepared by Motor Columbus, Consulting Engineers, Inc., CH-5401 Baden, Switzerland.

World Resources Institute. 1990. *World Resources 1990-91*. Oxford University Press, New York, NY.

Wright, F., C. Bartone, S. Arlosoroff. 1988. Integrated resource recovery: optimizing waste management. In: *Proceedings of the 5th International Solid Wastes Conference, International Solid Wastes and Public Cleansing Association*, Copenhagen, Denmark, pp. 619-624, 09/88.

Xianwen, C., Z. Yanhua. 1991. Landfill gas utilization in China. In: *Proceedings of the Third International Landfill Symposium*, Sardinia, Italy, October, pp. 1,747-1,753.

Yepes, G., T. Campbell. 1990 (draft). Assessment of Municipal Solid Waste Services in Latin America. Report in progress prepared for The World Bank, Technical Department, Infrastructure and Energy Division, Urban Water Unit, Latin America and the Caribbean Region. pp. 1-6, 20-26.

Young, L.Y., A.C. Frazer. 1987. *Geomicrobiology J., 5*:261.

Zehnder, A.J.B., et al. 1982. Microbiology of CH_4 bacteria. In: *Anaerobic Digestion* (D.E.

Hughes, ed.), Elsevier Biomedical Press B.V., Amsterdam, p. 45.

Zsuzsa, K.P. 1990. Possibilities for utilization of the energy content of the solid wastes of settlements. *Resour Conserva. Recycl., 4*:173-180.

[**Editor's Note on the Correlation Between Average Recovery Rate and Amount of Refuse**: In Figure 1, the correlation is greatly affected by a single point at the upper right hand corner. Without it the correlation for the remaining 20 landfills is only 0.15. This matter was discussed with Susan Thorneloe who sent me new data from more than 100 landfills, and they show a correlation similar to that reported here (for all points) and more robust. These results are still being analyzed and could not be completed for this chapter but will be published elsewhere.]

14. Fossil Fuel Industries

David A. Kirchgessner
United States Environmental Protection Agency
Office of Research and Development, National Risk---- Management
Research Laboratory, Research Triangle Park, North Carolina 27711

1. Introduction

This chapter focuses on methane emissions from the coal and natural gas industries. The petroleum industry is not addressed due to the lack of related quality data. Emission points are identified for each industry, and a discussion of factors affecting emissions is presented. A summary and discussion of available global and country-specific emission estimates are also provided. Where possible, assessments of data quality are offered.

Not included here are minor energy sources known or believed to emit some quantities of methane. These include, but are not necessarily limited to, coke production facilities, peat mining, shale oil mining operations, and the coalbed methane industry. This latter enterprise treats coal only as a methane source. It does not include methane that is captured and utilized from some coal mining operations.

2. Coal Mining Industry

Historically, coal use in the United States (U.S.) suffered a nearly catastrophic decline after World War II. It was not until the mid-1970s that bituminous coal production once again equaled the levels seen in the mid-1940s. Production in 1970 was about 547 million tonnes and by 1994 had grown to 938 million tonnes. Even more rapid growth can be expected in countries that have significant coal resources and large populations, and that are undergoing industrialization; India and China are examples. In the same time frame that the U.S. nearly doubled its coal production, India's production more than tripled from 74 million tonnes in 1970 to 250 million tonnes in 1994, and China's production nearly quadrupled from 327 million tonnes in 1970 to 1210 million tonnes in 1994.

The environmental effects of coal production range from the local to the global. Longwall underground mining results in varying amounts of subsidence at the surface that can cause extensive property damage and can be expensive to prevent. Most underground mines produce large quantities of waste water, and virtually all mines

Mohammad Aslam Khan Khalil   (Ed.)
Atmospheric Methane
© *Springer-Verlag Berlin Heidelberg 2000*

produce large amounts of coal and rock wastes. Surface mines produce scarring of the surface that can render the land useless, although the land can be restored to nearly original conditions at substantial expense. The combustion of all coals results in the release of large quantities of sulfur dioxide, nitrogen oxides, and particulate matter, and it is one of the largest sources of anthropogenic carbon dioxide emissions. Of particular interest here is that underground coal mining is also one of the largest global sources of anthropogenic methane emissions.

2.1 Sources of Methane Emissions in the Coal Industry

Methane is formed in coal during the process of coalification, with the quality and quantity of the gas created and retained being a function of the original organic matter composition and the conditions of burial. Generally, more methane is formed during coalification than can be stored within the coalbed itself, so excess methane migrates into, and can be stored in, the surrounding strata. This gas is retained by the coalbed and surrounding strata as long as they remain under pressure. Assuming that no geologic process breaches the reservoir first, mining releases this pressure, and the methane is allowed to escape.

Methane is emitted from three types of mines: underground, surface, and inactive or abandoned. Since methane in underground mines constitutes a safety hazard, methane levels in U.S. mines are monitored regularly by the Mine Safety and Health Administration (MSHA), and emissions are reasonably well known. Underground mines which have been abandoned or are temporarily inactive are monitored less frequently, if at all. Since emissions from surface mines are much smaller and are not a safety hazard, they are not measured on a regular basis.

Methane from underground mines can be released from the ventilation shaft, from gas drainage systems, and from coal crushing and handling operations (Boyer et al., 1990; Piccot et al., 1990). Methane can be released from all of the seams disturbed during mining. In an underground longwall mine, some studies suggest that this zone of disturbance may extend up to 160 meters into the roof rock and 40 meters below the seam being worked (Creedy, 1985). As a result, emissions from these mines are much greater than what would be expected from the coal alone. In areas where miners are working, methane levels must be kept below 0.5 percent. This is accomplished, in part, by sweeping the mine with large quantities of ventilation air. Although containing less than 1 percent methane, ventilation air contributes the largest amount to total emissions across the industry because the volume is so large.

Gas drainage systems are often employed to relieve some of the burden on the costly ventilation systems. Vertical wells can be drilled into the coal in advance of mining to drain methane from the coal and overlying strata. When a longwall miner passes under these wells and subsidence fractures the overlying strata, these wells become gob wells and drain the fractured area to prevent the released gas from venting through the mine workings. Horizontal boreholes and cross-measure boreholes can be drilled from within the mine into the coal and overlying strata, respectively. The methane is then conveyed to the surface through piping within the mine. In Europe methane recovered by gas drainage systems is most often used as an energy source, while in the U.S. most of the gas is still vented to the atmosphere. There is little published data on methane emissions from gas drainage systems. However, data obtained from industry representatives indicate that drainage well methane emissions may account for a significant fraction of the total emissions associated with longwall mines (Boyer et al., 1990; Kirchgessner et al., 1993a).

After coal leaves the mine it typically undergoes a series of operations collectively referred to as coal handling. They may include crushing, separation of impurities, size classification, drying, transportation, and storage. Coals desorb methane at different rates, but since coal is usually removed from a mine within a day of its being mined, some methane is usually left to be liberated from the coal during handling operations. To approximate these emissions the assumption is generally made that the fraction of *in-situ* coalbed methane remaining in the coal after mining is emitted completely to the atmosphere during post-mining operations. In fact, tight coals with a substantial amount of residual methane can retain some of the gas all the way to the combustion process where virtually all of it is burned. Researchers at British Coal have estimated that 40 percent of the methane in the mined coal is released after it leaves the mine (Creedy, 1993). The U.S. Environmental Protection Agency's Office of Research and Development (EPA/ORD) obtained a large number of run-of-mine samples from three mines (one in Alabama and two in Northern Appalachia) and conducted gas desorption analyses on them. Based on this limited evidence they also concluded that about 40 percent of the gas originally in the coal remained in the coal when it left the mine (Piccot et al., 1995). The amount of this gas that escapes to the atmosphere will be a function of the desorption characteristics of the coal, the amount of time between mining and combustion, and the treatment of the coal during that time.

Very few measurements have been taken at surface mines. Data from six surface mines in the U.S. gathered by EPA/ORD suggest that the primary sources of emissions are exposed coal surfaces, in particular the areas fractured by coal blasting. Methane emission measurements taken at the six mines using open-path Fourier Transform Infrared (FTIR) spectroscopy ranged from 0.1 to 1.8 million m^3/yr (Kirchgessner et al., 1993b; Piccot et al., 1995). In general, the strata overlying the coal do not appear to be a significant source of emissions but, as in underground mines, emissions may be contributed by underlying seams or faults. Measurements taken at inactive surface mines have shown that, while emissions can be significant at some sites, the category generally is not a large source.

Emissions from abandoned underground mines come from unsealed mine shafts or from vents installed to prevent the buildup of methane in the mines. Little research has been conducted on this source but EPA/ORD has made measurements at 20 abandoned U.S. mines. Emission rates vary widely from zero to over 21,240 m^3/day of relatively pure methane (Piccot et al., 1995). Total emissions for a mine, as well as the emission rate at any point in time, are functions of diurnal and, probably, seasonal cycles related to changes in temperature and barometric pressure, as well as longer term changes brought on by processes such as mine flooding and eventual depletion of available methane. Perhaps the largest problem in extrapolating these emission measurements to the population will be in separating abandoned mines from those which are temporarily inactive. Mines which are merely inactive but still emitting methane will then have to be added back into the general category of underground mines.

2.2 Factors Affecting Methane Emissions from Coal Mines

Numerous studies have examined the physical factors which control the production and release of methane by coal. These studies have been conducted to evaluate the potential of coalbed methane resources, to enhance the safety of underground mines, and to estimate global methane emissions. Generally, the studies address one of two topics: estimating the methane content of coal, or controlling the concentration of methane in the mine atmosphere and mine ventilation air. Most of these studies have focused on underground mines because these are the mines in which methane creates a safety hazard, and because deeper coals are more productive coalbed methane resources.

Studies of coalbed methane contents have identified pressure, coal rank, and moisture content as important determinants of coalbed methane content. Kim (1977) related gas content to coal temperature and pressure, and in turn to coal depth. After including coal analysis data to represent rank, Kim produced a diagram relating gas content to coal depth and rank. Although the validity of the rank relationship has been questioned, it generally appears to have been accepted by authors (Lambert et al., 1980; Murray, 1980; Ameri et al., 1981; Schwarzer and Byrer, 1983). Independently of Kim's work, Basic and Vukic (1989) established the relationship of methane content with depth in brown coals and lignite.

Several studies have recognized the decrease in methane adsorption on coal as moisture content increases in the lowest moisture regimes (Anderson and Hofer, 1965; Jolly et al., 1968; Joubert et al., 1974). Moisture content appears to reach a critical value above which further increases produce no significant change in methane content. Coals studied by Joubert et al. (1974) showed critical values ranging from 1 to 3 percent moisture.

Early investigations in the U.S. which attempt to identify correlates of methane emissions from coal mine ventilation air include those by Irani et al. (1972) and Kissel et al. (1973). Irani et al. developed a linear relationship between methane emissions and coal seam depth for mines located in five seams. Kissel et al. demonstrated a linear relationship between methane emissions and coalbed methane content for six mines. Although both studies suffer from a paucity of mines and/or seams in their analyses, Kissel et al. made the important observation that mine emissions greatly exceed the amount of methane associated with the mined coal seam alone. Emissions are produced not only by the mined coal, but also by the coal left behind, overlying and underlying seams, and nearby gas deposits. For the six mines studied, emissions per tonne mined exceeded coalbed methane per tonne mined by factors of 6 to 9.

In studies conducted by Boyer et al. (1990) and Kirchgessner et al. (1993a), regression equations were developed relating coal production rate and coalbed methane content to emissions from underground mines. These equations were developed to estimate global emissions from underground mines, using similar data and techniques. To develop these equations, multivariate regression analyses were performed using mine-specific data for the U.S. In the analysis performed by Kirchgessner et al., a database of 269 mine-specific emission measurements was used to produce an equation with an R^2 value of 0.59. This means that about 60 percent of the variation in methane emissions from the mines in the database can be explained by the independent variables included in the equation. In the analyses performed by Boyer et al., a database of about 60 mine-specific observations was used to produce an equation with an R^2 value of 0.35. Although not fully understood, one likely reason for the difference in R^2 values between the two studies is that the equation developed by Kirchgessner et al., was estimated using a database which contained over 4 times more individual mine measurements than the Boyer equation.

2.3. Summary of Country-Specific and Global Emission Estimates

Over the last 30 years there have been numerous attempts to estimate the global emissions of methane from coal mining operations. Those presented and discussed below represent the range of estimates that have been made and include those that are most commonly cited. This section summarizes these estimates, describes and compares the assumptions used in their development, and identifies key relationships that exist among them.

Table 1 presents a summary of global methane emission estimates developed by various researchers for coal mining operations. Estimates range from 7.9 to 64 teragrams (Tg)/yr. The 7.9 Tg/yr estimate is unrealistic. Although the specific assumptions used in developing this estimate are not clear, it appears to be based on the incorrect assumption that emissions from coal mines are equal to the amount of methane trapped in the coal removed from the mine (Bates and Witherspoon, 1952; Hitchcock and Wechsler, 1972). While this gas is released when the coal is fractured, the other release mechanisms described above contribute significantly to total mine emissions.

A review of the global estimates presented in Table 1 reveals that many of them are closely related; that is, the basis of many estimates can be traced back to key assumptions made by some of the earliest researchers. In general, estimates developed by Seiler (1984), Hitchcock and Wechsler (1972), Ehhalt (1974), Ehhalt and Schmidt (1978), Crutzen (1987), and Cicerone and Oremland (1988) were developed based in large part on methane emission factors developed by Bates and Witherspoon (1952) and Koyama (1963, 1964). Estimates by Boyer et al. (1990) and Kirchgessner et al. (1993a) are on a country-specific basis and were not developed based on other researchers' emission factors. Instead, new emissions relationships were developed based on measurement data contained in databases on coal properties and mine emission rates. The estimate of Fung et al. (1991) was developed differently from the other estimates discussed here. Fung et al. used a combination of global methane mass balances and atmospheric modeling techniques to infer a budget for all methane sources including coal mines. Several methane budget scenarios were constructed and tested to determine which was best able to reproduce the meridional gradient and seasonal variations of methane concentrations observed in the atmosphere. One budget scenario for all methane sources, including coal mines, was selected by Fung et al. because it was judged to reproduce the atmospheric record best. The coal mine estimate associated with this budget scenario is 35 Tg/yr. The estimate produced by the Coal Industry Advisory Board (1994), even though it is based on 1990 production and includes deep mining, surface mining, and post-mine handling emissions, yields a relatively low estimate of 24 Tg/yr because of the low emission factors employed.

Table 1. Estimates of global methane emissions from coal mining operations

Source	Year for Estimate	Emissions (Tg/yr)	Comments
Koyama (1963, 1964)	1960	20	Includes hard coal only.
Hitchcock and Wechsler (1972)	1967	7.9 to 27.7	Includes emissions from hard coals (6.3 to 22 Tg/yr) and lignite (1.6 to 5.7 Tg/yr). Based on Bates and Witherspoon, and Koyama.
Ehhalt and Schmidt (1978)	1967	7.9 to 27.7	Estimates taken from Hitchcock and Wechsler.
Seiler (1984)	1975	30	Results based on extrapolation of Koyama's results using updated production. Hard coal only.
Crutzen (1987)	(assume mid- 1980s)	34	Lignite not included. Based on Crutzen, and Ehhalt and Schmidt.
Cicerone and Oremland (1988)	(assume mid- 1980s)	25 to 45 (average 35)	Results based on analyses by Koyama, Ehhalt and Seiler. Hard coal only.
Boyer et al. (1990)	1987	33 to 64 (average 47.4)	Includes all coal types; mining and post-mining.
Fung et al. (1991)	(assume mid- 1980s)	35	Estimate developed by comparing model concentrations derived from published budgets with atmospheric methane measurements.

| Kirchgessner et al. (1993a) | 1989 | 45.6 | Includes all coal types; mining and post-mining. |
| CIAB (1994) | 1990 | 24 | Includes all coal types; mining and post-mining. |

Methodological and other differences among the estimates in Table 1 prevent their direct comparison. Some of these differences also prevent the estimates from being used to track the change in global emissions from coal mining over the years. First, the estimates are developed for different base years, 1960 through 1990, during which time coal production increased significantly. Second, the estimates fail to account for all the coal produced globally. Estimates developed by Koyama (1963, 1964), Seiler (1984), Crutzen (1987), and Cicerone and Oremland (1988) are known to account for emissions associated only with hard coal production and ignore brown coals and lignite. While lower rank coals contain much less methane than hard coal, their emissions on a global basis could be significant. Third, only the estimates of Boyer et al. and Kirchgessner et al. include estimates of post-mining operations. None of the estimates include emissions associated with abandoned or inactive mines.

Table 2. Summary of methane emission rates associated with global estimates

Source	Year	Emission rate (m^3 methane/tonne coal mined)
Bates and Witherspoon (1952)	Assume early 1950s	5.0 to 10.0
Koyama (1963)	1960	21.0
Hitchcock and Wechsler (1972)	1967	5.0 to 17.5
Seiler (1984)	1975	19.5
Crutzen (1987)	Assume mid-1980s	18.0 to 19.0
Boyer et al. (1990)	1987	14.2
Kirchgessner et al. (1993a)	1989	13.8

The fourth difference, and perhaps the most important to understand, is that very different emission factors have been employed by the various researchers. Where they can be determined with some degree of certainty, the emission factors used by the authors discussed above are shown in Table 2. It can be seen that the emission factors range from 5 to 21 m^3 of methane/tonne of coal mined. While it is not always explicit from their writing, it is clear that vastly different assumptions have been developed or accepted by different authors in choosing the factor. The lowest numbers appear to have been based on the assumption that the amount of methane liberated during the mining process is limited to the methane originally contained in the mined coal. As pointed out in Section 2.2, above, there are many more sources of methane in an underground mine than the mined coal alone. The derivation of the largest factors is not clear, and some of the later authors relied on these larger estimates. The factors employed by Boyer et al. and by Kirchgessner et al. are similar, and were independently derived from relationships between coal properties and measured mine emissions.

Table 3. Country-specific estimates of methane emissions from coal mining

Country	Source	Base Year	Total Emissions (Tg/yr)
China	CIAB (1994)[1]	1990	7.56
	Kirchgessner et al. (1993a)[2]	1989	9.30
	Boyer et al. (1990)[3]	1987	16.0
	EPA (1994)[2]	1990	9.5 - 16.6
	EPA (1996)[2]	1992	8.4 - 13.0
	Khalil et al. (1993)[4]	1988	7.0
U.S.S.R. (former)	CIAB (1994)	1990	4.83
	Kirchgessner et al. (1993a)	1989	7.90
	Boyer et al. (1990)	1987	7.70
	EPA (1994)	1990	4.8 - 6.0
	Andronova and Karol (1993)[2]	1988	1.4 - 11.0
U.S.	CIAB (1994)	1990	3.30
	Kirchgessner et al. (1993a)	1989	3.50
	Boyer et al. (1990)	1987	6.10
	EPA (1994)	1990	3.6 - 5.7
Germany	CIAB (1994)	1990	0.97
	Kirchgessner et al. (1993a)	1989	1.10
	Boyer et al. (1990)	1987	1.40
	EPA (1994)	1990	1.0 - 1.2
Poland	CIAB (1994)	1990	1.21
	Kirchgessner (1993a)	1989	3.60
	Boyer et al. (1990)	1987	3.30
	EPA (1994)	1990	0.6 - 1.5
	EPA (1995)[2]	1993	0.41
U.K.	CIAB (1994)	1990	0.76
	Kirchgessner et al. (1993a)	1989	1.30
	Boyer et al. (1990)	1987	1.50
	EPA (1994)	1990	0.6 - 0.9
	Mitchell (1993)[2]	NR[5]	1.7
	Creedy (1993)	1990/91	0.75

[1] Surface mining and post-mine handling included.
[2] Assume underground mining only.
[3] Post-mine handling included.
[4] Surface mining included.
[5] Not reported.

The disparities in the global estimates are also reflected in the individual country estimates produced by the authors mentioned above and others, and presented in Table 3. Considering that the base years for these estimates range between 1987 and 1993 one might have expected closer agreement, but it can be seen that, in the most extreme cases, the estimates for a given country vary by as much as a factor of 4. Again the databases available to the various authors, the emission factors they have chosen to use, the types of mining activities they have chosen to include, and perhaps even personal or institutional bias must be considered as possible explanations.

The types of analyses employed by Boyer et al. and Kirchgessner et al., in which mine- or basin-specific coal and coal mining characteristics are correlated to actual emissions, probably produce the most accurate bases for extrapolation. Without adequate databases for each country, however, the danger always exists that inaccurate extrapolations are being made. It is worth noting that the estimates produced by these authors are fairly consistently higher than the others. Unless there is a consistent bias towards higher emissions built into these analyses, which is not apparent, the higher global emissions shown in Table 1 may be closer to reality.

3. Natural Gas Industry

Natural gas has long been recognized as the environmentally preferred fossil fuel. It produces virtually no sulfur dioxide or particulate emissions, and far fewer nitrogen oxide and carbon monoxide emissions than other fossil fuels.

For this reason, and because it is widely available, relatively easily recovered, and readily usable, the global consumption of natural gas has approximately doubled since 1970. Among the industrialized nations, the U.S. is an exception to this pattern in that it consumes about 10 percent less natural gas today than in 1970. This has been attributed variously to an excessively restrictive regulatory structure (U.S. Department of Energy, 1991) and to a misconception of the future natural gas price structure stemming from an underestimate of available U.S. reserves in the 1970s (Hay et al., 1988). The U.S. National Energy Strategy produced in 1991 has recommended removing or revising excessive regulation inhibiting natural gas transactions (U.S. Department of Energy, 1991), and the Gas Research Institute has stated that domestic natural gas reserves are sufficient for the next several decades (Hay et al., 1988). Both of these factors should accelerate the slow rate of increase in U.S. domestic natural gas utilization which is already occurring.

Natural gas emits about half as much carbon dioxide per unit of energy output as coal, and about two-thirds as much as oil. Recognizing this, the Intergovernmental

Panel on Climate Change has formalized the recommendation to switch to natural gas as fuel where possible to achieve short term mitigation of the global climate change problem (Environmental Agency of Japan, 1990). It must first be demonstrated, however, that methane leakage from the increased production and utilization of natural gas would not nullify the benefit of decreased carbon dioxide production.

3.1 Sources of Methane Emissions in the Natural Gas Industry

The natural gas industry can be broadly divided into production, transmission, storage, and distribution sectors. Each of these sectors can contribute steady or fugitive methane emissions and intermittent emissions. Fugitive emissions result from normal operations and result primarily from leaking components such as valves, flanges, and seals. Intermittent emissions result from routine maintenance procedures, system upsets, and occasional large scale accidents.

Methane emissions from the production sector usually include those from well drilling, gas extraction, and field separation facilities. In this discussion gas processing plants are also included. Emissions from well drilling result primarily from occasional venting and flaring employed to prevent blowouts. During extraction, methane may be emitted by natural-gas-fired engines used for power generation, various wellhead components collectively referred to as the "Christmas tree," and occasional venting and flaring when gas volumes do not warrant recovery. Field separation may involve gas heating, gas or liquid separation, and gas dehydration. Principal sources of emissions are fugitive leaks, venting and flaring, natural-gas-powered devices, and combustion losses from heaters and dehydrators. Gas processing plants are usually located close to the production area and may be regarded as part of the production process. Gas plants are used to separate natural gas liquids from the gas stream and to fractionate the liquids into their components. The processes which are currently most commonly used in these plants are cryogenic expansion, refrigeration, and refrigerated absorption. Primary emissions sources from gas processing plants are fugitive losses, compressor exhaust, and venting and flaring.

Methane emissions associated with the transmission sector are produced by the pipelines, compressor stations, and metering and pressure regulation stations. Leaks from the pipelines are caused by corrosion, material and construction defects, miscellaneous leaks at valves, flanges, and fittings, and earth movement which can cause strains and cracks. Venting can occur at points in the pipeline where residual liquids collect and must be drained. Pneumatic devices powered by natural gas are found throughout the transmission sector and are typically vented to the atmosphere. Maintenance procedures such as pipe scraping result in emissions during launching and retrieving of the scraper. Dehydrators must receive periodic blowdowns and purges which are vented, and pipelines must occasionally be purged during installation, abandonment, replacement, repair, and emergency shutdown. Compressor stations produce fugitive emissions from the usual sources (e.g., flanges, seals), occasional unflared venting from system overpressure, and gas turbine start-up and operating emissions.

The primary sources of emissions from the distribution system, which delivers natural gas to the end users, are pipeline leaks. These leaks result, in varying degrees, from all of the same causes as leaks in transmission pipelines. Gas is intentionally vented after isolating segments of lines for repair, and is used to purge air from the pipeline after repair. Blow and purge operations on meters and regulators are typically vented to the atmosphere.

Injection facilities can be located at various points in the system, depending upon the facilities' function. Gas is frequently reinjected at the production site to maintain oil or gas reservoir pressure. Gas is also injected into underground reservoirs for storage. Normal operations at these facilities produce the usual fugitive emissions, releases during routine maintenance, and venting for overpressure protection of compressors, scrubber vessels, and wellhead injection stations.

The final category of emission sources (not discussed under the three-part industry breakdown above) is liquefied natural gas (LNG) facilities. Functions performed at an LNG facility include receiving, storage, and regasification. Equipment consists of unloading piping, pumps, insulated storage tanks for LNG, and heaters and compressors for regasification. During normal operation, fugitive releases occur but, because of the nature of these facilities, maintenance can be scheduled well in advance and the necessary controlled venting can be directed to the flare system. Pressure relief system releases are typically flared as well.

3.2 Summary of Country-Specific and Global Emission Estimates

Numerous estimates of methane emissions from the global gas industry are available dating back almost 30 years. As with the coal industry estimates reported above, they were originally produced in the interest of determining global balances of atmospheric trace gases, but more recently for assessing global climate change issues. Most of the global estimates shown in Table 4 have been drawn from the same sources as the coal industry estimates in Table 1.

Typically the estimates assume leakage rates of 1 to 4 percent and various base years. Ehhalt and Schmidt (1978) accepted Hitchcock and Wechsler's (1972) earlier and notably low estimate of 7 to 21 Tg/yr. An early base year provides part of the explanation but it is not clear how the assumed loss rates of 1 to 3 percent were applied. If a 1985 base year were used, these estimates would align well with the later ones. Estimates at the higher end of the typical range by Sheppard et al. (1982), Blake (1984), and Cicerone and Oremland (1988) are derived by adding assumed values for vented gas to the calculated values for gas leakage. Keeling (1973) arrived at a particularly high estimate of 40 to 70 Tg/yr by assuming a very high, and probably indefensible, global loss rate of 6 to 10 percent. While loss rates of this magnitude have been conjectured for the former U.S.S.R. and some of the eastern European countries, they have not been verified and almost certainly do not pertain to most countries.

Table 4. Estimates of global emissions from the natural gas industry.

Source	Base Year	Emissions (Tg/yr)	Assumed Loss Rates (%)
Hitchcock and Wechsler (1972)	1968	7-21	1-3
Keeling (1973)	1968	40-70	6-10
Ehhalt and Schmidt (1978)	1968	7-21	1-3
Sheppard et al. (1982)	1975	50	2 (leakage) + 25% for vented and flared
Blake (1984)	1975	50-60	2-3 (leakage) + 30 Tg for vented
Seiler (1984)	1975	19-29	2-3
Bolle et al. (1986)	NR[1]	35	3-4
Crutzen (1987)	NR	33	4
Cicerone and Oremland (1988)	Early 1980s	25-50	2.5 (leakage) + 14 Tg for vented and flared
Fung et al. (1991)	1986	40	NR

[1.] Not reported.

Sheppard et al. (1982) and subsequently Blake (1984) estimate emissions from venting and flaring at wellheads to be about 30 Tg/yr. Cicerone and Oremland (1988) provide a later, independent estimate of 14 Tg/yr. It can be inferred from their discussions that these estimates are for both gas and oil fields, and it is assumed that oil fields produce the majority of emissions. The estimates are not separated by industry, however, and such matters as flaring efficiencies and venting versus flaring practices by individual countries are not discussed. There are currently no known reliable data on global venting and flaring emissions from oil and gas fields.

While it is not always explicitly stated in the literature, the assumed leakage rates fall under the heading of what the gas industry refers to as "unaccounted for" gas (UAG). UAG is the difference between the volume of gas that a utility reports as purchased versus the volume sold, less any company use or interchange. It is a statistical figure attributable to diverse components including meter inaccuracies, gas theft, variations in temperature and pressure, billing cycle differences, and gas leakage or other actual losses. In the U.S. roughly 2 percent of gas marketed annually is classified as UAG (American Gas Association, 1986), while a recent study by EPA/ORD (Kirchgessner et al., 1997) has concluded that the actual rate of emissions was only about 1.4 percent in 1992. Used as a surrogate for actual gas losses, UAG should consistently result in an overestimate of actual emissions to the atmosphere and, for this reason, should not be the basis for emission estimates.

In recent years, coincident with the increased international interest in greenhouse gas emissions, a number of country estimates for losses from the gas industry have appeared in the literature and are summarized in Table 5. Without taking each entry in Table 5 point by point, it can be said generally that the authors have most commonly taken emission factors developed by others and applied them to the country of interest. Activity factors, such as production, consumption, or the number of pieces of specific equipment, have been taken from publicly available documents or, in some cases, proprietary industry information which cannot be readily verified. For this reason disparate estimates like those for the U.K. cannot be easily reconciled.

Table 5. Country-specific estimates of methane emissions from the gas industry

Source	Country	Base Year	Emissions (Tg/yr)	
Kirchgessner et al. (1997)	U.S.	1992	Production:	1.6
			Processing:	0.7
			Transmission:	2.2
			Distribution:	1.5
			Total:	6.0
Andronova and Karol (1993)	Former U.S.S.R.	1988	Maximum:	11.0
Mitchell (1993)	U.K.	NR[1]	Total:	1.9 - 10.8 median preferred
British Gas (1993)	U.K.	1991	Transmission:	0.02
			Distribution:	0.37
			Usage:	0.01
			Total:	0.4
Khalil et al. (1993)	China	1988	Total:	0.3
Mussig (1992)	Germany	1989	Total:	0.2
NGGIC (1994)	Australia	1988/89	Distribution:	0.13
Picard et al. (1992)	Canada	1989	Production, processing, and transmission in Alberta:	0.41
Guidotti and Castagnola (1994)	Italy	1990/92	Total:	0.14 - 0.24

Jensen (1993)	Denmark	1989	Production:	0.001
			Transmission:	0.0004
			Distribution:	0.016
			Total:	0.018[2]
	Norway	1989	Production:	0.009
	Sweden	1989	Transmission:	8 tonnes
			Distribution:	0.011
			Total:	0.012
	Finland	1989	Transmission:	0.003

[1] Not reported
[2] Totals may not be equal to sum of segments due to rounding.

A few of the estimates are extremely well documented and based on large data bases. Kirchgessner et al. (1997) conducted a 5-year study of the U.S. natural gas industry that was co-sponsored by EPA/ORD and the Gas Research Institute. An extensive sampling program was conducted at gas industry facilities to substantially expand the existing emissions data base, and total U.S. emissions were estimated using a statistically defensible methodology. Researchers from the Nordic Methane project participated in the U.S. program, and their methods, to the extent possible, mirror those applied in the U.S. The estimate developed by British Gas for the U.K. also appears to be supported by a substantial industry data base and a statistically robust methodology.

It seems clear that the data readily available even now do not allow a world estimate of methane emissions from the natural gas industry to be produced with much accuracy. Nearly half of the world's 1994 gas production occurred in non-OECD (Organization for Economic Co-operation and Development) countries (International Energy Agency, 1006), and information on emissions in these countries appears to be grossly under-represented in the literature. It is unlikely that this situation will change significantly until the number of high quality country studies available increases greatly.

References

Ameri, S., F. T. Al-Sandoon, and C. W. Byrer, Coalbed methane resource estimate of the Piceance Basin, *Report No. DOE/METC/TPR/82-6*, 44 pp., U.S. Department of Energy, Morgantown, West Virginia, 1981.

American Gas Association, Lost and unaccounted for gas, *Issue Brief 1986-28*, 4 pp., Arlington, Virginia, 1986.

Anderson, R.B., and L.J.E. Hofer, Activation energy of diffusion of gases into porous solids, *Fuel*, 44, 303-306, 1965.

Andronova, N.G., and I.L. Karol, The contribution of USSR sources to global methane emission, *Chemosphere*, 26, 111-126, 1993.

Banerjee, B.D., A.K. Singh, J. Kispotta, and B. B. Dhar, Trend of methane emission to the atmosphere from Indian coal mining, *Atmospheric Environment*, 28, 1351-1352, 1994.

Basic, A., and M. Vukic, Dependence of methane contents in brown coal and lignite seams on depth of occurrence and natural conditions, in *Proceedings of the 23rd International Conference of Safety in Mines Research Institutes*, pp. 282-288, U.S. Department of the Interior, Bureau of Mines, Washington, D.C.,1989.

Bates, D. R., and A. E. Witherspoon, The photo-chemistry of some minor constituents of the Earth's atmosphere (CO_2, CO, CH_4, N_2O), *Roy. Astronom. Soc. Monthly Not., 112,* 101-124, 1952.

Blake, D.R., Increasing concentrations of atmospheric methane, 1979-1983, Ph.D. thesis, University of California, Irvine, 1984.

Bolle, H.J., W. Seiler, and B. Bolin, Other greenhouse gases and aerosols, in *The Greenhouse Effect, Climate Change and Ecosystems,* edited by B. Bolin, pp. 157-203, John Wiley and Sons, New York, 1986.

Boyer, C.M., J.R. Kelafant, V.A. Kuuskraa, K.C. Manger, and D. Kruger, Methane emissions from coal mining: issues and opportunities for reduction, *Report No. EPA-400/9-90/008*, 143 pp., U.S. Environmental Protection Agency, Office of Air and Radiation, Washington, D.C., 1990.

British Gas, Methodology for determining emissions of methane from the onshore U.K. natural gas industry, Submitted to Watt Committee Working Party on Methane Emissions by British Gas, 31 pp., British Gas pic, London, 1993.

Cicerone, R.J., and R. Oremland, Biogeochemical aspects of atmospheric methane, *Global Biogeochem. Cycles,* 2, 299-327, 1988.

Coal Industry Advisory Board, *Global Methane and the Coal Industry,* 67 pp., Organization for Economic Co-operation and Development Publications, Paris, France, 1994.

Creedy, D.P., The origin and distribution of firedamp in some British coalfields, Ph.D. thesis, University College, Cardiff, Wales, 1985.

Creedy, D.P., Methane emissions from coal-related sources in Britain: development of a methodology, *Chemosphere,* 26, 419-440, 1993.

Crutzen, P.J., Role of the tropics in atmospheric chemistry, in *Geophysiology of Amazonia*, edited by R. Dickinson. pp. 107-130, John Wiley and Sons, New York, 1987.

Ehhalt, D.H., The atmospheric cycle of methane, *Tellus,* 26, 58-70, 1974.

Ehhalt, D.H., and U. Schmidt, Sources and sinks of atmospheric methane, *Pageoph*, 116, 452-463, 1978.

Environmental Agency of Japan, *International workshop on methane emissions from natural gas systems, coal mining, and waste management systems,* 709 pp., U.S. Agency for International Development and U.S. Environmental Protection Agency, Washington, D.C., 1990.

Fung, I., J. John, J. Lerner, E. Matthews, M. Prather, L.P. Steele, and P.J. Fraser, Three-dimensional model synthesis of the global methane cycle, *J. Geophys. Res.,* 96, 13033-13065, 1991.

Guidotti, G.R., and A.M. Castagnola, Methane sources and emissions in Italy, in *Proceedings of the Air and Waste Management Association International Specialty Conference: Global Climate Change-- Science, Policy, and Mitigation Strategies,* pp. 259-263, Phoenix, April 5-8,1994.

Hay, N.E., P.L. Wilkinson, and W.M. James, *Global climate change and emerging energy technologies for electric utilities: the role of natural gas,* 30 pp., American Gas Association, Arlington, Virginia, 1988.

Hitchcock, D.R., and A.E. Wechsler, Biological cycling of atmospheric trace gases, *Report No. NASW-2128*, 415 pp., National Aeronautic and Space Administration, Washington, D.C., 1972.

International Energy Agency, *Energy Statistics and Balances of Non-OECD Countries 1993-1994,* 574 pp., Organization of Economic Co-operation and Development, Paris, 1996.

Irani, M.C., E.D. Thomas, and T.G. Bobick, Methane emissions from U.S. coal mines, a survey, *Report No. IC 8558,* 18 pp., U.S. Department of Interior, Bureau of Mines, Pittsburgh, Pennsylvania, 1972.

Jensen, J., *Nordic Methane Project,* 98 pp., Nordic Gas Technology Center, Hørsholm, Denmark, 1993.

Jolly, D.C., L.H. Morris, and F.B. Hinsely, An investigation into the relationship between the methane sorption capacity of coal and gas pressure, *The Mining Engineer,* 127, 539-548, 1968.

Joubert, J.I., C.T. Grein, and B.Bienstock, Effect of moisture on the methane capacity of American coals, *Fuel,* 53, 186-191, 1974.

Keeling, C.D., Industrial production of carbon dioxide from fossil fuels and limestone, *Tellus,* 25, 174-198, 1973.

Khalil, M.A.K., M.J. Shearer, and R.A. Rasmussen, Methane sources in China: historical and current emissions, *Chemosphere,* 26, 127-142, 1993.

Kim, A.G., Estimating methane content of bituminous coalbeds from adsorption data, *Report No. RI 8245,* 24 pp., U.S. Department of the Interior, Bureau of Mines, Pittsburgh, Pennsylvania, 1977.

Kirchgessner, D.A., S.D. Piccot, and J.D. Winkler, Estimate of global methane emissions from coal mines, *Chemosphere,* 26, 453-472, 1993a.

Kirchgessner, D.A., S.D. Piccot, and A. Chadha, Estimation of methane emissions from a surface coal mine using open-path FTIR spectroscopy and modeling techniques, *Chemosphere,* 26, 23-44, 1993b.

Kirchgessner, D.A., R.A. Lott, R.M. Cowgill, M.R. Harrison, and T.M. Shires, Estimate of methane emissions from the U.S. natural gas industry, *Chemosphere,* 35, 1365-1390, 1997.

Kissel, F.N., C.M. McCulloch, and C.H. Elder, The direct method of determining methane content of coalbeds for ventilation design, *Report No. RI 7767,* 19 pp., U.S. Department of Interior, Bureau of Mines, Pittsburgh, Pennsylvania, 1973.

Koyama, T., Gaseous metabolism in lake sediments and paddy soils and the production of atmospheric methane and hydrogen, *J. Geophys. Res.,* 68, 3,971-3,973, 1963.

Koyama, T., Biogeochemical studies on lake sediments and paddy soils and the production of hydrogen and methane, in *Recent Researches in the Fields of Hydrosphere, Atmosphere, and Geochemistry,* edited by Y. Miyake and T. Koyama, pp. 143-177, Murucen, Tokyo, 1964.

Lambert, S.W., M.A. Trevits, and P.F. Steidl, Vertical borehole design and completion practices to remove methane gas from minable coalbeds, *Report No. DOE/CMTC/TR-80/2,* 163 pp., U.S. Department of Energy, Washington, D.C., 1980.

Mitchell, C., Methane emissions from the coal and natural gas industries in the UK, *Chemosphere,* 26, 441-446, 1993.

Murray, D.D. *Methane from Coalbeds-- a Significant Undeveloped Source of Natural Gas,* 37 pp., Colorado School of Mines Research Institute, Golden, Colorado, 1980.

Müssig, S., Possibilities for reduction of emissions-- in particular the greenhouse gases CO_2 and CH_4-- in the oil and gas industry, in *Proceedings of the European Petroleum Conference,* pp. 237-246, Cannes, November 16-18, 1992, (Published by the Society of Petroleum Engineers, *Paper No. SPE 25041), 1992.*

National Greenhouse Gas Inventory Committee, *Australian Methodology for the Estimation of Greenhouse Gas Emissions and Sinks: Workbook for Fugitive Fuel Emissions (Fuel Production, Transmission, Storage and Distribution) Workbook 2,* 50 pp., Department of Environment, Sport and Territories, Canberra, Australia, 1994.

Picard, D.J., B.D. Ross, and D.W.H. Koon, *A Detailed Inventory of CH_4 and VOC Emissions* *from the Upstream Oil and Gas Operations in Alberta: Volume 1-- Overview of Emissions,* 65 pp., Canadian Petroleum Association, Calgary, Alberta, 1992.

Piccot, S.D., A. Chadha, J. DeWaters, T. Lynch, P. Marsosudiro, W. Tax, S. Walata, and J.D. Winkler, Evaluation of significant anthropogenic sources of radiatively important trace gases, *Report No. EPA-600/8-90-079; NTIS PB91-127753*, 198 pp., U.S. Environmental Protection Agency, Office of Research and Development, Research Triangle Park, North Carolina, November 1990.

Piccot, S.D., S.S. Masemore, E.S. Ringler, and D.A. Kirchgessner, Developing improved methane emission estimates for coal mining operations, in *Proceedings: The 1995 Symposium on Greenhouse Gas Emissions and Mitigation Research, Report No. EPA/600/R-96/072; NTIS PB96-187752*, pp. 2/53-64, Washington, D.C., June 27-29, 1995.

Schwarzer, R.R., and C.W. Byrer, Variation in the quantity of methane adsorbed by selected coals as a function of coal petrology and coal chemistry, *Report No. DE-AC21-80MC14219, V.1,* 102 pp., U.S. Department of Energy, Morgantown, West Virginia, 1983.

Seiler, W., Contribution of biological processes to the global budget of CH_4 in the atmosphere, in *Current Perspectives in Microbial Ecology*, edited by M.J. Klug and C.A. Reddy, pp. 468-477, American Society for Microbiology, Washington, D.C., 1984.

Sheppard, J.C., H. Westberg, J.F. Hopper, and K. Ganesan, Inventory of global methane sources and their production rates, *J. Geophys. Res., 87, 1305-1312, 1982.*

U.S. Department of Energy, National energy strategy, *Report No. DOE/S0082P, 217 pp.,* Washington, D.C., 1991.

U.S. Environmental Protection Agency, Assessment of the potential for economic development and utilization of coalbed methane in Czechoslovakia, *Report No. EPA/430/R-92/1008,* 91 pp., Office of Air and Radiation, Washington, D.C., 1992.

U.S. Environmental Protection Agency, International anthropogenic methane emissions: estimates for 1990: report to congress, edited by M.J. Adler, *Report No. EPA/230-R-93-010,* 310 pp., Office of Policy Planning and Evaluation, Washington, D.C., January 1994.

U.S. Environmental Protection Agency, Reducing methane emissions from coal mines in Poland: A handbook for expanding coalbed methane recovery and utilization in the Upper Silesian Coal Basin, *Report No. EPA/430-R-95-003,* 141 pp., U.S Office of Air and Radiation, Washington, D.C., April 1995.

U.S. Environmental Protection Agency, Reducing methane emissions from coal mines in China: the potential for coalbed methane development, *Report No. EPA/430-R-96-005,* 173 pp., Office of Air and Radiation, Washington, D.C., 1996.

15. Geological Sources of Methane

A. G. Judd
Centre for Marine & Atmospheric Science
School of the Environment, University of Sunderland
Sunderland SR2 7BW, UK

1. Introduction

Methane is one of many gases produced by geological processes. Others include: CO_2, H_2, H_2S, N_2, SO_2, H_2O (as steam or water vapour), and the petroleum gases (ethane, propane, butane and pentane). This chapter considers geological sources of methane, and natural pathways which enable it to enter the atmosphere. For the purposes of this chapter 'geological' sources are taken to include sediments of all ages: ancient and modern, including those being actively deposited at the present day. Reference is also made to releases associated with geological resource extraction, however the most significant of these (coal and petroleum) are discussed in specific chapters elsewhere in this volume.

Methane may be formed in various geological environments (see Table 1) which can be divided into three categories: biogenic / bacterial, thermogenic and abiogenic.

Biogenic or bacterial methane: is derived by the degradation of organic matter by bacteria. This occurs in anoxic conditions at shallow depths (mainly in the top few metres or tens of metres).

Thermogenic methane: is derived by the breakdown of complex organic molecules at the elevated temperatures and pressures found at greater depths (generally 1 to 4km., but this is dependent upon geothermal gradients). This category also includes methane derived by the breakdown of crude oils.

Abiogenic methane: is derived by inorganic processes deep within the Earth's crust or the underlying mantle.

The processes of methanogenesis are described elsewhere in this volume and by Floodgate and Judd (1992).

Methane from these sources may be distinguished by carbon isotope ratios (see Whiticar, this volume). In most cases the processes involved have been in continuous operation for extended periods of geological time; in the cases of near-surface processes, conditions have remained essentially unchanged since the end of the last (Weichselian or Wisconsinian) glacial maximum. Deeper processes may have been on-going continuously for considerably longer.

Mohammad Aslam Khan Khalil   (Ed.)
Atmospheric Methane
© *Springer-Verlag Berlin Heidelberg 2000*

Table 1. Geological origins of methane

Source type	Category	Escape pathways	Time from source to escape	^{13}C‰
microbial degradation of organic matter in sediments - current activity	biogenic	diffusion, bioturbation, seepage	days → years	-60 to -70‰
microbial degradation of organic matter in sediments - *past activity*	biogenic*	seepage, diffusion, mud volcanoes	years → tens of millions of years	-60 to -70‰
thermal degradation of organic matter in sediments / sedimentary rocks	thermogenic*	seepage, diffusion, mud volcanoes	millions → hundreds of millions of years	-20 to -52‰
abiogenic sources	abiogenic*	volcanic, geothermal, hydrothermal activity, etc.	→ billions of years	-5 to -45‰

* 'fossil' methane
Sources: MacDonald, 1993, Schjoell, 1988, Kadko et al., 1995; Whiticar (this volume).

2. Sources

2.1 Organic-Rich Sediments - Bacterial Methane

At the present day bacterial or 'biogenic' methane is derived by the degradation of organic matter under anoxic conditions at shallow depths within sediments in shallow sediments in lakes, estuaries and coastal embayments, fjords and basins (enclosed or with narrow communications to the sea), and marine areas. Activity continues, albeit at reduced rates, away from land, in deeper water and with depth in the sediment on the continental shelf, in the deeper waters of the continental slope and at the base of the slope. Accumulations of 'fossil' bacterial methane, generated by the same processes in similar conditions and environments during the geological past, have been trapped within sediments and sedimentary rocks. These are both widespread and large (Rice and Claypool, 1981).

Conditions for methanogenesis are to be found mainly in fine-grained sediments (clays and muds) in which organic carbon is concentrated, and from which the expulsion of evolved gas is impeded by low permeability. In certain situations, such as within rapidly-accumulating sediments, considerable accumulations of methane may occur. Coarse-grained sediments (sands and gravels), through which fluids may flow relatively easily, rarely contain sufficient organic carbon for methanogenesis, and any methane formed is likely to be rapidly oxidised if porewaters are oxygenated. However, methane which has migrated from fine-grained sediments may accumulate in coarse-grained sediments that are 'capped' by impermeable, finer-grained sediments. The emission of methane from these accumulations may occur if critical pore fluid pressures are exceeded as sediment accumulates, or if conditions are disturbed for example during earthquake loading or by anthropogenic activities (civil engineering construction, drilling etc.).

Methane generation and accumulation do occur on land, however more data are available from offshore where seismic reflection surveys can be used to identify gas accumulations (Judd and Hovland, 1992). Examples have been reported from numerous sedimentary environments; the examples cited in Table 2 are representative of the numerous reports.

Table 2. Gas (methane) in recent sediments: example locations (many other examples are reported in the literature; Hovland and Judd (1988) provides a review).

Estuaries	Various around the UK	Taylor, 1992
Coastal lagoons	Cape Lookout Bight, Nth Carolina, USA	Martens & Klump, 1980
Fjords, estuaries & coastal indentations	Saanich Inlet, BC, Canada Chesapeake Bay, USA Ria de Muros, Spain	McCartney & Barry, 1965 Hill et al., 1992 Acosta, 1984
Harbours	Suva, Fiji	Hochstein, 1970
Deltas	Mississippi, USA Yangtze, China	Prior & Coleman, 1982 Butenko et al., 1985
Continental shelf	Norton Sound, Bering Sea Kattegat, Denmark	Nelson et al., 1979 Jørgensen et al., 1990
Continental slope	Gulf of Cadiz, offshore Spain	Baraza and Ercilla, 1996
Accretionary wedges subduction zones	Makran accretionary prism, offshore Pakistan	von Rad et al., 1996

2.2 Hydrocarbon-Bearing Sediments

'Thermogenic' methane is generated principally from organic matter buried to great depths in sedimentary rocks. Under the combined influences of heat and pressure, residual organically-derived molecules (the residue remaining after bacterial activity has been terminated) are 'cracked' to form the petroleum hydrocarbons of which methane is the simplest. Methane is also produced from woody material during the process of coal formation, and during the thermal alteration of oil. (MacDonald, 1983; Schoell 1988). Once generated, methane is removed from the source rocks by the processes of primary migration. Then, during secondary migration, it will rise by the bulk flow of bubbles driven by buoyancy, by diffusion, or in solution in pore waters. Ultimately secondary migration will bring the methane to the surface (or seabed), however significant proportions become trapped in petroleum reservoirs *en route*. In this case further upward progress (tertiary migration) occurs either by the leakage of the seal at the top of the reservoir, or by spillage when the reservoir becomes over full.

The processes of thermogenic generation and migration have been in progress over a time-scale measured in tens of millions of years. Hunt (1979) estimated that there is 1.4×10^{18} to 2.8×10^{19} g of carbon present as dispersed gas in sedimentary basins, the distribution of which is illustrated in Figure 1. Migration may be continuous or periodic, individual migration events being triggered when critical conditions are obtained, for example, during earthquake activity. Surface (or seabed) fluxes of petroleum hydrocarbons (of which methane has the smallest molecular size, is most soluble in water and is therefore the most mobile) occur in all areas underlain by petroliferous sediments, as is discussed below (section 3.1).

2.3 Volcanic and Related Emissions

Various gases are known to emanate from centres of igneous activity (see Figure 2) and, in many circumstances, these gases include methane. Because of the infrequency of volcanic events it is unrealistic to imply a steady input of methane to the atmosphere. According to Lacroix (1993), estimates of contributions made by individual eruptions vary from 0.003 Tg CH_4 (Mount Etna in 1974) to 1.94 Tg CH_4 (Santorini in 1470 BC). An average of 0.13 Tg CH_4.year^{-1} (conservative, 0.02; liberal 2.33) for the years 1800 to 1969 was presented by Leavitt (1982), input whilst Lacroix (1993) estimated the annual as 3.5 ± 2.7 Tg CH_4. However, these estimates considered only volcanic events on land.

Submarine volcanoes, like their counterparts on land, probably release substantial quantities of gas, including methane, during eruptions, and steadily release small quantities of gas for extended periods after (or between) eruptions. For example, Upstill-Goddard et al. (1996) reported that the waters at Deception Island, an active volcano, are strongly supersaturated with methane (~460% at the sea surface, ~600-800% at 100-150 m). The number of submarine volcanoes (and seamounts) is very large: in the South Pacific Ocean there are an estimated 8 per 10,000 km^2 (Hekinian, 1984), so there may be about 130,000 in the Pacific Ocean alone.

2.4 Geothermal and Hydrothermal Sources

Hydrothermal emanations occur wherever water from the surface or porewaters are heated or superheated in areas with a high geothermal gradient. They commonly contain a variety of low molecular weight hydrocarbons, including methane (Des Marais et al., 1988). Hydrothermal activity occurs both at active plate margins (e.g., the ocean spreading centres) and other areas of volcanic or igneous activity on land (e.g., Italy, Yellowstone Park, Wyoming, U.S.A.) and offshore (e.g., the White Island geothermal field, New Zealand; Guaymas Basin, Gulf of California). Hydrothermal gases include He, H_2, H_2S and CO_2 but methane is a common component (see Table 3).

Fig. 1. Major sedimentary basins of the world

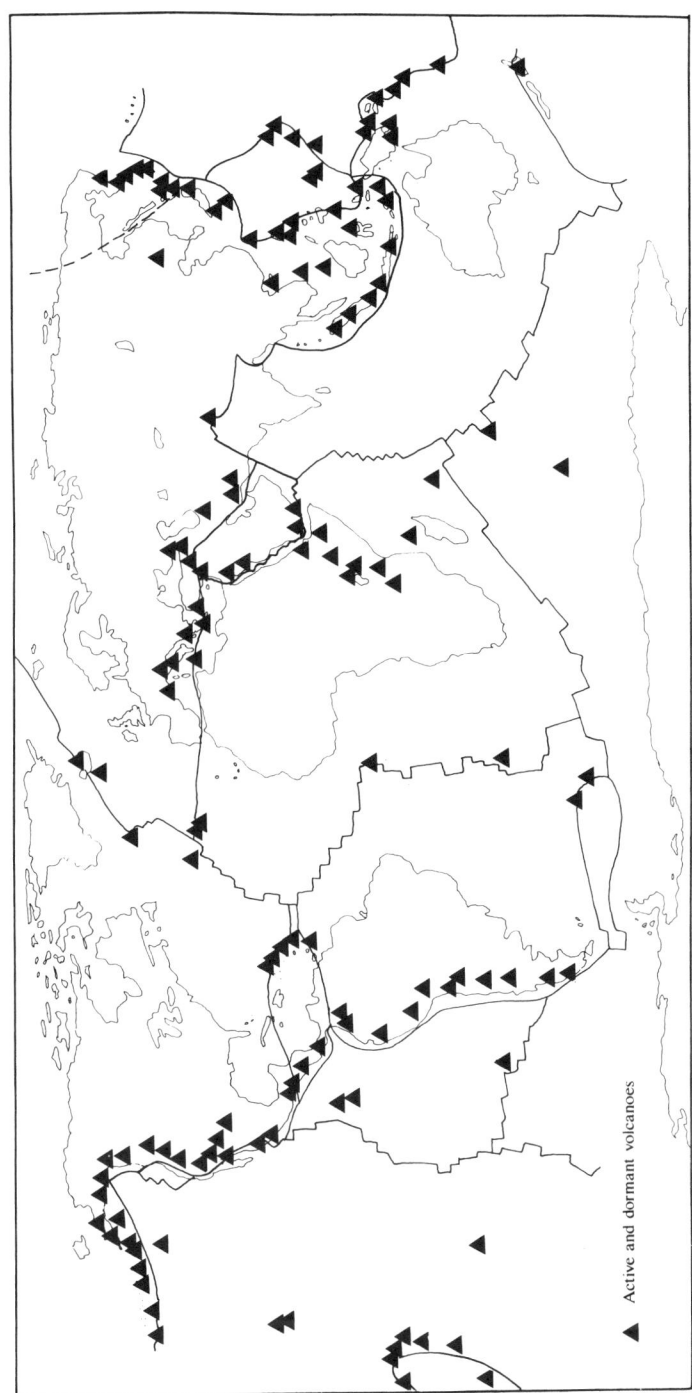

Fig. 2. Active and dormant volcanoes

Table 3. Methane and other gases from geothermal / hydrothermal sources

Location	Bay of Plenty, New Zealand	Cerro Prieto, California Norte, Mexico	Offshore Milos, Greece
CH_4	Present	3.7 - 4.8%	0.1 - 9.7%
CO_2	Dominant		54.9 - 91.9%
Other gases	N_2, NH_3, H_2S		H_2O 0 - 3.0% H_2S 0 - 8.1%
Source reference	Glasby, 1971	Des Marais et al., 1988	Dando et al., 1995

According to Welhan (1988), the principal sources of methane in hydrothermal fluids are:

(1)   thermogenic;
(2)   biogenic;
(3)   outgassing from the mantle (see below); and
(4)   inorganic synthesis

Hydrothermal activity is common along the ocean spreading centres. These extend over a distance of some 55,000 km. Methane has been found to be a volatile component of hydrothermal fluids escaping into the marine environment (Lilley et al., 1982; Vidal et al., 1982: Welhan and Craig, 1983; Horibe et al., 1986; Jean-Baptise et al., 1990). Vent fluids have concentrations of methane 10^5 to 10^7 times higher than the ocean bottom waters into which they are discharged (de Angelis et al., 1993). Plumes of fluid rise, and extend laterally from the vent fields; 'steady state' emission plumes rise hundreds of metres above the seabed, whereas 'event plumes' may rise as much as a kilometre. Within these plumes anomalous concentrations of methane have been detected; de Angelis et al. (1993) reported high methane levels >15 km. from the source. Welhan and Craig (1983) estimated the total annual emissions of methane from the spreading centres to be 1.6 x 10^8 m^3 (about 0.1 Tg CH4.year-1); however, this may be a conservative estimate (Malahoff, 1985). Kadko et al. (1995) considered the methane to be of magmatic origin because the ratios of methane to other gases (^{3}He, H_2 and CO_2) are similar to those of MORB (Mid-Ocean Ridge Basalt) glass.

Lacroix (1993) estimated that hydrothermal events on land provided 2.3 ± 1.4 Tg CH_4 yr^{-1}; there is no reliable global estimate for the contributions of all (land and marine) hydrothermal and geothermal sources.

2.5 Crystalline Basement Sources

Sherwood et al. (1988) reported that methane is *"ubiquitous and discharging freely"* from boreholes drilled for metalliferous ore exploration on the Canadian Shield where it represents up to 80% of the free gas phase. They also referred to similar occurrences from the Baltic Shield and the Kola Peninsula, U.S.S.R. Söderberg and Flodén (1990) described submarine gas (mainly methane) seeps associated with tectonic lineaments in crystalline rocks in the Stockholm Archipelago.

The origin of these gases is unclear; they may have been derived from organic precursors, by migration in gaseous form, or from migrating methane-bearing connate waters (Parnell, 1988; Sherwood et al., 1988; Parnell and Swainbank, 1990). It is probable that flux rates are not high; however, rocks of crystalline basement cover extensive areas of the Earth's surface. No estimate of the global emission of methane from this source is offered. This source is discussed in greater detail by Lacroix (1993).

2.6 'Deep Earth' Gas

Gold and Soter (1980, 1982) considered that large volumes of methane, remnants of the primitive volatile constituents of the Earth, may be emanating from the mantle, particularly along deep-seated faults. Gas derived from a deep crustal or mantle source has been reported by several authors (for example, Lawrence and Taviani, 1988; see also Lacroix, 1993), although alternative derivations may be offered. For example, gas that seeps from the ophiolites in the Philippines may be of mantle origin, although Abrajano et al. (1988) concluded that it may equally be derived by the reduction of water and carbon during the low temperature serpentinization of the ophiolites. Seepage gases from this site contain up to 55% methane.

Some authors (Cicerone and Oremland, 1988, for example) vigorously discount the validity of the abiogenic genesis of "primordial" methane by the outgassing of the mantle. There has, as yet, been no definite evidence to irrefutably support the claims of Gold and his associates, nor has their claim been totally disproved. However there is extensive evidence (reviewed by Lacroix, 1993) of methane derived from deep crustal and mantle sources; this includes some of the methane sources described in the previous three sections.

3. Release Pathways

3.1 Natural Gas Seeps

Natural seepages of oil and gas are known to be widespread in both land and marine environments (Link, 1952; Landes, 1973; Wilson et al., 1974; Kvenvolden and Harbaugh, 1983; Hovland and Judd, 1988; Clarke and Cleverly, 1991), see Figure 3. However, much of the data relating to natural seeps are proprietary data held by the oil industry. Indeed petroleum seeps gave the first indications of the presence of petroleum in most of the world's petroleum provinces (Link, 1952) and are still used in petroleum exploration (Philip, 1987).

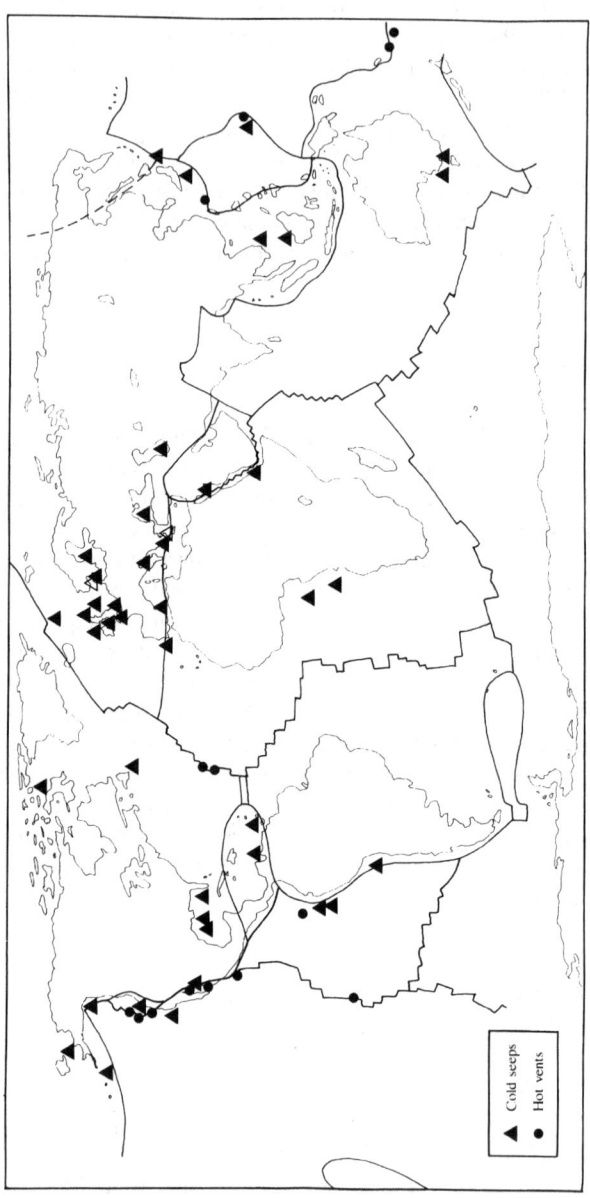

Fig. 3. Natural gas seepages: cold seeps and hot vents.

Seeps may occur in any environment in which methane is generated. Seeps are most commonly associated with organic-rich sediments in which thermogenic or bacterial methane is generated. However, Clayton et al. (1993) reported the natural emission of methane from coal-bearing rocks in Alabama, USA; Selley (1992) claimed that the reports of petroleum (presumably including methane) from the Upper Carboniferous Coal Measures of Great Britain were so numerous that it *"would have been the study of a lifetime"* to document them. Oil seeps attract more attention than gas seeps as they are sources of pollution. On land oil seeps are easier to observe than gas seeps, however it is possible that some gas seepage occurs wherever oil is seeping (Clarke and Cleverly, 1991).

Gas seepages on land have been reported (for example, by Simoneit et al., 1979), and it is probable that atmospheric concentrations of methane are higher over areas characterised by methane-bearing rocks. Data presented by Stadnik et al. (1986) show that the measured atmospheric methane levels in eleven regions of the former U.S.S.R. were consistently higher in petroliferous areas (1.97 to 6.6 ppm; mean 3.47 ppm) compared to the regional background (1.15 to 2.9 ppm; mean 1.85). During hydrocarbon prospecting programmes seasonal surveys of the methane concentration of snow accumulations in the former U.S.S.R. have shown that thermogenic methane passes through the underlying permafrost, particularly in fissured and faulted areas (Glotov et al., 1985; Bordkov et al., 1988; and Vyshemirskiy et al., 1989). Andronova and Karol (1993) estimated this flux at 0.2 to 0.9 Tg CH_4 yr^{-1} from the permafrost regions of the former U.S.S.R.

A significant proportion of the oceans' continental shelves consist of sedimentary basins. Estimates of the areal extent of marine petroliferous sedimentary basins vary, for example, 27 x 10^6 km^2 (Earney, 1980) and 35 x 10^6 km^2 (Trotsyuk and Avilov, 1988). The presence of shallow gas, mainly methane (either bacterial or thermogenic), is not uncommon in such areas. Similarly, there is ample evidence of the escape of gas through the seabed (Hovland and Judd, 1988). This takes the form of direct evidence (visual and acoustic) of gas bubbles emissions from the seabed. Indirect evidence includes topographic features formed by gas expulsion (pockmarks) or the mobilisation of gassy sediments (mud diapirs), benthic communities supported by chemosynthetic bacteria, and methane-derived carbonates.

Gas seeps, identified on echo sounder, sonar, and seismic records, have been reported from many parts of the world (Hovland and Judd, 1988). However, a relatively small proportion of the world's continental shelves has been surveyed in detail and most of the relevant data are held in confidence by the petroleum industry. Where information is available, it is evident that seepage areas may produce considerable volumes of methane at the seabed (Judd and Hovland, 1992).

Measurements of methane flux rates from seepages are not numerous, Clarke and Cleverly (1991), who noted that seepage rates could be measured most easily under water, reported measurements of 28 gassy seeps that were found to produce between 10 m^3 and 3 x 10^6 m^3 (6.8 x 10^{-9} to 2.0 x 10^{-3} Tg) CH_4 per year.

The most prolific area of marine petroleum seepage in the world is probably the Santa Barbara Channel, California, U.S.A., where tar, crude oil, and gas emanate from petroleum-bearing rocks overlain by only a thin veneer of surficial sediments. Gas seepage rates off Coal Oil Point may be as high as 400 g CH_4 m^{-2} yr^{-1} (Fischer and Stevenson, 1973).

Estimates of the numbers of seeps in two areas reach 19,000 and 8,600 respectively (Watkins and Worzel, 1978; Addy and Worzel, 1979). Flow rates of 1 ml min^{-1} to 50 1 min^{-1} were reported, so flux rates may be in the order of 0.01 to 56 g CH$_4$ m^{-2} yr-1 at the seabed.

Global estimates of seepage methane entering the atmosphere have been undertaken by various authors. Ehhalt and Schmidt (1978) used the Wilson et al. (1974) study of the world-wide distribution of oil seepages for their calculation of the oceanic flux of methane. Trotsyuk and Avilov (1988) measured the disseminated flux of methane in the Black Sea and extrapolated world wide. Lacroix (1993) calculated hydrocarbon reservoir depletion and migration rates and, by estimating the rate of removal by oxidation, etc., derived an estimate of the rate of emission to the atmosphere. Cranston (1994) considered methane release by coastal sediments, marine sediments and gas hydrates. The resultant figures are presented in Table 4.

Table 4. Estimates of methane emissions from natural seepages.

Author(s)	Estimate Tg CH$_4$.yr^{-1}	Notes
Andronova and Karol 1993	0.2 to 0.9[a]	perma-frost regions of the former U.S.S.R.
Ehhalt & Schmidt, 1978	1.3 to 16.6	world's oceans
Trotsyuk & Avilov, 1988	1.9	world's continental shelves
Hovland et al., 1993	8 to 65 (at seabed)	world's continental shelves
Lacroix, 1993	17 14	world's oceans
Cranston, 1994	1 to 10	world's ocean and marine sediments
Judd et al., 1997	0.12 to 3.5	United Kingdom continental shelf
see Appendix	0.4 to 12.2[b]	continental shelves world-wide
a + b	0.6 to 13.1	total (conservative)

Hovland et al. (1993) considered that the importance of seabed seeps had been under-estimated. They presented a crude estimate that the total flux on the world's continental shelves is between 8 and 65 Tg CH$_4$ yr^{-1} at the seabed. Judd et al. (1997) undertook a regional review of seismic data to estimate the distribution of gas seeps on the United Kingdom's continental shelf. They estimated the number to be approximately 173,000, even though there are published reports of only eight. Having estimated seabed flux rates and losses to solution as bubbles rise to the sea surface, Judd et al. (1997) concluded that the UK continental shelf contributes between 0.12 and 3.5 Tg CH$_4$ year^{-1} to the atmosphere. If this estimate is extrapolated (using the method of Hovland et al., 1993 - see Appendix) to include continental shelves world-wide (possibly a conservative first approximation considering that the North Sea was considered by Landes, 1973, to have a 'low' seepage potential) the resultant estimate is 12.5 Tg CH_4 year^{-1}. To this should be added seepage from onshore sedimentary basins; these are represented on Table 4 only by the releases through the permafrost regions of the former U.S.S.R. (0.2 to 0.9 Tg CH_4 year^{-1}).

3.2. Mud Volcanoes

Mud volcanoes are similar in shape to magmatic volcanoes, but the material ejected consists of gas, water, mud and rock fragments which originate from sedimentary rocks at depth. These features occur in various parts of the world (see Figure 4), particularly in zones of tectonic convergence. The lateral tectonic forces cause an increase in pore fluid pressure, so sediments containing significant concentrations of gas (e.g., petroleum-bearing sediments) may be mobilised. They force their way vertically through overlying strata to emerge, often explosively, at the land surface or the seabed. Reviews have been presented by Gansser (1960), Ridd (1970), Higgins and Saunders (1974), and Hedberg (1980).

It is estimated that there are over 900 mud volcanoes on land, and at least a further 220 (probably more than 670) on the seabed (L.I. Dimitrov, pers. comm., 1997). Over 200 of these lie in Azerbaijan, the most prolific mud volcano region. These range in size from 5 to 500 m in height with basal diameters of up to 3.5 km. and craters up to 500 m across (Jakubov et al., 1971). One individual mud volcano, Solakhai, has produced 5,800 x 10^6 m^3 of mud breccia which covers an area of 5800 ha to an average depth of 100 m. (Jakubov et al., 1971). The nature of the gases emitted varies from place to place. They may be CO_2-rich, but they are generally methane-dominated (Reitsema, 1979). The composition of mud volcano gases from Azerbaijan is shown in Table 5).

Table 5.   Composition of Azerbaijan mud volcano gases

Gas	Concentration Range
Methane	85 - 99%
Carbon dioxide	<10%
Carbonic acid	<11.4%
Higher hydrocarbon gases	<3.4%
Hydrogen sulphide	<1%
Nitrogen	< trace
Hydrogen	< trace
Inert gases	< trace

source (Jakubov et al., 1971)

The volumes of gas emitted are dependent upon both the size and periodicity of major eruptions, and the rate of emission during periods of quiescence. The nature of the gases varies across the mud volcano belt; ranges are shown in Table 5.

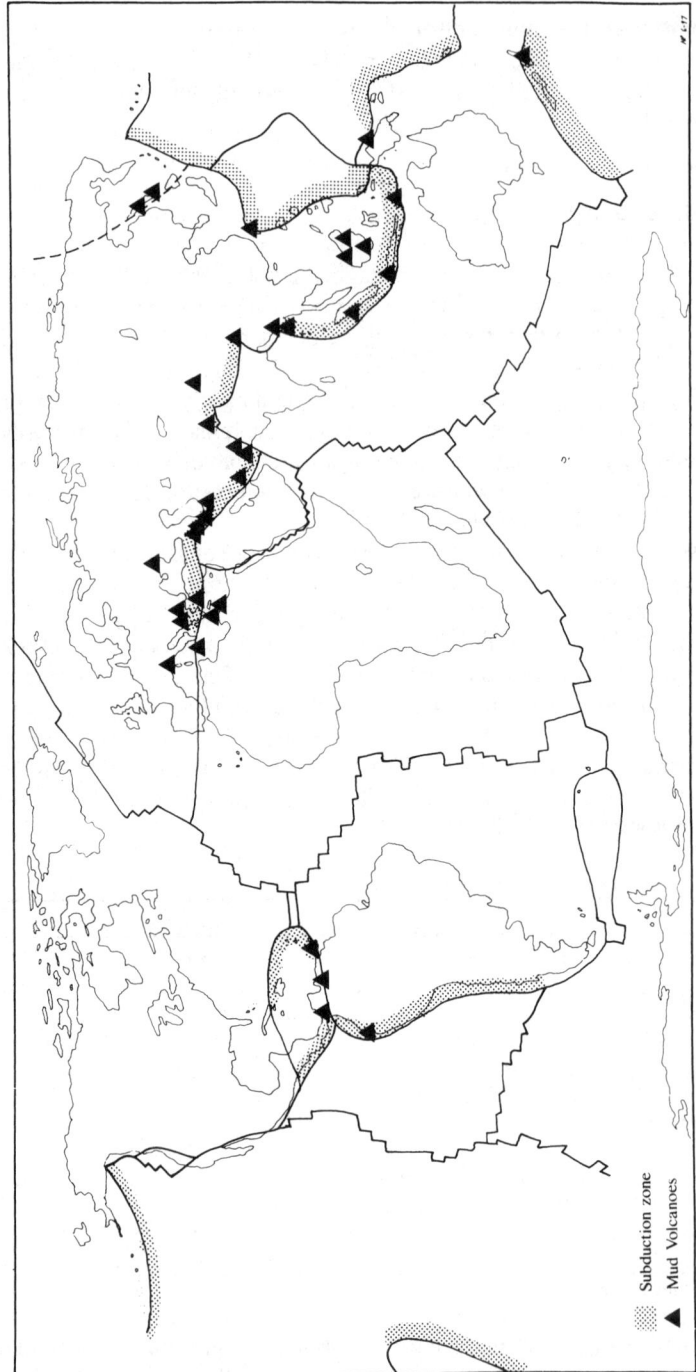

Fig. 4. Global distribution of mud volcanoes (various sources, including L.I. Dimitrov, personal communication, 1997)

Eruptive events are periodic and may be short-lived, some lasting only a few hours; however, in these events enormous volumes of gas may be emitted. Gentle emanations may persist through periods of quiescence which last many years. Sokolov et al. (1969) reported that during quiet activity in the Baku region, groups of mud volcanoes emit 1-3,000 m^3 gas or more in a day and that eruptive phases may produce "*some hundred millions of m^3 in 1-2 days*"; sometimes the gases ignite with flames reaching as high as 350 m.

Accurate global estimates of hydrocarbon gas emissions are not available, but Jakubov et al. (1971) and Guliev and Feizullayev (1996) both estimated that 250 x 10^{6}m^3.yr^{-1} are emitted as a result of eruptions, and a further 20 x 10^{6}m^3.yr^{-1} continuous emanations. Together this amounts to approximately 0.18 Tg CH^4.yr^{-1} from Ajerbaijan alone. Dimitrov (personal communication, 1997) suggests that the global contribution of mud volcanoes lies in the range 9.7 to 12.9 Tg CH$_4$.yr^{-1} (0.7 to 3.9 Tg CH$_4$.yr^{-1} from continuous emanations, the remainder from eruptions). Diffuse fluid flow (unmeasured) through sediments surrounding mud volcanoes may result in even larger fluxes than these channelled flows.

3.3. Gas Hydrates

Gas hydrates, crystalline ice-like compounds in which gas (principally methane) molecules are trapped in a rigid, cage-like lattice of water. The gas sequestered by or trapped beneath these hydrates has evidently been derived elsewhere in the sediments, and may be of thermogenic or bacterial origin. They occur in zones that are strictly controlled by the pressure and temperature regimes, and by the availability of gas (see Figure 5).

They are extensive in or near the permafrost regions on land and offshore, and under large areas of the world's deep oceans (generally in water depths of >500 m where the sediment temperature is low and the pressure is high). These occurrences represent an enormous reservoir of methane; this was estimated by Kvenvolden (1993) to total 2 x 10^{16} m^3 (1.4 x 10^7 Tg) methane (of which 1.4 x 10^{13} to 3.4 x 10^{16} m^3 is in permafrost regions, and 3.1 x 10^{15} to 7.6 x 10^{18} m^3 in deep oceanic sediments). However, Ginsburg and Soloviev (1994, 1995) considered a figure of 10^{15} m^3 (7 x 10^5 Tg) methane to be more realistic.

In many cases the hydrates also represent an impermeable layer under which additional gas may be trapped (Dillon and Paull, 1983).

Evidence from various areas indicates that gas from hydrate locations escapes to the seabed and possibly to the atmosphere, for example: Sassen and MacDonald (1994) reported the occurrence of gas hydrates and associated gas seep plumes at the seabed in the Gulf of Mexico; Carson et al. (1993) reported methane derived carbonates, bacterial mats and a cold seep community from which methane bubbles were emanating on the Cascadia accretionary wedge; Paull et al. (1995), reporting a similar occurrence on the Blake Ridge, noted that acoustic evidence showed a plume of gas bubbles rising 320 m above the seabed. Destabilisation and emission to the atmosphere may also occur on land. Kvenvolden (1988) estimated that 3 Tg CH$_4$.yr^{-1} from gas hydrates enters the atmosphere.

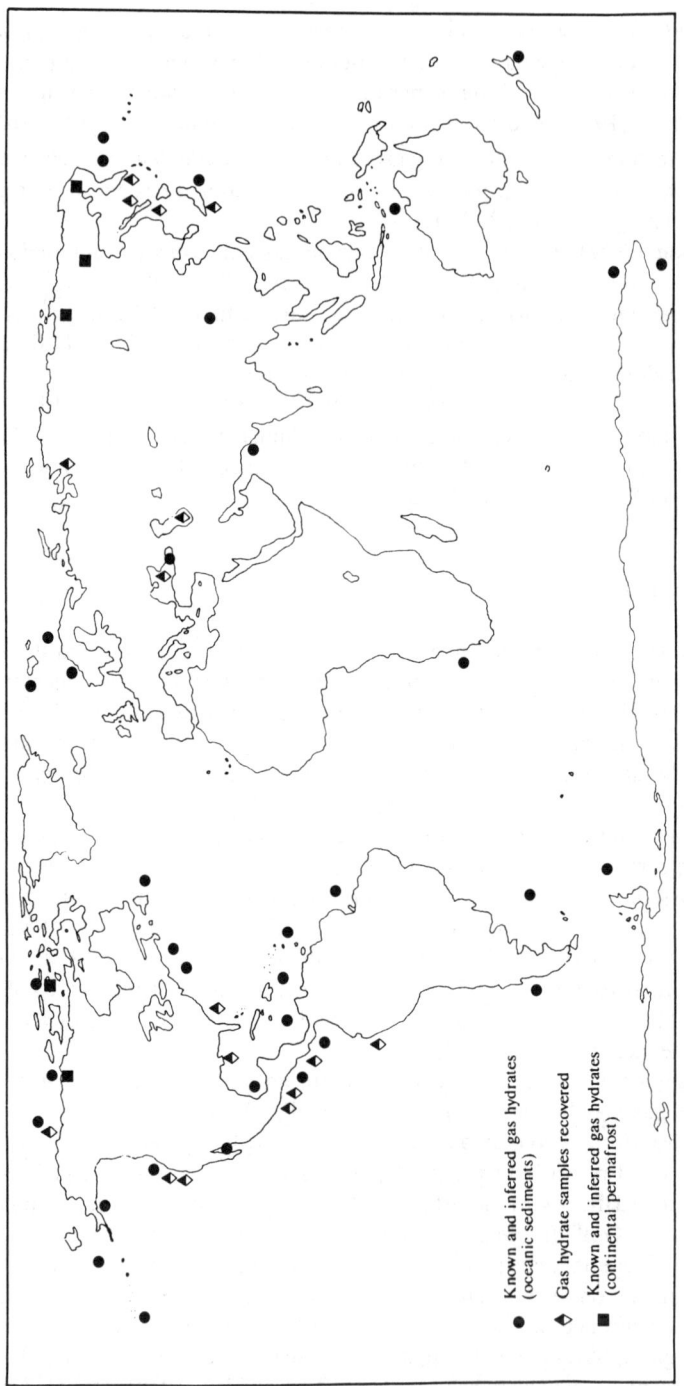

Fig. 5 Distribution of gas hydrates (primarily after Kvenvolden, 1993; and Ginsburg and Soloviev, 1994)

3.4. Natural Coal Seam Fires

Bustin and Matthews (1985) reported that methane comprises 0.3 to 4.0% of the total gaseous emissions from natural coal seam fires. The global contribution made by these fires is probably less than 1 Tg CH_4 yr^{-1}.

4. Releases Associated with Geological Resource Extraction

The mining of coal and the exploitation of petroleum account for the most significant emissions of methane (see Beck et al., 1993); however, the extraction of certain other natural resources produce minor amounts of methane.

4.1. Groundwater Abstraction

Zor'kin et al. (1985) estimated that there is 10^{16} to $10^{17}m^3$ methane dissolved in the Earth's waters (the "hydrosphere"). Groundwater may be a source of atmospheric methane (Aravena et al., 1989; Geyh and Softner, 1989; Kimmelmann et al., 1989), and Coleman et al. (1988) showed that methane may be released during groundwater abstraction. Lacroix (1993) estimated that 1.1 ± 0.8 Tg CH_4 yr^{-1} is derived from groundwater on the basis of average methane contents of groundwater and global rates of groundwater usage.

4.2. Peat Mining

Using the methane content of peat presented by Glotov et al. (1985) and an estimated global production rate of 23×10^6 t yr^{-1}, Lacroix (1993) suggested that peat mining may be responsible for 2 ± 1 Tg CH_4 yr^{-1}. However, it is acknowledged that this estimate is only a crude approximation and that further work is required.

4.3. Mining of Other Geological Resources

Methane is a common gaseous component of igneous, sedimentary, and metamorphic rocks, and emissions of methane from uranium, salt, and metal sulphide mines have been reported (Sokolov et al., 1972; Hyman, 1987; Lacroix, 1993). At present it is not possible to provide a realistic estimate of the methane emissions from these mining activities; however, one example indicated that mining may represent a significant source. According to Sokolov et al. (1972) 0.3 Tg CH_4 yr^{-1} is emitted from the Witwatersrand gold mines of South Africa.

5. Discussion

The contributions of the geological sources described in this chapter are summarised in Table 6. In many cases inadequate data have prevented reliable estimation, and in several cases the estimates presented are crude. Further work is therefore required. The significance of these sources lies not only in the volumes of methane emitted but also in the isotopic composition, the variability and unpredictability of the flux rates, and the fact that they often provide methane at isolated locations.

Table 6. Contributions to atmospheric methane levels from geological sources.

	Natural methane reservoirs	Source
organic-rich sediments (bacterial methane)	n/a	
hydrocarbon-bearing sediments (thermogenic methane)	$(1.4 \times 10^6$ to 2.8×10^{17} Tg carbon)	Hunt, 1979
gas hydrates	1.4×10^7 Tg CH_4	Kvenvolden, 1993
	Natural Release Pathways Tg $CH_4 . yr^{-1}$	
natural gas seeps	0.6 to 13.1	
mud volcanoes	9.7 to 12.9	
hydrothermal / geothermal activity	0.9 to 3.7	Lacroix, 1993 (land events only)
volcanic activity	0.8 to 6.2	Lacroix, 1993 (land events only)
natural coal seam fires	<1	Lacroix, 1993
	Geological Resource Extraction Tg $CH_4 . yr^{-1}$	
Coal mining	35 to 48	Beck et al., 1993
Oil / gas abstraction	20 to 50	Beck et al., 1993
Groundwater abstraction	0.3 to 1.9	Lacroix, 1993
Peat mining	1.0 to 3.0	Lacroix, 1993
other	n/a	
Total: Natural Release Pathways	**12.0 to 36.9**	
Total: Geological Resource Extraction	**56.3 to 102.9**	

It is suggested that significant volumes of methane may enter the world's seas and oceans from natural gas seepages, hydrothermal vents and gas hydrates. It is generally considered that almost all is consumed in the water. The proportion reaching the atmosphere is dependent primarily upon the water depth and temperature, and on bubble sizes. Judd et al. (1997) estimated that 75% of methane bubbles reached the sea surface from a North Sea seep in a water depth of 175 m.. Cranston et al. (1994) reported observing bubbles breaking the sea surface above seeps supplied by gas hydrates in a water depth of >700 m in the Sea of Okhotsk. In contrast de Angelis et al. (1993) considered that the majority of the methane entering the seawater from the Endeavour segment of the Juan de Fuca Ridge is oxidised in the seawater (microbial oxidation makes contributions to organic carbon production, having a significant effect on local deep sea productivity).

The significance of these submarine geological sources may lie not only in the contribution they make to methane concentrations of the hydrosphere, but also in the fact that they are 'point sources' of methane. Wernecke et al. (1994) demonstrated that elevated seawater methane concentrations existed within only 40 m of a gas seepage in the Gullfaks field (North Sea). Regional studies which rely on limited numbers of measurements may provide inaccurate pictures of seawater methane distribution if such sources are present but not identified.

Geological methane sources, both on land and in the oceans, can be expected to be active over long periods of geological time. However, emission rates may vary considerably on a human time scale, for example, with cycles of volcanic activity, the periodicity of mud volcano eruptions, the triggering of gas escapes from petroleum reservoirs by seismic activity, or the destabilisation of gas hydrates. Consequently it is difficult to assess the true contribution they make to atmospheric levels

Deep geological sources are responsible for a supply of methane which rises towards the Earth's surface (or the seabed) over extended periods of geological time regardless of surface conditions. When permafrost, ice or gas hydrates impede emissions to the atmosphere, accumulations may build up. At the present day these are exemplified by gas trapped beneath gas hydrates seen as BSRs (Bottom Simulating Reflections) on seismic reflection profiles. During periods of climate amelioration, such as at the end of the last glacial period, the melting of permafrost, ice and gas hydrates will allow these accumulations to escape. Judd et al. (1994) described evidence of such an event: an anomalously large pockmark formed when gas trapped beneath seabed permafrost was released as cold Arctic waters were flushed from the North Sea by warmer Atlantic waters. This occurred about 13,000 years ago. Such events may have provided significant positive feedback to global warming.

Models combining ^{13}C and ^{14}C contents of atmospheric methane and its source imply that fossil or ^{14}C-depleted sources constitute about 20% (Whalen et al., 1989) to 30% (Lowe et al., 1988) of the total atmospheric budget. The budget of Cicerone and Oremland (1988) proposes a total methane source strength of 540 Tg CH_4 yr^{-1} of ^{14}C-depleted methane which cannot be accounted for. As the majority of methane from the sources described here (i.e., excluding only the bacterial methane from sediments geologically recent sediments) the evidence presented in here suggests that the geological sources may account for a significant proportion of the 'missing' fossil methane.

Acknowledgements. In preparing some parts of this chapter I have relied upon material previously published by the late Dr. A.V. Lacroix (Lacroix, 1993; and sections of the chapter on '*Minor Sources*' in the first edition of this volume). I gratefully acknowledge the contribution this represents. I thank Gabriel Ginsburg, Nick Owens, and Lyoubomir Dimitrov for the constructive comments on this chapter.

Appendix

Global contributions of gas seeps on continental shelves to atmospheric methane (following the method of Hovland et al., 1993)

Assumption 1: contributions from the UKCS lie in the range:

$$0.12 \text{ to } 3.5 \text{ Tg CH}_4 \text{ yr}^{-1} \text{ (Judd et al., 1997)}$$

Assumption 2: UKCS area $= 0.62 \times 10^6 \text{ km}^2$
therefore, the UKCS flux is estimated at:
$$0.12 / 0.62 \times 10^6 = \textbf{0.19} \times \textbf{10}^{-6} \textbf{ Tg CH}_4 \textbf{ km}^{-2} \textbf{ yr}^{-1}$$
to $3.5 / 0.62 \times 10^6 = \textbf{5.6} \times \textbf{10}^{-6} \textbf{ Tg CH}_4 \textbf{ km}^{-2} \textbf{ yr}^{-1}$
N.B. the UKCS includes, but is not exclusively underlain by, sedimentary basins.

Assumption 3: The area of the world's continental shelves (including adjacent seas) $= 27.4 \times 10^6 \text{ km}^2$ (Trotsyuk and Avilov, 1988)

Areas of high, medium and low seepage potential were estimated by Wilson et al. (1973) — adjusted so that the total equals $27.4 \times 10^6 \text{ km}^2$ — as follows:

High Medium Low
$1.7   15.7   10.0 \times 10^6 \text{ km}^2$

Contributions from these areas are assumed to equate to the UKCS flux estimate as follows:
Low: zero
Medium at 1/33 of UKCS flux rate:
$= [(0.19 \text{ to } 5.6 / 33) \times 10^{-6}] \times [15.7 \times 10^6]$
$= 0.09 \text{ to } 2.66 \text{ Tg CH}_4 \text{ km}^{-2} \text{ yr}^{-1}$
High: UKCS flux rate
$= [0.19 \text{ to } 5.6 \times 10^{-6}] \times [1.7 \times 10^6]$
$= 0.32 \text{ to } 9.52 \text{ Tg CH}_4 \text{ km}^{-2} \text{ yr}^{-1}$
Therefore, the global flux estimate is estimated as:
$0.4 \text{ to } 12.2 \text{ Tg CH}_4 \text{ km}^{-2} \text{ yr}^{-1}$

References

Abrajano, T. A., N. C. Sturchi, J. K. Bohlke, G. L. Lyon, R. J. Poreda, C. M. Stevens. 1988. Methane-hydrogen gas seeps, Zambales Ophiolites, Philippines: deep or shallow origin? In: Origins of Methane in the Earth (M. Schoell, ed.), *Chem. Geol., 72*:211-222.

Acosta, J. 1984. Occurrence of acoustic masking in sediments in two areas of continental shelf of Spain: Ria de Muros (NW) and Gulf of Cadiz (SW). *Marine Geol. 58*: 427-434.

Addy, S. K., J. L. Worzel. 1979. Gas seeps and sub-surface structure off Panama City, Florida. *Am. Assoc. Petrol. Geol. (Bull.), 63*:668-675.

Andronova, N. G., I. L. Karol. 1993. The contribution of USSR sources of global methane emission. *Chemosphere, 26:* (1-4): 111-126.

Aravena, R., L. I. Wassenaar, J. F. Baricek. 1989. Investigating carbon sources for methane and dissolved organic carbon in a regional confined aquifer using ^{14}C, radiocarbon. *Proc. 14th Internat. Radiocarbon Conf., 31*:170-171.

Baraza, J., G. Ercilla, 1996. Gas-charged sediments and large pockmark-like features on the Gulf of Cadiz slope (SW Spain). *Marine Geology 13:*253-261.

Beck, L. L., S. D. Piccot, D. A. Kirchgessner. Industrial sources. In *Atmospheric Methane: Sources, Sinks, and Role in Global Change,* Ch. 17, NATO ASI Series, edited by M.A.K. Khalil, 1993, Springer-Verlag, pp. 399-431.

Bordkov, Yu K., V. I. Yefimov, J. B. Timkia. 1988. Result of a gas-biochemical survey of snow cover for direct exploration for hydrocarbon deposits in the Venisey-Khatanga Downwarp (S.S.R). *Petrol. Geol., 22:*203-205.

Bustin, R. M., W. H. Matthews, 1985. In situ gasification of coal, a natural example: additional data on the Aldridge Creek coal fire, south-eastern British Columbia. *Can. J. Earth Sci., 22*: 1858-1864.

Butenko, J., J. D. Milliman, Y.-C Ye. 1985. Geomorphology, shallow structure, and geological hazards in the East China Sea. *Cont. Shelf Res., 4*: 121-141.

Carson, B., G. Westbrook, R. Musgrave. 1993. Cascadia Margin science operator report leg 146, *JOIDES Journal*, June 1993. Ocean Drilling Program, Texas A & M University: 11-16.

Cicerone, R. J., R. S. Oremland. 1988. Biogeochemical aspects of atmospheric methane. *Global Biogeochem. Cycles, 2*:299-327.

Clarke, R. H., R. W. Cleverly. 1991. Petroleum seepage and post-accumulation migration. In England, W. A., A. J. Fleet, *Petroleum Migration.* Geol. Soc. Sp. Publ. No. 59 (Geological Society of London, Bath), 265-271.

Clayton, J. L., J. S. Leventhal, D. R. Rice, J. C. Pashin, D. Mosher, P. Czepiel. 1993 Atmospheric methane flux from coals - preliminary investigations of coal mines and geologic structure in the Black Warrior Basin, Alabama. In Howell, D. G. (Ed.) *The Future of Energy Gases. United States Geological Survey Prof. Paper 1570*: 471-492.

Coleman, D. D., C. Liu, K. M. Riley. 1988. Microbial methane in the shallow paleozoic sediments and glacial deposits of Illinois, USA. *Chem. Geol., 71*:23-40.

Cranston, R. E. 1994. Marine sediments as a source of atmospheric methane. *Bull. Geol. Soc. Denmark, 41*: 101-109.

Cranston, R. E., G. D. Ginsburg, V. A. Soloviev, T. D. Lorenson. 1994. Gas venting and hydrate deposits in the Okhotsk Sea. *Bull. Geol. Soc. Denmark, 41*: 80-85.

Dando, P. R., J. A. Hughes, Y. Leahy, S. J. Niven, L. J. Taylor, C. Smith. 1995. Gas venting from submarine hydrothermal areas around the island of Milos, Hellenic volcanic arc. *Continental Shelf Research, 15*: 913-929.

de Angelis, M. A., M. D. Lilley, E. J. Olson, J. A. Baross. 1993. Methane oxidation in deep-sea hydrothermal plumes of the Endeavour Segment of the Juan de Fuca Ridge. *Deep Sea Res. 40*: 1169-1186.

Des Marais, D. J., M. L. Stallard, N. L. Nehring, A. H. Truesdell. 1988. Carbon isotope geochemistry of hydrocarbons in the Cerro Prieto geothermal field, Baja California Norte, Mexico. *Chem. Geol. 71*:159 - 167.

Dillon, W. P., C. K. Paull. 1983. Marine gas hydrates - II: geophysical evidence. In: *Natural Gas Hydrates: Properties, Occurrence and Recovery* (J. L. Cox, ed.), Butterworth, Boston, p 73-90.

Earney, F. C. F. 1980. *Petroleum, and Hard Minerals from the Sea*. V. H. Winston and Sons, New York, 291p.

Ehhalt, D. H., Schmidt, U. 1978. Sources and sinks of atmospheric methane. *Pure Appl. Geophys., 116:*452-464.

Fischer, P. J., A. J. Stevenson. 1973. Natural hydrocarbon seeps along the northern shelf of the Santa Barbara Channel, California. Paper 1728, *Offshore Technol. Conf.*, Houston, Texas.

Floodgate, G. D., A. G. Judd. 1992. The origins of shallow gas. *Continental Shelf Res., 12:*1,145-1,156.

Gansser, A. 1960. Über Schlammvulkane und Saldome (Mud volcanoes and salt domes). *Naturf Gesell Zürich Vierteljahrssch, 105*:1-46.

Geyh, M. A., B. Softner. 1989. Groundwater analysis of environmental carbon and other isotopes from the Jakarta Basin Aquifer, Indonesia. *Radiocarbon, 31*:919-925.

Ginsburg, G. D., V. A. Soloviev. 1994. *Submarine Gas Hydrates* VNIIOkeangeologia, St. Petersburg, 199p. (in Russian)

Ginsburg, G. D., V. A. Soloviev. 1995. Submarine gas hydrate estimation: theoretical and empirical approaches. *Offshore Technology Conference*, Paper *OTC 7693*, Houston, Texas, USA, May 1995.

Glasby, G. P. 1971. Direct observations of columnar scattering associated with geothermal gas bubbling in the Bay of Plenty, New Zealand. *N. Z. J. Mar. Freshwater Res., 5*:483-496.

Glotov, V., V. V. Ivanov, N. A. Shilo. 1985. Migration of hydrocarbons through permafrost rock. *Trans. (Doklady) U.S.S.R. Acad. Sci., Earth Sci. Sect., 285*:192-194.

Gold, T., S. Soter. 1980. The deep earth gas hypothesis. *Sci. Am., 242*:154-161.

Gold, T., S. Soter. 1982. Abiogenic methane and the origin of petroleum. *Energy Exploration and Exploitation, 1*:89-104.

Guliev, I. S., A. A. Feizullayev. 1996. Geochemistry of hydrocarbon seepages in Azerbaijan. IN Schumacher, D., M. A. Abrams (Eds.) *Hydrocarbon migration and its near-surface expression.* AAPG Mem 66: 63-70.

Hedberg, H. D. 1980. Methane generation and petroleum migration. In: *Problems of Petroleum Migration*. Am. Assoc. Petrol. Geol., Studies in Geology No. 10 (W.H. Roberts, III, R. J. Cordell, eds), pp 179-206.

Hekinian, R. 1984. Undersea volcanoes. *Sci. Am., 251*:46-55.

Higgins, G. E., J. B. Saunders. 1974. Mud volcanoes - their nature and origin. *Verhandl Naturforschung Gellschaft, Basel, 84*:101-154.

Hill, J. M., J. P. Halka, R. Conkwright, K. Koczot, S. Coleman., 1992. Distribution and effects of shallow gas on bulk estuarine sediment properties. *Cont Shelf Res. 12*: 1219-1229.

Hochstein, M. P., 1970 Seismic measurements in Suva harbour (Fiji). New Zealand *J. of Geol. & Geophys., 13*:269-281.

Horibe, Y., K. Kim, H. Craig. 1986. Hydrothermal methane plumes in the Mariana back-arc spreading centre. *Nature, 324*:131-133.

Hovland, M., A. G. Judd. 1988. *Seabed pockmarks and seepages: impact on geology, biology and the marine environment.* Graham & Trotman, London, 293p.

Hovland, M., A. G. Judd, R. A. Burke. 1993. The global flux of methane from shallow submarine sediments. *Chemosphere, 26*:559-578.

Hunt, J. M. 1979. *Petroleum Geochemistry and Geology*, W. H. Freeman, San Francisco.

Hyman, D. M. 1987. A review of mechanisms of gas outbursts in coal. *U.S. Dept. Of the Interior, Bureau of Mines Info. Circ. 9155* 11p.

Jakubov, A. A., A. A. Ali-Zade, M. M. Zeinalov. 1971. *Mud volcanoes of the Azerbaijan SSR*. Akademija Navk Azerbaijan SSSR, 257p. (in Russian with English summary)

Jean-Baptiste, P., S. Belviso, G. Alaux,, B. C. Nguyen, N. Mihalopoulos. 1990. 3He and methane in the Gulf or Arden. *Geochim. et Cosmochim. Acta, 54*:111-116.

Jørgensen, N. O., T. Laier, B. Burchardt, T. Cederberg. 1990. Shallow hydrocarbon gas in the northern Jutland-Kattegat region, Denmark. *Bull. Geol. Soc. Denmark 38*: 69-76.

Judd, A. G., M. Hovland. 1992. The evidence of shallow gas in marine sediments. *Cont Shelf Res. 12*: 1081- 1096.

Judd, A. G., D. Long, M. Sankey. 1994. Pockmark formation and activity, U. K. block 15/25, North Sea. *Bull. Geol. Soc. Denmark, 41*:34-49.

Judd, A. G., G. Davies, J. Wilson, R. Holmes, G. Baron, I. Bryden. 1997. Contributions to atmospheric methane by natural seepages on the UK continental shelf. *Marine Geol., 137*:165-189.

Kadko, D., J. Baross, J. Alt. 1995. The magnitude and global implications of hydrothermal flux. In S. E. Humphries, R. A. Zierenberg, L. S. Mullineaux, R. E. Thomsen. *Seafloor Hydrothermal Systems: physical, chemical, biological and geological interactions*. Geophysical Monograph 91, American Geophysical Union.

Kimmelmann, A. A., Aldo de Cunha, S. Reboucas, M. M. Freitas Santiago, 1989. 14-C analysis of groundwater from the Botucatu Aquifer system in Brazil. *Radiocarbon, 31*:926-933.

Kvenvolden, K. A. 1988. Methane hydrate: a major reserve of carbon in the shallow geosphere. *Chem. Geol., 71*:41-51.

Kvenvolden, K. A. 1993. Gas hydrates as a potential energy resource - a review of their methane content. IN Howell, D. G. (Ed.) *The Future of Energy Gases*. United States Geological Survey Prof. Paper 1570, pp555-561.

Kvenvolden, K. A., J. W. Harbaugh. 1983. Reassessment of the rates at which oil from natural sources enters the marine environment. *Marine Env. Res., 10*:223-243.

Lacroix, A. V. 1993. Unaccounted for sources of fossil and isotopically-enriched methane and their contribution to the emissions inventory: A review and synthesis. *Chemosphere, 26*: 507-558.

Landes, K. K. 1973. Mother nature as oil polluter. *Am. Assoc. Petrol. Geol. (Bull), 53*:2,431-2,479.

Lawrence, J. R., M. Taviani. 1988. Extreme hydrogen, oxygen and carbon isotope anomalies in the pore water and carbonates of the sediments and basalts from the Norwegian Sea: methane and hydrogen from the mantle. *Geochim, et Cosmochim, Acta, 52*:2,077-2,083

Leavitt, S. W. 1982. Annual volcanic carbon dioxide emission: an estimate from eruption chronologies. *Environ. Geol., 4*:15-21.

Liley M. D., J. A. Baross, L. I. Gordon. 1982. Dissolved hydrogen and methane in Saanich Inlet, British Columbia. *Deep Sea Res., 28*:1,471-1,484.

Link, W. K. 1952. Significance of oil and gas seeps in world oil exploration. *Am. Assoc. Petrol. Geol. (Bull), 36*:1,505-1,540.

Lowe, D. C., C. A. M. Brenninkmeiher, M. R. Manning, R. Sparks, G. Wallace. 1988. Radiocarbon determination of atmospheric methane at Baring Head, New Zealand. *Nature, 332*:522-525.

MacDonald, G. J. 1983. The many origins of natural gas. *J. Petrol. Geol., 5*:341-362.

Malahoff, A. 1985. Hydrothermal vents and polymetallic sulfides of the Galapagos and Gorda/Juan de Fuca Ridge systems and of submarine volcanoes. In: *Hydrothermal Vents of the Eastern pacific: An Overview* (M. L. Jones, ed.), Bull. Biol. Soc., Washington, 6:19-41.

Martens, C. S., J. V. Klump. 1980. Biogeochemical cycling in an organic rich coastal marine basin. 1. Methane sediment water exchange processes. *Geochem. Et. Cosmochim. Acta 44*: 471-490.

302   Judd

McCartney, B. S., B. McK. Barry. 1965. Echo sounding on probable gas bubbles from the bottom of Saanich Inlet, British Columbia. *Deep Sea Res. 12*: 285-294.

Nelson, C. H., D. R. Thor, M. W. Sandstrom, K. A. Kvenvolden. 1979. Modern biogenic gas-generated craters (sea-floor 'pockmarks') on the Bering Sea shelf, Alaska. *Geol. Soc. Am. Bull, 90*: 1144-1152.

Parnell, J. 1988. Migration of biogenic hydrocarbons into granites - a review of hydrocarbons in British plutons. *Mar & Petrol. Geol., 5*:385-395.

Parnell, J., I. Swainbank. 1990. Pb - Pb dating of hydrocarbon migration into a bitumen-bearing ore deposit, North Wales (United Kingdom). *Geology, 18*:1,028-1,030.

Paull, C. K., W. Ussler III, W. S. Borowski, F. N. Spiess, 1995. Methane-rich plumes on the Carolina continental rise: Association with gas hydrates. *Geology, 23*: 89-92.

Philip, R. P. 1987. Surface prospecting methods for hydrocarbon accumulations. In: *Advances in Petroleum Geochemistry*, Vol. II (J. Brooks and D. Welte, eds.), Academic Press, London, p 209-253.

Prior, D. B., J. M. Coleman, 1982. Active slides and flows in underconsolidated marine sediments on the soles of the Mississippi delta. IN Saxov, S., J. K. Nieuwenhuis (Eds.) *Marine Slides and Other Mass Movements*, Plenum Press, New York, 21-49.

Reitsema, R. H. 1979. Gases of mud volcanoes in the Copper River Basin, Alaska. *Geochim, et Cosmochim. Acta, 43*: 183-187.

Rice, D. R., G. C. Claypool. 1981. Generation accumulation and resource potential of biogenic gas. *Am. Assoc. Petrol. Geol. (Bull.), 65*:5-25.

Ridd, M. F. 1970. Mud volcanoes in New Zealand. *Am. Assoc. Petrol. Geol. (Bull.) 54*:601-616.

Sassen, R., I. R. MacDonald, 1994. Evidence of structure H hydrate, Gulf of Mexico continental slope. *Organic Geochemistry, 22*: 1029-1032.

Schoell, M. 1988. Multiple origins of methane in the earth. *Chem. Geol., 71*:1-10.

Selley, R. C. 1992. Petroleum seepages and impregnations in Great Britain. *Marine & Petrol. Geol. 9*: 226-244.

Sherwood, B., P. Fritz, S. K. Frape, J. A. Macko, S. M. Weise, J. A. Welhan. 1988. Methane occurrences in the Canadian Shield. *Chem. Geol., 71*:223-23.

Simoneit, B. R. T., P. T. Crisp, B. G. Rohrback, B. M. Didyk. 1979. Chilean paraffin dirt - II. Natural gas seepage at an active site and its geochemical consequences. In: *Physics and Chemistry of the Earth*, Vol. 12: Advances in Geochemistry 1979 (A. G. Douglas and J. R. Maxwell, eds.), (Proc. 9th Internat. Meeting on Organic Geochem., Newcastle, Sept. 1979) p 171-176.

Söderberg, P., T. Flodén. 1992. Gas seepages, gas eruptions and degassing structures in the seafloor along the Strömma tectonic lineament in the crystalline Stockholm Archipelago, east Sweden. *Cont Shelf Res. 12*: 1157-1172.

Sokolov, V. A., Z. A. Buniat-Zade, A. A. Geodekian, F. G. Dadashev. 1969. The origin of gases of mud volcanoes and the regularities of the powerful eruptions. In: *Advances in Organic Chemistry - 1969* (P. Schenk and I. Havemar, eds.), Pergamon Press, Oxford, p 473-484.

Sokolov, V. T., V. Tichomolova, O. A. Cheremisinov. 1972. The composition and distribution of gaseous hydrocarbons and dependence on depths, as a consequence of their generation and migration. In: *Advances in Geochemistry - 1971* (H.R. Gaertner and H. Wehner, eds.), Pergamon Press, Oxford, p 479-486.

Stadnik, Ye. V., I. Ya. Sklyarenko, I. S. Guliyev, A. A. Feyzullayev. 1986. Methane distribution in the atmosphere above tectonically different regions. *Trans. (Doklady) U.S.S.R. Acad. Sci., Earth Sci. Sect., 289*: 190-192.

Taylor, D. I. 1992. Nearshore shallow gas around the U. K. coast. *Cont Shelf Res. 12:* 1135-1144.

Trotsyuk, V. Y., V. I. Avilov. 1988. Disseminated flux of hydrocarbon gases from the sea bottom and a method of measuring it. *Trans. (Doklady) U.S.S.R. Acad. Sci., Earth Sci. Sect., 291*:218-220.

Upstill-Goddard, R. C., A. P. Reed, N. J. P.Owens. 1996. Simultaneous high-precision measurements of methane and nitrous oxide in water and seawater by single phase equilibration gas chromatography. *Deep-Sea Research, 43*: 1669-1682.

Vidal, F. V., J. A. Welhan, V. N. V. Vidal. 1982. Stable isotopes of helium, nitrogen and carbon in a coastal submarine hydrothermal system. *J. Volcano Geother. Res., 12*:101-110.

von Rad, U., H. Rösch, U. Berner, M. Geyh, V. Marchig, H. Schulz. 1996. Authigenic carbonates derived from oxidized methane vented from the Makran accretionary prism off Pakistan. *Marine Geology 136:* pp55-77.

Vyshemirskiy, V. S., R. S. Khakimzyanova, V. F. Shugurov. 1989. A gas survey of snow cover in the Kuznetsk Basin. Trans. *(Doklady) U.S.S.R. Acad. Sci., Earth Sci. Sect., 309*:172-174.

Watkins, J. S., J. L. Worzel. 1978. Serendipity gas seep area, South Texas offshore. *Am. Assoc. Petrol. Geol. (Bull.), 62*:1,067-1.074.

Welhan, J. A. 1988. Origins of methane in hydrothermal systems. *Chem. Geol., 71*:183-198.

Welhan, J. A., H. Craig. 1983. Methane hydrogen and helium in hydrothermal fluids at 21°N on the East Pacific Rise. In: *Hydrothermal Processes at Seafloor Spreading Centres* (Rana *et al.*, eds.), Plenum Press, New York, p 391-409.

Wernecke, G., G. Flöser, S. Korn, C. Weitkamp, W. Michaelis. 1994. First measurements of the methane concentration in the North Sea with a new in-situ device. . *Bull. Geol. Soc. Denmark, 41*: 5-11.

Whalen, M., M. Tanaka, B. Henry, B. Deck, J. Zeglen, J. S. Vogel, J. Southon, A. Shemesh, R. Fairbanks, W. Broecker. 1989. Carbon-14 in methane sources and in atmospheric methane: the contribution from fossil carbon. *Science, 245*:286-290.

Whiticar, M. J. (this volume) Can stable isotopes be used to constrain atmospheric methane budgets?

Wilson, R. D., P. H. Monaghan, A. Osanik, L. C. Price, M. A. Rogers. 1974. Natural marine oil seepage. *Science, 184*:857-865.

Zor'kin, L. M., F. G. Dadashev, A. A. Dadashev, Krylova. 1985. Peculiarities of the isotopic concentration of methane from petrogas-condensate and gas condensate deposits of Azerbaijan. *Dan SSR, 280*:1,225-1,228.

16. Methane in the Global Environment

Donald J. Wuebbles, Katharine A. S. Hayhoe, and Rao Kotamarthi
University of Illinois at Urbana-Champaign
Department of Atmospheric Sciences, 105 S. Gregory St., Urbana, IL 61801

1. Introduction

The concentration of methane in the atmosphere has increased dramatically over the last few centuries, from 0.7 ppmv to more than 1.7 ppmv, and continues to increase. This increasing concentration of methane (CH_4) in the atmosphere is of particular concern because of the potential effects that it can have on global atmospheric chemistry and climate. Given its relatively long atmospheric lifetime, methane emissions do not appear, in general, to have an appreciable effect on local or regional air pollution. However, methane chemistry does have an important influence on the global atmosphere, affecting the amount of ozone (O_3) in both the troposphere and stratosphere, the amount of hydroxyl (OH) in the troposphere, and the amount of water vapor (H_2O) in the stratosphere. Methane oxidation is also an important source of atmospheric carbon monoxide (CO) and formaldehyde (CH_2O). Methane is the most abundant reactive trace gas in the troposphere. In addition, methane is a greenhouse gas, and its increasing concentrations are of special interest to concerns about climate change.

This chapter discusses the effects of methane on atmospheric chemistry and climate. Discussion focuses on three major topics: effects on tropospheric chemistry, effects on stratospheric and upper atmospheric chemistry, and effects on climate.

2. Effects on Tropospheric Chemistry

2.1 Oxidizing Capacity of the Atmosphere

Photochemistry in the troposphere generates oxidants that are important to destroying many important gases emitted into the atmosphere. This self-cleansing feature of the atmosphere is called the oxidizing capacity. Several chemical species determine the oxidizing capacity of the troposphere. In descending order of reactivity, the most important oxidants are hydroxyl, ozone, hydroperoxyl radical (HO_2), and organic peroxy radicals (RO_2). H_2O_2 is an effective oxidant in aqueous media (i.e. cloud and rain water). The most reactive oxidant is hydroxyl; it is responsible for the oxidation of the majority of the gases emitted into the atmosphere, and is the primary scavenger for methane, most of the higher hydrocarbons (referred to as non-methane

Mohammad Aslam Khan Khalil   (Ed.)
Atmospheric Methane
© *Springer-Verlag Berlin Heidelberg 2000*

hydrocarbons or NMHCs), carbon monoxide, methylchloroform (CH_3CCl_3), methyl chloride (CH_3Cl), methyl bromide (CH_3Br), hydrogen sulfide (H_2S), and sulfur dioxide (SO_2).Therefore, the atmospheric hydroxyl concentration determines the lifetime and hence the atmospheric abundance of these compounds. The reactions of these gases with hydroxyl generally result in the formation of compounds that can be removed by wet deposition. As will be discussed further later, ozone is the primary driver of photochemical processes that recycle gases because its photolysis controls hydroxyl formation. Hydroperoxyl radicals oxidize nitric oxide (NO) to nitrogen dioxide (NO_2) which allows ozone formation to occur via nitrogen dioxide photolysis. Nitric oxide is often the limiting species in this process.

The concentrations of tropospheric oxidants is highly variable (e.g., Browell et al., 1996; Kleinman et al., 1996; Kley et al., 1996; Crossley, 1995; Wennberg et al., 1994). Extensive measurements of ozone are currently available (WMO, 1995), with a few scattered measurements of H_2O_2 (Heikes et al., 1996; Daum et al., 1996). Measurements of OH are gradually becoming available both from surface sites and remote upper troposphere (JAS 95, JGR 83).

The main factors that regulate the oxidizing capacity of the atmosphere are:

1. Species such as methane, higher hydrocarbons, carbon monoxide, and nitrogen oxides ($NO_x = NO + NO_2$) undergo chemical reactions that produce and destroy important oxidizing species.
2. Stratospheric ozone controls the penetration of ultraviolet radiation which determines photochemical activity in the troposphere.
3. Climate characteristics such as atmospheric temperature, cloudiness and humidity also impact photochemical activity in the troposphere.

The underlying phenomenon that links all of these processes is photochemistry. If the oxidizing capacity of the troposphere is perturbed then this would produce a change in the levels of trace species (e.g., see Thompson, 1992, or Logan et al., 1981). As an example, a decrease in oxidizing capacity implies that trace species would have longer residence times, permitting transport of pollutants over long distances and resulting in perturbations over remote regions.

As the most abundant organic species in the atmosphere, methane plays an influential role in determining the tropospheric oxidizing capacity. An important series of reactions are initiated via the reaction of methane with hydroxyl. This consumption of methane is so effective that 80-90% of the methane destruction occurs in the troposphere (Cicerone and Oremland, 1988). A rise in the background level of methane — due to growing emissions — can reduce hydroxyl which would result in a further increase in the methane concentration. Therefore, a positive feedback exists in this chemical cycle which could lead to an overall decrease in the oxidizing capacity of the troposphere. The relationship between methane and hydroxyl will be discussed further in the next section.

Is the oxidizing capacity of the atmosphere changing? Isaksen (1988) concludes that although changes have been observed in key atmospheric species, no direct evidence exists from the trace gas budgets that the oxidizing capacity of the atmosphere has changed on a global scale. Basically, there do not exist sufficient measurements of the tropospheric concentrations of hydroxyl, ozone, and the other oxidants to determine if the oxidizing capacity is changing. Model results produced by Lu and Khalil (1991) and Pinto and Khalil (1991) suggest that mean hydroxyl concentrations may have changed only slightly even though climatic conditions and trace gas (e.g., methane) levels vary tremendously between ice ages, interglacial epochs, and the present time. In both studies, the authors found that increases in hydroxyl destruction due to rising methane and carbon monoxide levels are offset by increases in the production processes. As a result of the increasing atmospheric emissions and concentrations of methane, carbon monoxide, and several other gases, it is likely that changes in the oxidizing capacity of the atmosphere have occurred but the magnitude of the perturbation will be regionally dependent and remains poorly understood.

Studies by Prinn et al. (1992, 1995), based on CH_3CCl_3 measurements from the seven ALE/GAGE surface stations, suggest that these measurements could be used as a proxy for the global oxidizing capacity, by inferring global average OH mixing ratios from known CH_3CCl_3 loss by activity with OH and known sources. The most recent evaluation of this data suggests a small trend of 0.1 % per year in global average OH concentrations.

Generally, changes in oxidizing capacity would be expected to vary in accordance with a location's proximity to major pollution sources. Also, Thompson et al. (1989) and Crutzen and Zimmermann (1991) note that due to high concentrations of hydroxyl and ozone in the tropics, this region may be very important to future changes in global oxidant levels.

2.2 Methane Oxidation by OH

On a global scale, methane oxidation is one of the major reaction pathways affecting atmospheric concentrations of hydroxyl. Depending on nitric oxide levels, methane oxidation can be either a production or destruction process for odd-hydrogen (OH + HO_2). Thus, different chemically coherent (Thompson et al., 1989) regions can be distinguished on the basis of concentrations of nitrogen oxides. Polluted (high NO_x) environments where odd hydrogen is produced include the temperate zone of the Northern Hemisphere and planetary boundary layer of the tropics during the dry season. Unpolluted (low NO_x) environments where odd hydrogen is destroyed include marine areas, the free troposphere over the tropics, and most of the Southern Hemisphere (e.g., Fishman et al., 1979; Crutzen, 1988; Cicerone and Oremland, 1988; WMO, 1991, 1995). Under warm, humid conditions, nitrogen oxide levels must be substantially higher. In general, these conclusions apply to the boundary layer. Most current modeling studies such as those presented here suggest an increase in methane will increase ozone throughout most of rest of the troposphere.

2.2.1 High NO_x Areas.
To illustrate this qualitative description, we will consider the details of the methane oxidation cycle. In areas characterized by high concentrations of nitrogen oxides, methane oxidation primarily occurs via the following mechanism:

$$CH_4 + OH = CH_3 + H_2O \tag{1}$$
$$CH_3 + O_2 + M = CH_3O_2 + M \tag{2}$$

$$CH_3O_2 + NO = CH_3O + NO_2 \tag{3}$$
$$CH_3O + O_2 = CH_2O + HO_2 \tag{4}$$
$$HO_2 + NO = OH + NO_2 \tag{5}$$
(2x) $\quad NO_2 + hv = NO + O \ (\lambda \leq 400 \ nm) \tag{6}$
(2x) $\quad O + O_2 + M = O_3 + M \tag{7}$

Net: $\quad CH_4 + 4 O_2 = CH_2O + H_2O + 2 O_3 \tag{8}$

Initially, methane is attacked by hydroxyl to form a highly reactive methyl (CH_3) radical. Note that nitric oxide is oxidized to nitrogen dioxide twice in the mechanism via the reduction of a peroxy-radical to an oxy-radical. One step provides a pathway for methylperoxyl radical (CH_3O_2) to form methoxy radical (CH_3O) which in turn produces formaldehyde (CH_2O). The other step reduces hydroperoxyl to hydroxyl. These reactions generate nitrogen dioxide which then photolyzes to provide the oxygen atom needed for ozone production. Summing the reactions algebraically, we find that the products of the net reaction are ozone and formaldehyde (CH_2O). Note that M in the reaction sequence above is any molecule, typically N_2 or O_2.

There are three reaction pathways for the oxidation of formaldehyde to carbon monoxide. The first is the direct photolysis of formaldehyde to carbon monoxide and molecular hydrogen:

$$CH_2O + hv = CO + H_2 \ (\lambda \leq 350 \ nm) \tag{9}$$

This reaction is independent of the nitric oxide levels and occurs at wavelengths shorter than 350 nanometers (nm). The second also begins with photolysis but it has a different product channel:

$$CH_2O + hv \ = CHO + H \ (\lambda \leq 350 \ nm) \tag{10}$$
$$CHO + O_2 = CO + HO_2 \tag{11}$$
$$H + O_2 + M = HO_2 + M \tag{12}$$
(2x) $\quad HO_2 + NO = OH + NO_2 \tag{13}$
(2x) $\quad NO_2 + hv \ = NO + O \ (\lambda \leq 400 \ nm) \tag{14}$
(2x) $\quad O + O_2 + M = O_3 + M \tag{15}$

Net: $\quad CH_2O + 4 O_2 = CO + 2 OH + 2 O_3 \tag{16}$

The third pathway is initiated by hydroxyl attack:

$$CH_2O + OH = CHO + H_2O \tag{17}$$
$$CHO + O_2 = CO + HO_2 \tag{18}$$
$$HO_2 + NO = OH + NO_2 \tag{19}$$
$$NO_2 + hv \ = NO + O \ (\lambda \geq 400 \ nm) \tag{20}$$
$$O + O_2 + M = O_3 + M \tag{21}$$

Net: $\quad CH_2O + 2 O_2 = CO + H_2O + O_3 \tag{22}$

In the troposphere, Crutzen (1988) calculated that the averaged relative fractions of these three formaldehyde oxidation pathways are roughly 50-60 %, 20-25 %, and 20-30 %, respectively. Provided that sufficient formaldehyde survives atmospheric removal processes such as wet deposition to undergo further reaction, odd hydrogen and carbon monoxide are formed as end products of methane oxidation. Figure 1 provides a graphical summary of the methane oxidation pathway in high NO_x regions. The odd hydrogen species are circled in order to emphasize that they are produced and consumed. Also, note the variety of loss processes that impact formaldehyde: photolysis, chemical reaction with hydroxyl, and heterogeneous removal.

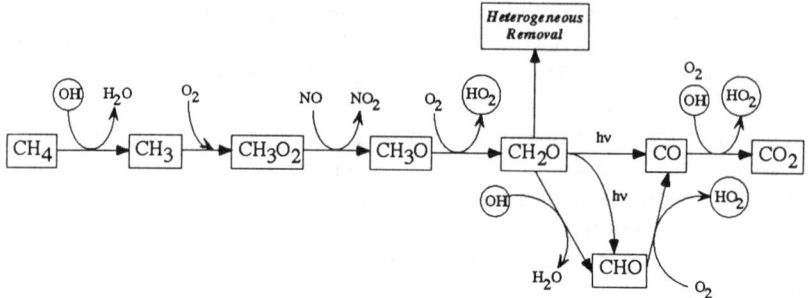

Fig. 1. Methane oxidation pathway in the presence of sufficient nitrogen oxides. Odd hydrogen species are circled in order to indicate steps involving production or destruction

2.2.2 Low NO_x Areas.   The photochemistry of methane in the unpolluted troposphere can be divided similarly into several reaction pathways. In these areas characterized by low levels of nitrogen oxides, methane oxidation occurs primarily by the following reaction sequence:

$$CH_4 + OH = CH_3 + H_2O \tag{23}$$
$$CH_3 + O_2 + M = CH_3O_2 + M \tag{24}$$
$$CH_3O_2 + HO_2 = CH_3O_2H + O_2 \tag{25}$$
$$CH_3O_2H + OH = CH_2O + OH + H_2O \tag{26}$$

Net: $$CH_4 + OH + HO_2 = CH_2O + 2 H_2O \tag{27}$$

Here we observe that hydroperoxyl reacts with methylperoxyl to form methyl peroxide (CH_3O_2H). In turn, methyl peroxide reacts with hydroxyl to produce formaldehyde. The methyl peroxide formed in this sequence also participates in a catalytic subcycle that consumes odd hydrogen:

$$CH_3O_2 + HO_2 = CH_3O_2H + O_2 \tag{25}$$
$$CH_3O_2H + OH = CH_3O_2 + H_2O \tag{28}$$

Net: $$OH + HO_2 = H_2O + O_2 \tag{29}$$

From this subcycle, we see that methyl peroxide also can produce methylperoxyl radical. According to DeMore et al. (1997), the recommended branching ratios for the $CH_3O_2H + OH$ reaction are 70% ($CH_3O_2 + H_2O$) and 30% ($CH_2O + OH + H_2O$).

In addition to reaction (9), there are two more reaction pathways. One involves formaldehyde photolysis with a different product channel:

$$CH_2O + h\nu = CHO + H \ (\lambda \leq 350 \ nm) \tag{30}$$

$$CHO + O_2 = CO + HO_2 \tag{31}$$

$$H + O_2 + M = HO_2 + M \tag{32}$$

$$HO_2 + O_3 = OH + 2 \ O_2 \tag{33}$$

Net: $\quad CH_2O + 2 \ O_3 = CO + 2 \ O_2 + 2 \ OH \tag{34}$

The other pathway is initiated by hydroxyl attack on formaldehyde:

$$CH_2O + OH = CHO + H_2O \tag{35}$$

$$CHO + O_2 = CO + HO_2 \tag{36}$$

$$HO_2 + O_3 = OH + 2 \ O_2 \tag{37}$$

Net: $\quad CH_2O + O_3 = CO + H_2O + O_2 \tag{38}$

The branching ratios for these three pathways vary with altitude. In the troposphere (surface to 10 km), we calculate with our two-dimensional model of the global atmosphere that the range of relative fractions for these three formaldehyde oxidation paths are approximately 38-54%, 23-28%, and 39-18%, respectively.

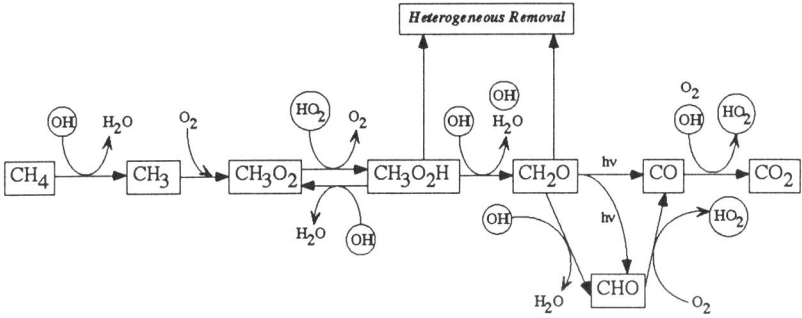

Fig. 2. Methane oxidation pathway in the case of insufficient levels of nitrogen oxides. Odd hydrogen species are circled in order to indicate steps involving production or destruction.

As a result of these mechanisms there is a net destruction of odd hydrogen by the oxidation of methane in low NO_x regions. Figure 2 provides a graphical summary of the methane oxidation pathway in low NO_x regions. The odd hydrogen species are circled in order to emphasize that they are produced and consumed. Another interesting feature is the subcycle - reactions (25) and (28) - involving methylperoxyl and methyl peroxide. Two odd hydrogen species are destroyed each time the cycle turns over. Also, note that methyl peroxide and formaldehyde are affected by several loss processes. In particular, formaldehyde can be destroyed by photolysis, chemical reaction with hydroxyl, and heterogeneous removal.

2.3 Carbon Monoxide Production from Methane Oxidation

The methane oxidation cycle is an important source of carbon monoxide, accounting for roughly a quarter of the carbon monoxide in the troposphere. Carbon monoxide concentrations are a great deal more variable than methane, due to its relatively short atmospheric lifetime (approximately 1-3 months) and because of the variety of natural and anthropogenic sources that contribute to its budget. These sources include fossil fuel combustion, biomass burning in the tropics, and the oxidation of natural hydrocarbons (those emitted by vegetation, e.g. isoprene). As in the case of methane, the carbon monoxide oxidation cycle also depends on the levels of nitric oxide present in the atmosphere. For polluted regions (high NO_x):

$$CO + OH = CO_2 + H \tag{39}$$
$$H + O_2 + M = HO_2 + M \tag{40}$$
$$HO_2 + NO = OH + NO_2 \tag{41}$$
$$NO_2 + h\nu = NO + O \; (\lambda \leq 400 \; nm) \tag{42}$$
$$O + O_2 + M = O_3 + M \tag{43}$$
$$\text{Net:} \quad CO + 2 \, O_2 = CO_2 + O_3 \tag{44}$$

As before, we note that hydroperoxyl oxidizes nitric oxide to form nitrogen dioxide which in turn is photolyzed to produce the oxygen atom required for ozone formation. When nitrogen oxide levels are low, there is no net consumption of odd hydrogen:

$$CO + OH = CO_2 + H \tag{45}$$
$$H + O_2 + M = HO_2 + M \tag{46}$$
$$HO_2 + O_3 = OH + 2 \, O_2 \tag{47}$$
$$\text{Net:} \quad CO + O_3 = CO_2 + O_2 \tag{48}$$

Regardless of the nitric oxide levels, one important result of carbon monoxide oxidation is the production of carbon dioxide (CO_2), a very important greenhouse gas.

Based on the reaction sequences for methane, formaldehyde, and carbon monoxide in polluted as well as unpolluted environments, the effect of the complete oxidation of one mole of methane on odd hydrogen is +0.4 to 0.5 moles in the high NO_x case and -3.5 to -3.9 in the low NO_x case. Correspondingly the change in ozone for complete oxidation of methane is +3.6 to 3.8 moles for the high NO_x case and -1.7 to -1.8 moles for the low NO_x case. According to our calculations, the catalytic subcycle involving reactions (25) and (28) has a sizeable impact on the amount of odd hydrogen loss in the low NO_x case. Crutzen (1988) included this cycle in his results whereas lower effects determined by Cicerone and Oremland (1988) did not include its effect. Our analyses of the resulting changes in odd-hydrogen and ozone are substantially in agreement with Crutzen's for both cases. The resultant destruction of odd hydrogen depends on the hydroxyl concentration, the methyl peroxide chemical reaction pathways, and the heterogeneous removal rates of important intermediate species.

These reaction mechanisms are extremely important for the photochemistry of the unpolluted troposphere because methane and carbon monoxide are the main reaction partners of hydroxyl. Observational data from ground-based stations show that the global level of methane is increasing (e.g., Rasmussen and Khalil, 1981; Khalil and Rasmussen, 1983; Steele et al., 1987; Blake and Rowland, 1988; scientific reviews can be found in WMO, 1989, 1995; IPCC, 1990,1996). The rate of increase for methane during the last decade is approximately 15-17 ppbv (~1%) per year (Khalil et al., 1989; Khalil and Rasmussen, 1990; Wallace and Livingston, 1990). Polar ice cores provide a record of the atmospheric concentration of methane over much longer time scales (i.e., present time to 100,000 years ago). Based on measurements from these cores, the mixing ratio of methane started to increase about 200 years ago (Craig and Chou, 1982; Khalil and Rasmussen, 1983, 1989). Determining the trend for carbon monoxide is difficult because it has a short tropospheric lifetime and the ground-based observational network is inadequate (Khalil and Rasmussen, 1984; WMO, 1991). Due to the likelihood of local contamination, the data are noisy and the trends may not be globally representative (Zander et al., 1989).

The major source of CO in the atmosphere is combustion processes, ranging from automobile exhaust to biomass burning and oxidation of CH_4 and NMHCs. An earlier analysis by Khalil and Rasmussen (1988) indicates that hemispheric and global average concentrations of carbon monoxide show small increasing trends of about 1 ppbv per year. Their investigation also supports Zander et al.'s statement that the rate of change of carbon monoxide varies between different locations. Khalil and Rasmussen conclude that their data evaluation may indicate an increase in the global trend of carbon monoxide, but accurate determinations of the rate of change will require systematic data form a much longer period of time. More recent analyses suggest that during the 1970s and 1980s, CO in the atmosphere was increasing at a rate of 1.0% year in the northern hemisphere (WMO, 1995). Marine boundary layer measurements for the period June 1990 to June 1993 indicated that CO levels were decreasing at a rate of approximately 5.6% year in both the northern and southern hemispheres (Novelli et al., 1994). Larger decreases were found from the end of 1991 to 1993. The decreases over the period 1987 to 1992 were found to be in the range of 1-2% (Khalil and Rasmussen, 1994). Khalil and Rasmussen (1994) attributed much of this decrease to reduction in biomass burning in the tropics, starting from 1987. Novelli et al. (1994) estimated that changes in emissions from biomass burning and increased pollution controls in the northern hemisphere could account for only 20-30% of the change in CO. Thus, the negative trends in the CO, particularly the accelerated decreases during the 1990-1993 period, have not been satisfactorily explained. A possibility exists that the enhanced decreases during the 1990-1993 could have resulted from changes in the tropospheric oxidant levels. Large reductions were observed in column ozone during the period, as a result of the increased sulfate loading in the lower stratosphere from the Mount Pinatubo explosion.

Another key consideration is the nonlinear nature of the coupling between methane, carbon monoxide, hydroxyl, ozone, and nitrogen oxides. A greatly simplified form (Thompson et al., 1989) of the photochemical steady-state equation for hydroxyl concentration is:

$$[OH] = \frac{2k[O(^1D)]\,[H_2O]}{k[CO] + k[CH_4] + k[NMHC]}$$

where k is the chemical reaction rate constant and [X] is the species concentration.

In the unpolluted troposphere, an increase in carbon monoxide, methane, and non-methane hydrocarbons decreases hydroxyl. At higher concentrations of nitrogen oxides, ozone formation increases with methane and carbon monoxide and feeds back to form hydroxyl via the reaction between excited oxygen atoms ($O(^1D)$) and water vapor; the numerator and denominator both increase in this case. As the nitrogen oxide levels get even higher, hydroxyl again decreases as carbon monoxide is added. At any given time a perturbation in emissions of one trace gas produces a change that depends on the ambient levels of other species. This implies that a particular species may not be perturbed in the future the same way that it has been in the past. Therefore, we see that predictions of ozone and hydroxyl become very difficult when emission rates change.

2.4 Modeling Atmospheric OH Concentrations

Interest in the chemical cycles of trace species has elicited much research into the effects of methane perturbations. Several studies (e.g., Khalil and Rasmussen, 1985; Levine et al., 1985; Thompson and Cicerone, 1986a; Thompson and Kavanaugh, 1986; Hough, 1991) have concluded that the increasing mixing ratio of methane may be caused by both its increasing emission rate and the downward perturbation of the tropospheric hydroxyl concentration. Khalil and Rasmussen (1985) conclude that some depletion of hydroxyl -- on the order of 20% -- is likely to have occurred over the past 100-200 years. They suggest that most of the increase in methane (about 70%) is due to the increase in emissions from anthropogenic sources. Model calculations by Pinto and Khalil (1991) and Lu and Khalil (1991) also suggest that changes in OH have only played a minor role in explaining the increases in methane over the last few centuries.

Anthropogenic emissions of methane, nitrogen oxides, carbon monoxide, and non-methane hydrocarbons in the Northern Hemisphere appear to have increased substantially over the last several decades, with corresponding increases in their concentrations (Isaksen, 1988; Penkett, 1988; WMO, 1989, 1991; IPCC, 1990, 1995). Liu et al. (1988) argues that there should have been a decrease of hydroxyl levels in the Southern Hemisphere due to increases in methane and carbon monoxide. They attribute this to low nitrogen oxide emissions. It is not possible to imply such trends of hydroxyl in the Northern Hemisphere. Unfortunately, only a few measurements of hydroxyl exist and they are tainted by uncertainty. Existing hydroxyl measurements only provide an indication of local hydroxyl levels and have not yet provided useful information about global distributions and their trends.

Model studies of the relationship of hydroxyl to other species provide the best alternative to understanding its atmospheric behavior provided that the chemistry is well understood. Isaksen (1988) determined that the hydroxyl distribution calculated in atmospheric models agrees reasonably well with indirect determinations obtained from measurements of trace species oxidized by hydroxyl. Thompson et al. (1989), using a one-dimensional model of global tropospheric chemistry, show that increasing carbon monoxide and methane emissions, while holding nitrogen oxide levels constant, will decrease hydroxyl and increase ozone in all remote regions they have analyzed. Isaksen and Hov (1987) use their two-dimensional model to show that hydroxyl levels exhibit a strong seasonal and latitudinal variation. Based on their calculated hydroxyl distribution, the oxidation rate also changes dramatically with season and latitude. The largest changes are evident at latitudes greater than 50¡ in the winter hemisphere. Over the year, hydroxyl levels change by an order of magnitude and even methane varies noticeably. Isaksen and Hov estimate that global hydroxyl levels may have changed only slightly in recent decades. Lu and Khalil (1991) and Pinto and Khalil (1991) reached similar conclusions. However, it is difficult to judge the extent of the perturbation until the distributions and trends of other relevant constituents (e.g., nitrogen oxides) are better understood. Ehhalt et al. (1991) generally find a consistency between their limited hydroxyl measurement database and regional model calculations.

Estimates of future trends in hydroxyl concentrations have been obtained from one-dimensional and two-dimensional atmospheric models. Thompson et al. (1989) studied future atmospheres characterized by higher methane levels and increased anthropogenic emissions of carbon monoxide and nitric oxide with a one-dimensional model. Specifically, they found that most remote regions will lose hydroxyl as a result of increasing carbon monoxide and methane if nitrogen oxide levels remain about the same. They state that this should be the case in the mid- and higher-latitude Southern Hemisphere, which is mostly ocean, and perhaps over large regions of the tropics where 72% of the 60N-60S total of hydroxyl is found. In addition, near areas where biomass burning takes place, emissions of carbon monoxide, nitric oxide, methane, and non-methane hydrocarbons may suppress hydroxyl.

In a similar effort, Thompson et al. (1990) used a one-dimensional model to calculate future changes in tropospheric ozone and hydroxyl due to carbon monoxide, methane, and nitrogen oxide emissions for chemically coherent regions during the years from 1985 to 2035. One of the scenarios studied was an increase in methane mixing ratio of 0.8% per year and carbon monoxide mixing ratio of 0.5% per year. They found that initially there was an increase in hydroxyl for a few years in the urban mid-latitude region followed by a decrease, resulting in only a few percent change by the end of the simulation (see Figure 3). In all the lower nitrogen oxide regions, hydroxyl losses were dominant and regions with the smallest ozone increases had the greatest loss of hydroxyl. They found that urban and nonurban regions may respond quite differently to chemical and climatic perturbations which is due to differences in nitrogen oxide levels. Also, the effects of large-scale stratospheric ozone change or other chemical changes in the troposphere may counteract the influences of increased methane and carbon monoxide emissions. Isaksen et al.

(1987) used their two-dimensional tropospheric model, which has a domain from the surface up to 14 km, to examine a scenario for a 1.5% per year increase in the emission of methane for the years 1950--2010. They found that this gave rise to an average increase in ozone of 0.45% per year while the hydroxyl radical concentration dropped by about 0.4% annually.

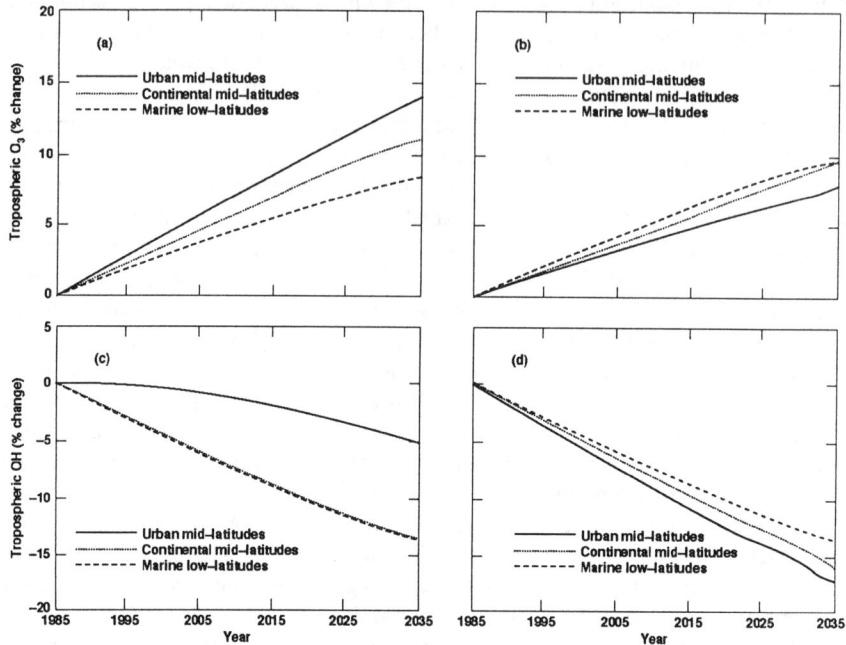

Fig. 3. Calculated percentage change in tropospheric ozone (a and b) and hydroxyl (c and d) during the years 1985 to 2035 for an increase in the methane mixing ratio of 0.8% per year. Parts (a) and (c) depict urban as well as continental mid-latitudes and marine low-latitude regions. Parts (b) and (d) are for marine as well as southern hemisphere mid-latitudes and continental low-latitudes (Thompson et al., 1990).

As an indication of the possible importance of the relationship between methane and hydroxyl, Figure 4 shows calculated changes in global mean concentration of methane as a function of changes in the methane flux (relative to the present emissions). In these model calculations, only the methane flux was allowed to change. The importance of hydroxyl feedback is demonstrated by the non-linear response in the calculated methane concentration for a change in the surface flux. As described above, the actual response would depend on the changes occurring in emissions and concentrations of other atmospheric constituents as well.

2.5 Tropospheric Ozone

2.5.1 Review of Tropospheric Ozone Sources and Sinks. Approximately 10% of the ozone in the atmosphere is located in the troposphere. The downward transport

of ozone from the stratosphere traditionally was thought to be the major source of tropospheric ozone (Crutzen, 1988, and references therein). It is now generally regarded that the net tropospheric photochemical production of ozone is of similar magnitude to the downward transport source (e.g., Fishman et al., 1979; Fishman, 1985; WMO, 1985; Isaksen, 1988; Penkett, 1988; Hough and Derwent, 1990; WMO, 1995). Further support is lent to this notion from model calculations cited by Isaksen (1988) and Penkett (1988) that show the column-integrated photochemical production of ozone is of the same order of magnitude as the stratospheric ozone flux.

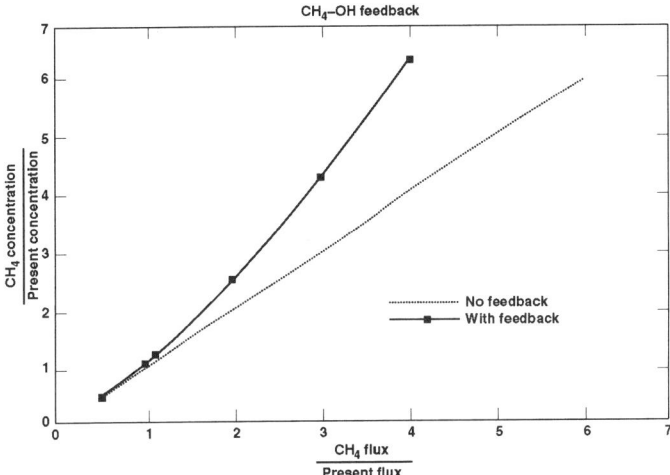

Fig. 4. Calculated changes in average global mean concentration of CH_4 as a function of changes in fluxes. The dotted line represents no feedback between OH and CH_4 (WMO, 1991).

by volume, pptv (Fishman et al., 1979; WMO, 1985). Oceans or regions of the world characterized by low nitric oxide concentrations are probably a net photochemical sink of odd oxygen (Liu et al., 1983; WMO, 1985).

Nitrogen dioxide photolysis is the only known photochemical mechanism for producing ozone in the troposphere. This implies that the generation rate is roughly proportional to the concentration of nitric oxide. High concentrations of nitrogen oxides over the continental boundary layer signify that this region is likely a net source of ozone. Increases in nitric oxide emissions may lead to further ozone increases, especially in the tropics. However, the magnitude of odd oxygen (O + $O(^1D)$ + O_3) production is ultimately limited by the supply of carbon monoxide, methane, and nonmethane hydrocarbons. The oxidation of one mole of carbon monoxide molecule can form one mole of ozone. In contrast, the complete oxidation of a mole of methane can produce 3 to 4 moles of ozone. Because there are insufficient nitrogen oxides present in the background troposphere, only about 10% of the potential tropospheric ozone production is being realized.

Ozone can be removed from the troposphere by dry deposition and photochemical mechanisms. Dry deposition occurs when an atmospheric constituent is transferred directly to a surface such as leaves or soils. Photochemical ozone destruction in the troposphere occurs through a variety of processes. The two primary mechanisms are photolysis and reaction with hydroperoxyl. According to the studies performed by Fishman et al. (1979) and Hough and Derwent (1990), these two reactions account for most (i.e., 70-90%) of the destruction of ozone in the troposphere. Ozone also can react with hydroxyl but this chemical sink plays a minor role based on the calculations in Fishman's paper. In addition, the rate of loss of ozone is almost independent of nitrogen oxide levels for concentrations below 200 parts per trillion

Ozone is the primary driver of the photochemistry that recycles gases emitted from natural and anthropogenic sources because its photolysis regulates hydroxyl formation. Initially, ozone is photolyzed by ultraviolet light:

$$O_3 + h\nu = O(^1D) + O_2 \quad (\lambda \leq 330) \tag{49}$$

Most of the excited oxygen atoms are collisionally deactivated by nitrogen and oxygen (collectively represented as M) back to the ground state:

$$O(^1D) + M = O + M \tag{50}$$

The remainder react with water vapor to form hydroxyl:

$$O(^1D) + H_2O = 2\ OH \tag{51}$$

This reaction constitutes the primary tropospheric source of hydroxyl radicals. Due to this interconnection, the present composition of the earth's atmosphere would be totally different if ozone were not present because hydroxyl removes many trace gases from the troposphere.

2.5.2 Methane Oxidation and Tropospheric Ozone Formation. The methane oxidation pathways regulate the tropospheric ozone budget. For this analysis, we will sum the net reactions and take the formaldehyde branching ratios into account to obtain a stoichiometric relation involving methane and carbon dioxide. In regions characterized by high nitric oxide concentrations, the net methane oxidation reaction is:

$$CH_4 + 7.2\text{-}7.5\ O_2 = CO_2 + 0.4\text{-}0.5\ OH + 3.6\text{-}3.8\ O_3 \\ + 1.2\text{-}1.4\ H_2O + 0.5\text{-}0.6\ H_2 \tag{52}$$

We see that complete methane oxidation results in net production of ozone in the polluted troposphere. For areas with low nitric oxide levels, the net methane oxidation reaction is:

$$CH_4 + 1.5\text{-}1.7\ OH + 2.2\ HO_2 + 1.7\text{-}1.8\ O_3 = CO_2 + 3.3\text{-}3.6\ H_2O + 3\ O_2 \\ + 0.4\text{-}0.6\ H_2 \tag{53}$$

In this case, complete methane oxidation results in net destruction of approximately 1.7-1.8 moles of ozone. The effect of methane on tropospheric ozone has been summarized in Table 1.

The formation of tropospheric ozone occurs primarily through smog formation-type mechanisms that involve peroxy radicals, including hydroperoxyl, methylperoxyl, or any complex organic peroxy radical. Recall that formation of tropospheric ozone occurs because nitric oxide reacts with peroxy radicals (illustrated with hydroperoxyl) to form nitrogen dioxide:

$$HO_2 + NO = OH + NO_2 \qquad (41)$$

Then, nitrogen dioxide can photolyze to produce an oxygen atom and nitric oxide. The oxygen atom then combines with molecular oxygen to form ozone:

$$NO_2 + hn = NO + O \quad (\lambda \leq 400 \text{ nm}) \qquad (42)$$

$$O + O_2 + M = O_3 + M \qquad (43)$$

The nitric oxide molecule can react again with peroxy radicals to repeat this process. Thus, nitrogen oxides act as a catalyst to produce ozone in the troposphere. At the low nitric oxide levels characteristic of the remote troposphere, peroxy radicals react directly with ozone to destroy it:

$$HO_2 + O_3 = OH + 2 O_2 \qquad (47)$$

Current data show that the rate constant for reaction (41) is about 4000 times greater than that for reaction (47). Therefore, the formation cycle dominates the destruction cycle if the concentration ratio of nitric oxide to ozone exceeds 1:4000. For example, model results cited by Fishman et al. (1979), Crutzen (1988) and Cicerone and Oremland (1988) indicate that the nitric oxide concentration must be greater than 5-10 pptv in the lower troposphere and above 20 pptv in the upper troposphere for this to happen.

Over the last several years many researchers have focused their attention on global changes in tropospheric ozone. The studies performed to date show that the effects of nitrogen oxide levels are crucial, particularly where their concentrations are low: hydroxyl concentrations tend to decrease, hydroperoxyl and hydrogen peroxide increase, and ozone appears to be insensitive to methane changes. Isaksen (1988) and Penkett (1988) have arrived at several conclusions based on their assessment:

Thompson et al. (1989) concluded from their modelling study that nitric oxide, carbon monoxide, and methane emission increases will suppress hydroxyl and increase ozone. These trends may be opposed by stratospheric ozone depletion and climate change. Stratospheric ozone depletion would tend to decrease ozone (except where nitrogen oxide levels are high) and increase hydroxyl through enhanced ultraviolet photolysis. Increased levels of water vapor (one possible result of climate change) also would decrease ozone and increase hydroxyl. Liu et al. (1988) have come to the conclusion that tropospheric ozone has increased in the Northern Hemisphere. Crutzen (1988) also suspects that ozone and hydroxyl concentrations are

Table 1. Change in radiative forcing (Wm^{-2}) due to increasing methane composed to total change due to all greenhouse gases as derived by IPCC (1990) for several future scenarios. The percentage effect of methane on the total change in radiative forcing is also shown only direct methane effects are included.

Scenarios		1765Ð2025	1765Ð2050	1990Ð2025	1990Ð2050
BaU	CH$_4$	0.72	0.90	0.30	0.48
	All	4.59	6.49	2.14	4.04
	%	15.7	13.9	14.0	11.9
B	CH$_4$	0.56	0.65	0.14	0.23
	All	3.8	4.87	1.35	2.42
	%	14.7	13.3	10.4	9.5
C	CH$_4$	0.51	0.53	0.09	0.11
	All	3.63	4.49	1.18	2.04
	%	14.0	11.8	7.6	5.4
D	CH$_4$	0.47	0.43	0.05	0.01
	All	3.52	3.99	1.07	1.54
	%	13.3	10.8	4.7	0.6

1. Nitrogen oxides, carbon monoxide, methane, and nonmethane hydrocarbons are the major ozone precursors. Nitrogen oxides are usually the limiting species due to their short lifetime and widely scattered sources.
2. Ozone produced from anthropogenic nitrogen oxide emissions is a significant contributor to the budget in the Northern Hemisphere and is the dominant term in the budget of the planetary boundary layer over industrialized areas.
3. Ozone concentrations measured in the boundary layer over industrialized areas during the last forty years exhibit increasing trends that range from 20-100%. Northern Hemispheric observations made at remote stations and in the free troposphere also show an increase of 0.5-2% per year during the past several decades. An increase greater than a factor of two is obtained by comparing observations made during the end of the nineteenth century at a station near Paris to present-day observations made at stations with similar ambient conditions (Bojkov, 1986; Volz and Kley, 1988).
4. The observed seasonal variations of surface ozone and peroxyacetyl nitrate tend to indicate that the spring ozone maximum in the Northern Hemisphere could be influenced significantly by photochemical production during the winter and spring.

decreasing in clean atmospheric environments and increasing at midlatitudes in the Northern Hemisphere. He cites a previous study that indicates average surface ozone concentrations at remote locations are changing in the expected directions. Following the Pinatubo volcanic eruption in 1991, there are indications that a change in tropospheric ozone may have resulted from the decline in methane discussed earlier (Bekki et al., 1994).

Oltmans and Levy (1994) analyzed surface ozone data from a network of 14 stations collected over several years. A small positive annual average trend in ozone mixing ratios were obtained mid and high-latitude northern hemisphere stations; no trend for a southern hemisphere tropical station and a negative trend near south pole. Because of the large day-to-day and seasonal variations, the significance of these trends is difficult to judge and needs to be further verified. Logan (1994) analyzed vertical ozone sonde profile data. Beginning from 1970 ozone amounts for the entire troposphere increases by 2% over Europe and less than 1% over eastern USA.

Long-term ozonesonde balloon measurement programs are in operation at a small number of ground stations (Tiao et al., 1986) located at northern midlatitudes (there is also one in the southern hemisphere). Ozone soundings at these sites mostly began in the early 1970s although some started in the late 1960s (Tiao et al., 1986). By analyzing the ozone time series from the beginning of the data sets and averaging over all the stations, a statistically significant increasing trend in the tropospheric ozone concentrations at northern midlatitudes is observed (Logan, 1985; WMO, 1985, 1989, 1991, 1995; Tiao et al., 1986; Lacis et al., 1990; Miller et al., 1992). The ozone increase from these measurements is largest in the lower troposphere, with ozone amounts increasing about 8 % per decade in the lowest kilometer of the atmosphere. However, the limitations (e.g. sparsity of the stations, sign of the tropospheric trend differs for each station) are such that these data sets do not necessarily provide conclusive evidence for a global tropospheric ozone increase.

2.6 Formaldehyde Production from Methane Oxidation

Regardless of the nitric oxide levels, methane oxidation is responsible for much of the formaldehyde in the atmosphere. Based on the estimate of Lowe and Schmidt (1983), the formaldehyde production rate is on the order of 10^{14} grams per year. The other major photochemical precursors of formaldehyde consist of nonmethane hydrocarbons and higher aldehydes. The nonmethane hydrocarbons are emitted from natural and anthropogenic sources while the higher aldehydes are produced by in situ photochemistry. There are appreciable anthropogenic sources of formaldehyde as well (e.g., automobile exhaust).

In order to analyze the photochemical mechanisms that produce formaldehyde, we must distinguish between different regions on the basis of nitrogen oxide levels. As we can see from the methane oxidation mechanism under low nitrogen oxide conditions, methyl peroxide reacts with hydroxyl to produce formaldehyde:

$$CH_3O_2H + OH = CH_2O + OH + H_2O \qquad (26)$$

Since methyl peroxide has a long lifetime against photolysis, heterogeneous processes (which occur on aerosols) can serve as a loss mechanism for this species in unpolluted environments. This implies that formaldehyde production is controlled by aerosol concentrations in remote regions. However, in polluted environments formaldehyde formation does not involve methyl peroxide. In this case, it would be regulated by nitric oxide concentrations because methylperoxyl is reduced to methoxyl. Methoxyl then reacts with oxygen to produce formaldehyde:

$$CH_3O_2 + NO = CH_3O + NO_2 \qquad (54)$$

$$CH_3O + O_2 = CH_2O + HO_2 \qquad (55)$$

Once it is generated, formaldehyde is destroyed rather quickly under direct insolation. During the daytime, photolysis is the dominant loss process for formaldehyde in the remote troposphere while in polluted atmospheres reaction with hydroxyl and hydroperoxyl will be comparable to photolytic degradation. At night, the only appreciable formaldehyde consumption is by reaction with nitrate radical (NO_3) but this is slow in comparison to the daytime reactions with hydroxyl and hydroperoxyl. In summary, formaldehyde is an important intermediate in the removal processes of hydrocarbons as well as the general chemical reactivity of the troposphere.

3. Effects on Upper Atmospheric Chemistry

Although about 85 % of the total emissions of methane is consumed by reaction with tropospheric hydroxyl (Cicerone and Oremland, 1988), the remaining methane flux, on average about 60 teragrams of methane per year (Tg CH_4 / year), enters the stratosphere. In the stratosphere and above, the reaction with OH continues to be the dominant sink, but reactions with chlorine atoms and excited oxygen atoms are also important. Reaction with chlorine atoms account for about 9 % of the methane loss (Brenninkmeijer et al., 1995). Recent studies (e.g., Burnett and Burnett, 1995) also indicate that the increased oxidation of methane resulting from increasing concentrations of chlorine in the stratosphere has led to enhanced production of stratospheric OH.

3.1 Stratospheric Ozone

3.1.1. A Review of Stratospheric Ozone Formation and Depletion. In order to put the role of methane in stratospheric chemistry into context, it is useful to first discuss the importance of ozone and the changes occurring in its distribution. Changes in the distribution and amount of ozone in the global troposphere and stratosphere have received much attention. Much of the concern about ozone has centered on the importance of ozone as an absorber of ultraviolet radiation; its concentrations determine the amount of ultraviolet radiation reaching the Earth's surface. Absorption of solar radiation by ozone also explains the increase in temperature with altitude found in the stratosphere. Finally, ozone is also a greenhouse gas and can influence climate.

Approximately 90 % of the ozone in the atmosphere is contained in the stratosphere. In the stratosphere, the production of ozone begins with the photodissociation of oxygen (O_2) at ultraviolet wavelengths less than 242 nm. This reaction produces two ground-state oxygen atoms that can react with oxygen to produce ozone. Since an oxygen atom is essentially the same as having an ozone, it is common to refer to the sum of the concentrations of ozone and oxygen atoms (both ground state and excited state) as odd-oxygen. The primary destruction of odd-oxygen in the stratosphere comes from catalytic mechanisms involving various free radical species. Nitrogen oxides, chlorine oxides, and hydrogen oxides participate in catalytic reactions that destroy odd-oxygen.

The chlorine and bromine catalytic mechanisms are particularly efficient. Because of the growing levels of reactive chlorine in the stratosphere resulting from emissions of trichlorofluoromethane ($CFCl_3$), difluorodichloromethane (CF_2Cl_2), along with bromine-containing halons and other halocarbons, these mechanisms have been the subject of much study due to their recent as well as future potential effects on concentrations of stratospheric ozone. These catalytic cycles can turn over thousands of times before the catalyst is converted to a less reactive form. Because of this cycling, relatively small concentrations of reactive chlorine or bromine can have a significant impact on the amount and distribution of ozone in the stratosphere. The total amount of chlorine in the current stratosphere is about 3 parts per billion by volume (ppbv), much of which is in the form of less reactive compounds like hydrochloric acid (HCl).

Methane plays an important role in the chlorine chemistry of the stratosphere, serving both as a source and a sink in key reactions affecting reactive chlorine. The direct reaction of methane with a chlorine atom is the primary source of hydrochloric acid, the primary chlorine reservoir species. However, hydroxyl produced through the oxidation of methane in the stratosphere can react with the hydrochloric acid to return the chlorine atom, thus reinitiating the chlorine catalytic mechanism. As mentioned earlier, measurements of OH in the stratosphere (Burnett and Burnett, 1995) suggest that there has been an increase in OH, largely as a result of the increasing levels of stratospheric chlorine.

In addition to being involved in the reaction taking reactive chlorine to the less reactive hydrochloric acid, methane has several other effects on stratospheric ozone. Hydrogen oxides produced from the dissociation of methane can react catalytically with ozone, particularly in the upper stratosphere. In the lower stratosphere, the primary effect of these hydrogen oxides is to react with nitrogen oxides and reactive chlorine, reducing the effectiveness of the ozone destruction catalytic cycles involving nitrogen oxides and chlorine oxides. The hydrogen oxides at these altitudes can also react catalytically to destroy ozone.

Measurements of ozone from ground-based stations and from satellites indicate that concentrations of ozone in the stratosphere have been decreasing. Ozone at 3 millibars (mbars), about 40 km altitude, has been decreasing globally by 3-4% per decade, in good agreement with the model calculations of the expected effects from chlorofluorocarbons (CFCs) and other trace gas emissions (WMO, 1989, 1991, 1995; DeLuisi et al., 1989; Wuebbles et al., 1991). Satellite and surface-based measurements of the total ozone column indicate that stratospheric ozone is decreasing throughout much of the world, with the effects increasing with latitude in both hemispheres

(WMO, 1989, 1991, 1995; Stolarski et al., 1991). The ozonesonde and the SAGE satellite data sets (WMO, 1991, 1995; Miller et al., 1992) indicate that a significant fraction of this ozone change is occurring in the lower stratosphere and also indicate that ozone in the lower stratosphere is decreasing at a faster rate than can be explained by current theory. Part of this lower stratospheric ozone decrease can be explained by the dilution of the Antarctic ozone hole after its late springtime breakup, while recent studies (WMO, 1991, 1995) indicate that heterogeneous chemistry on stratospheric sulfate aerosols explains much, but not all, of the lower stratospheric ozone decrease. The role of increasing atmospheric methane in the changes occurring in the ozone distribution is not well understood.

Beginning in the late 1970s, a special phenomenon began to occur in the springtime over Antarctica, referred to as the Antarctic ozone "hole" (Solomon, 1988). A large decrease in total ozone is occurring over Antarctica beginning in early spring. Decreases in total ozone column of more than 60% as compared to historical values have been observed by both ground-based and satellite techniques. Measurements made in 1987 indicated that more than 95% of the ozone over Antarctica at altitudes from 15 to 20 km had disappeared during September and October (WMO, 1989). During the 1990s, the Antarctic ozone holes have characteristically been large.

Measurements also indicate that the unique meteorology during the winter and spring over Antarctica sets up special conditions producing a relatively isolated air mass (the polar vortex). Polar stratospheric clouds form if the temperatures are cold enough in the lower stratosphere, a situation which often occurs within the vortex over Antarctica. Heterogeneous reactions can occur between atmospheric gases and the particles composing these clouds. Measurements indicate that reactions of hydrochloric acid and chlorine nitrate ($ClONO_2$) on these particles can release reactive chlorine once the sun appears in early spring. Thus, the reactions on the cloud particles allow chlorine to be in a very reactive state with respect to ozone. The ozone hole ends in late spring with the breakup of the vortex. The weight of scientific evidence strongly indicates that man-made chlorinated (produced from CFCs) and brominated chemicals are primarily responsible for the substantial decreases of stratospheric ozone over Antarctica in springtime (WMO, 1989, 1991, 1995). Similar processes to those affecting the Antarctic region have also been measured in the Northern Hemisphere (WMO, 1995), although the larger variability in temperature and atmospheric dynamics in the Northern Hemisphere has resulted in smaller ozone decreases due to these processes.

3.1.2 Calculated Effects of Methane on Stratospheric Ozone. As indicated above, the increasing atmospheric concentrations of methane, carbon dioxide, carbon monoxide, nitrous oxide, and various chlorinated and brominated compounds are all thought to be affecting the distribution of ozone in the troposphere and stratosphere. There have been a number of research studies using numerical models to examine the combined effects on ozone from the increases occurring in concentrations of methane and the other gases listed above (Wuebbles et al., 1983, 1991; WMO, 1985, 1988, 1989, 1995; Stordal and Isaksen, 1987). When combined with the effects of the other trace gas emissions, it is difficult to evaluate the role of methane in the observed and projected ozone trends. For this reason, it is useful to examine studies that have only considered the effects of increasing methane on ozone.

Numerical models of atmospheric chemical and physical processes generally calculate that increasing methane concentrations result in a net ozone production in the troposphere and lower stratosphere and net ozone destruction in the upper stratosphere (Owens et al., 1982, 1985; WMO, 1985, 1991, 1995; Isaksen and Stordal, 1986). The net effect from these calculations has been that methane by itself causes a net increase in ozone. For a doubling of the methane concentration (early papers went from 1.6 to 3.2 ppmv, while recent analyses assume 1.7 to 3.4 ppmv), published effects on the calculated change in total ozone range from +0.3 % (Prather, in WMO, 1985) to +4.3 % (Owens et al., 1985). With radiative feedback effects included, the published model results tend to be in the upper end of this range (Owens et al., 1985; WMO, 1985; Isaksen and Stordal, 1986).

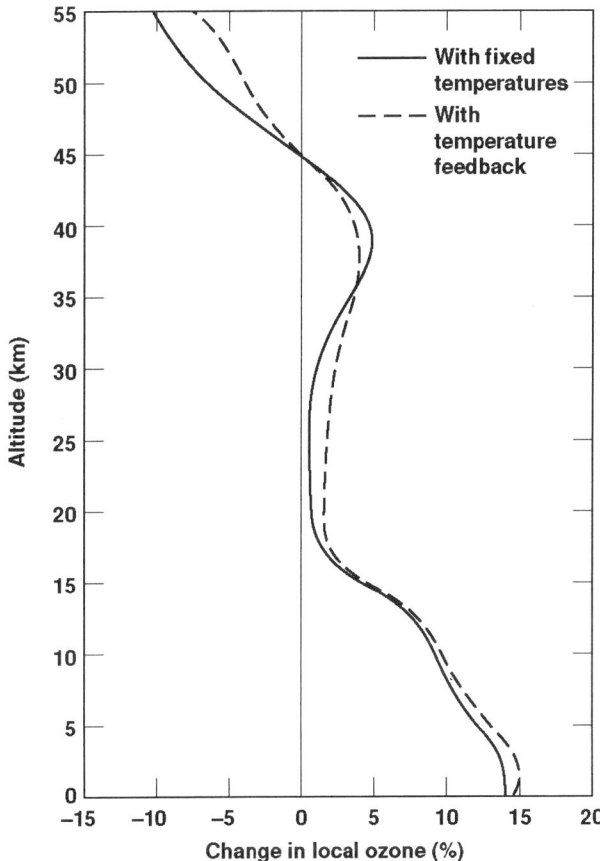

Fig. 5. Calculated percentage change in local ozone for a doubling in the concentration of atmospheric methane from 1.6 to 3.6 ppmv. Profiles obtained with the 1985 version of the Lawrence Livermore National Laboratory one-dimensional model (WMO, 1985).

As an example, the derived change in ozone from our earlier one-dimensional model as reported in WMO (1985) is shown in Figure 5. The calculated change in total ozone for a doubling of the methane concentration from this model (with temperature feedback) was +2.9 %. Calculated changes in ozone with altitude from other published studies (e.g., Owens et al., 1985; Isaksen and Stordal, 1986) are very similar. More recent analyses based on our two-dimensional chemical-transport model (e.g, Wuebbles et al., 1991; Wuebbles and Kinnison, 1996) are presented here. Figure 6 gives the changes in total ozone as a function of latitude and season, while Figure 7 shows the changes in ozone as a function of latitude and altitude for July. These two-dimensional model calculations gives a 3.4% increase in globally averaged total ozone from a doubling of methane.

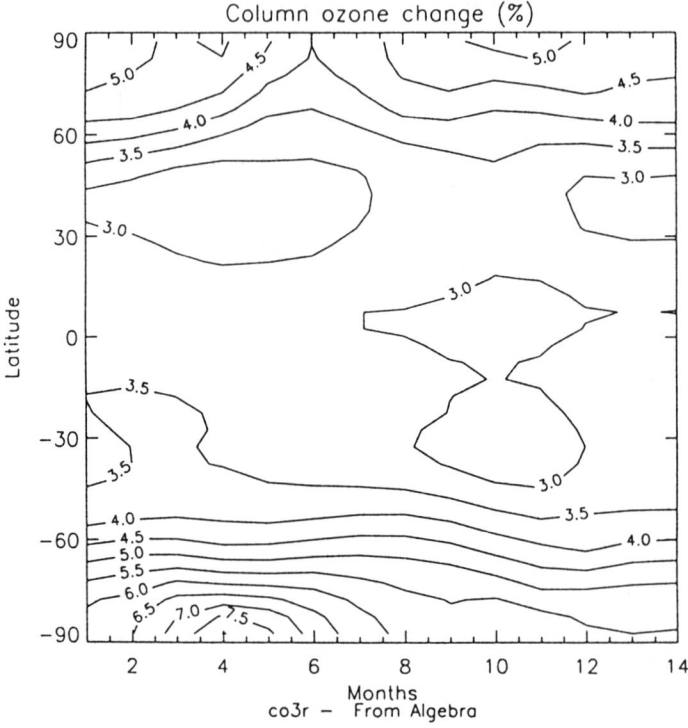

Fig. 6. Calculated percentage change of total ozone for a doubling in the concentration of atmospheric methane from 1.7 to 3.4 ppmv. Based on the 1997 version of our two-dimensional model as described in the text.

The calculations of increasing methane give a small percentage increase in the lower stratospheric ozone, a larger increase near 40 km altitude, and a decrease in ozone above 45 km. As mentioned earlier, the hydrogen oxides produced by methane oxidation affect the efficiency of the nitrogen oxide and chlorine oxide catalytic ozone destruction mechanisms. However, the effect of methane in the lower stratosphere will depend on the efficiency of the nitrogen oxide catalytic cycle; if the amount of reactive odd nitrogen is reduced, then the additional hydrogen oxides from methane could destroy ozone in this region. With the additional heterogeneous reactions involving, N_2O_5, $BrONO_2$ and $ClONO_2$ reduce the efficiency of the nitrogen oxide catalytic cycle and the increased HOx from double methane has a larger effect on the lower stratosphere than compared to previous calculations (Wuebbles and Tamaresis, 1993). In the upper stratosphere, the additional hydrogen oxides react catalytically with ozone, leading to the decrease in ozone determined at these altitudes in the model calculations.

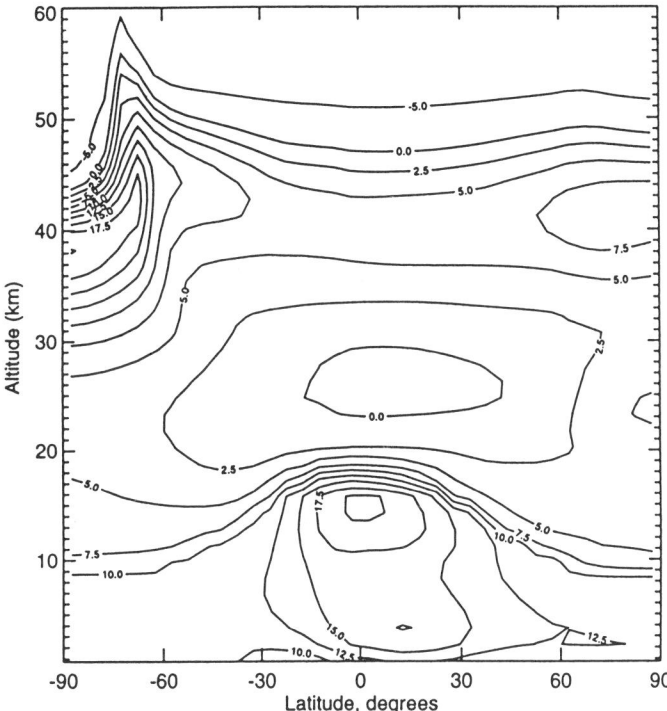

Fig. 7. Calculated percentage change of local ozone in July for a doubling in the concentration of atmospheric methane from 1.7 to 3.4 ppmv. Based on the 1997 version of our two-dimensional model.

Figure 8, which is adapted from WMO (1991), compares the calculated globally and annually averaged increase in tropospheric and lower stratospheric ozone from four different models for a doubling of the methane surface flux. While several of the models agree reasonably well with each other, there is more than a factor of two difference in the calculated response of ozone between some of the models. In these model calculations, the globally-averaged increase in tropospheric ozone for the methane increase ranges from 10.4 to 15.2 % (WMO, 1991). Even larger differences are found in the calculated changes in tropospheric hydroxyl from these models, with globally averaged hydroxyl decreases ranging from -10.2 to -17.7 % for a doubling of methane (WMO, 1991). The differences in these results are indicative of the large remaining uncertainties in modeling of tropospheric chemical and physical processes.

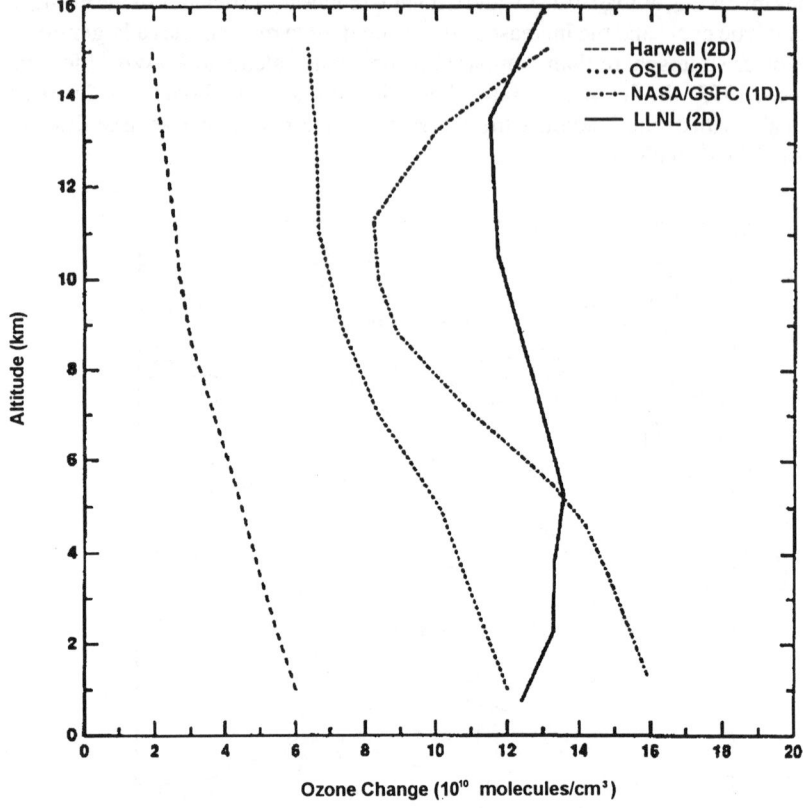

Fig. 8. Global annual average height profiles of O_3 changes for doubled CH_4 surface flux based on model calculations. Symbols in parentheses after model names indicate whether it is one-dimensional (1D) or two-dimensional (2D). (Adapted from WMO, 1991. LLNL curve from Wuebbles and Tamaresis, 1993.)

The current version of our two-dimensional model has an improved representation of the troposphere. This version of the model has a representation of convective mixing. Besides CH_4, several other hydrocarbons (NMHCs) including C_2H_6, C_3H_8, C_2H_4, C_3H_6, isoprene and their degradation products are represented. The surface emissions of isoprene were obtained from the GEIA inventory (Guenther et al., 1995) and the rest are set according to the estimates of Strand and Hov (1993). The importance of including hydrocarbons in a model in evaluating a $2xCH_4$ scenario is shown in Figure 9. The change in column ozone resulting from a no NMHC calculation and doubling CH_4 to that compared to a simulation with hydrocarbons is an increased response ranging from 3-8 % over that of case including NMHCs in the calculated change in column ozone. The effect on the Southern Hemisphere is negligible as there are no significant non-NMHC emissions from this region. Thus, the model response is damped for a doubling of methane surface concentration when non-methane hydrocarbons are included in the model. This results from the additional sources of reactive oxides available in the lower troposphere from hydrocarbons sources and the impact of doubling CH_4 in this region, mainly through reaction 5 is now reduced. The current model gives a feedback factor, defined as the % change in HO for a 1% change in CH_4 (WMO, 1995) of 0.32%, which is with in the range of calculated in WMO (1995).

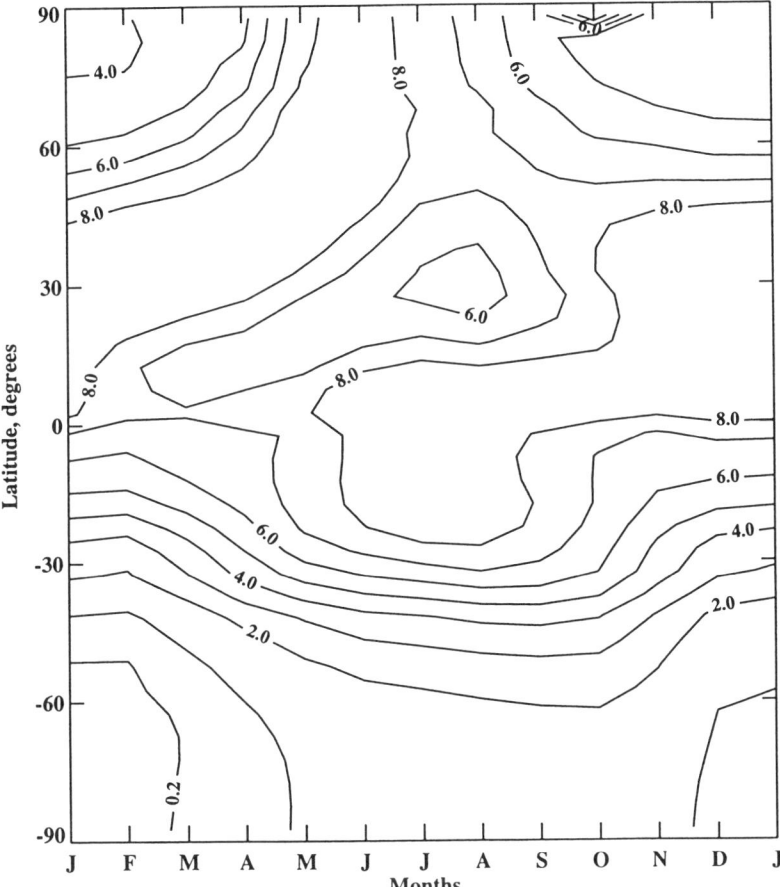

Fig. 9. Percent change in column ozone for a case with no non-methane hydrocarbons in the model to a case with non-methane hydrocarbons as a result of doubling the surface mixing ratio of methane from 1.7 to 3.4 ppmv. Based on the 1997 version of our two-dimensional model.

Figure 10 shows the change in tropospheric ozone at two latitudes (40 N and S) and two seasons (summer and winter) calculated with our two-dimensional model for a doubling of CH_4 surface mixing ratios. These results, along with those from Figure 6. suggest that significant differences may be expected as a function of season and latitude in the effect on ozone from increasing methane.

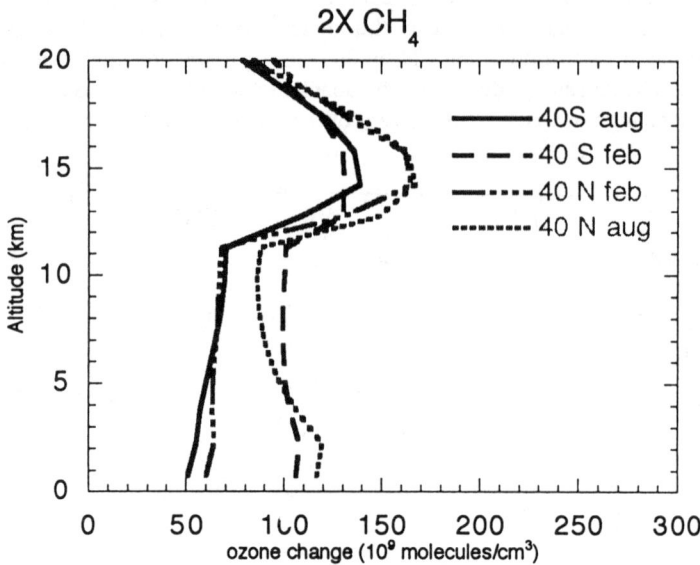

Fig. 10. Vertical profiles of ozone increases at 40°N latitude and 40°S latitude during February and August for doubled CH_4 emission (based on WMO, 1991).

3.2 Stratospheric Water Vapor

Concentrations of water vapor in the atmosphere vary from as much as 15,000 ppmv near the surface in the tropics to as low as 3 ppmv in the lower stratosphere (Ellsaesser et al., 1980; WMO, 1985; Parameswaran and Krishna Murthy, 1990; Schwab et al., 1990). The spatial distribution of water vapor in the troposphere is primarily determined by evaporation, condensation and transport processes. Human activities are currently thought to have little impact on tropospheric water vapor concentrations. Increased water vapor concentrations as a result of global warming is a well recognized climatic feedback process; increasing temperatures allow more water vapor to remain in the atmosphere, but, since water vapor is one of the most important greenhouse gases, the added water vapor further enhances the greenhouse radiative forcing.

Very little of the tropospheric water vapor penetrates into the stratosphere. The mechanisms limiting the transport of tropospheric water vapor into the stratosphere is still not well understood. As a consequence, it is not known how water vapor concentrations in the lower stratosphere will respond to climate change effects on tropospheric water vapor concentrations. Concentrations of water vapor increase with altitude in the stratosphere, from 3 ppmv in the lower stratosphere to about 6 ppmv in the upper stratosphere. This increase in concentration with altitude occurs as a result of the oxidation of methane.

The methane oxidation reactions roughly produce two moles of water vapor for each mole of methane that is destroyed. Stratospheric water vapor concentrations should increase as concentrations of methane increase. Since methane concentrations have increased from about 0.7 ppmv in the pre-industrial atmosphere to the current concentration of 1.7 ppmv, this implies that upper stratospheric water vapor concentrations have increased by roughly 2 ppmv over this time period. Actually the increase in water vapor should be somewhat less than this due to methane reactivity with chlorine and oxygen atoms. Both modeling and data analysis studies (e.g., Le Texier et al., 1988; Hansen and Robinson, 1989) are in agreement with this conclusion, indicating that the overall stratospheric water vapor yield from methane is somewhat less than two.

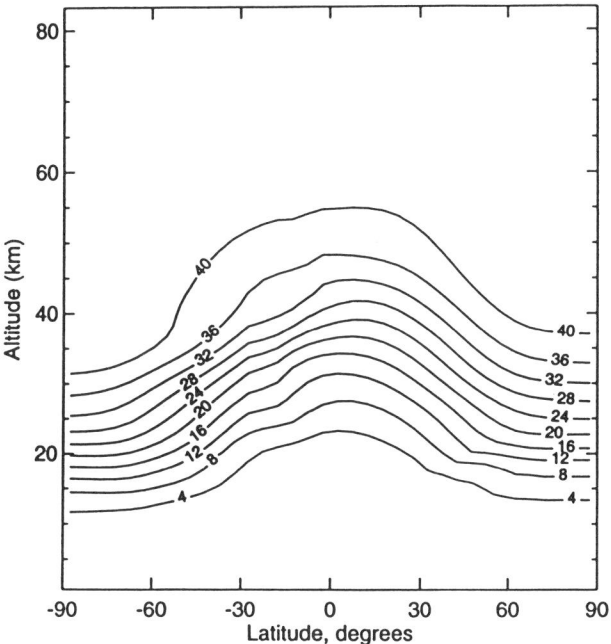

Fig. 11. Calculated change in stratospheric water vapor due to a increase in methane from 1.7 to 3.4 ppmv using our two-dimensional chemical transport model.

Figure 11 shows the increase in stratospheric water vapor calculated with our two-dimensional model for a doubling in the surface mixing ratio of methane. The calculated concentration of stratospheric water vapor at 50 km increases by approximately 40 % or about 3 ppmv for a doubling of the methane surface mixing ratio from 1.7 to 3.4 ppmv.

3.3 Polar Stratospheric Clouds

As discussed above, the occurrence of polar stratospheric clouds is thought to be an important element in the formation of the Antarctic ozone hole. Increases in stratospheric water vapor concentrations from methane oxidation could contribute to the formation of stratospheric clouds in the polar lower stratosphere, and to increased effectiveness of the heterogeneous processes determining the destruction of ozone presently occurring in the Antarctic (the springtime *ozone hole*) and, to a lesser degree, in the arctic (Blake and Rowland, 1988; Ramanathan, 1988a, b; WMO, 1989, 1995). Cooler stratospheric temperatures as a result of increasing concentrations of carbon dioxide and other greenhouse gases could also enhance the formation of polar stratospheric clouds (Ramanathan, 1988a,b; Shine, 1989; Blanchet, 1989). The implications of these processes are still highly uncertain; however, increasing concentrations of methane may be partially responsible for the apparent increase in the frequency of polar stratospheric clouds during the last several decades.

3.4 Water Vapor in the Mesosphere and Above

Increasing methane concentrations should also be leading to increasing concentrations of water vapor in the mesosphere. Increasing concentrations of methane and carbon dioxide should also lead to cooler temperatures in the stratosphere, mesosphere, and thermosphere (WMO, 1985, 1989; Brasseur and Hitchman, 1988; Roble and Dickinson, 1989). With the increase in water vapor and the cooler temperatures, there is the potential for increased occurrences of noctilucent clouds near the mesopause (Thomas et al., 1989). Roble and Dickinson (1989) also point out that other changes in mesospheric and thermospheric composition should occur due to both the cooler temperatures and direct chemical effects from increased concentrations of methane and carbon dioxide. In addition, exospheric hydrogen will increase with increasing methane (Ehhalt, 1986).

4. Effects on Climate

4.1 Direct Effects

The concern that human activities may be affecting global climate has largely centered around carbon dioxide because of its importance as a greenhouse gas and also because of the rapid rate at which its atmospheric concentration has been increasing. However, research over the last decade has shown that other greenhouse gases are contributing about half of the overall increase in the greenhouse radiative forcing on climate. Methane is a greenhouse gas. The increasing concentration of methane is a major contributor to the increase in the greenhouse effect. In addition to its direct radiative forcing effect on climate, methane can also influence climate indirectly through chemical interactions; these will be discussed below.

Like other greenhouse gases, methane absorbs infrared radiation (also called longwave or terrestrial radiation) emitted by the relatively warm planetary surface and emits radiation to space at the colder atmospheric temperatures, leading to a net trapping of infrared radiation within the atmosphere. This is called the greenhouse effect. The balance between the absorbed solar radiation and the emitted infrared radiation determines the net radiative forcing on climate.

Although its atmospheric abundance is less than 0.5 % that of carbon dioxide, methane is an important greenhouse gas. Donner and Ramanathan (1980) calculated that the presence of methane at current levels causes the globally averaged surface temperature to be about 1.3 K higher than it would be without methane. On a molar basis, an additional mole of methane in the current atmosphere is about 21 times more effective at affecting climate than an additional mole of carbon dioxide (Ramanathan et al., 1985, 1987; IPCC, 1990). Correspondingly, on a mass basis, an additional kilogram of methane is about 58 times more effective as a greenhouse gas than a kilogram of carbon dioxide. It should be noted, however, that the actual lifetime of carbon dioxide in the atmosphere is much longer than the atmospheric lifetime of methane (IPCC, 1990, 1996). But it should also be recognized that the dissociation of methane eventually produces carbon dioxide, leading to additional climatic forcing from the original emission of methane.

The strongest bands for absorption by methane in the infrared are in the short wavelength edge of the window region. The most important infrared spectral feature of methane is the 7.66 mm (1306 cm^{-1}) absorption band. Due to saturation of the line cores for methane and emission from the pressure-broadened Lorentz line wings, radiative forcing from methane increases approximately as the square root of its concentration (IPCC, 1990; Wigley, 1987). Overlap with absorption by water vapor and other species (particularly nitrous oxide) also affects the efficiency of methane absorption.

Wang et al. (1991) show that the greenhouse radiative forcing for methane has different effects on climate than carbon dioxide, and that methane needs to be accounted for explicitly when attempting to predict the climate response to increasing concentrations of greenhouse gases. However, few calculations with three-dimensional global climate models (GCMs) have yet considered the explicit effects of methane on climate. Hansen et al. (1988) and Wang et al. (1991) have explicitly considered the effects of methane in GCM studies, while many other studies have included methane only through accounting for its radiative forcing through use of an increase in carbon dioxide as a proxy. The vast majority of the climate modeling studies that have included methane directly have been done with radiative-convective models and other models with simplified treatments of climatic processes.

Increasing concentrations of methane are thought to be a significant fraction of the increase in radiative forcing from greenhouse gases over the last two centuries. IPCC (1990) calculated that the direct radiative effect of the increase in methane since the mid-1700s has accounted for an increase in radiative forcing of 0.42 Wm^{-2} (climate models indicate that the 4 Wm^{-2} associated with a doubling of CO_2 from 300 to 600 ppmv would give approximately a 1.5 to 4.5 K increase in surface temperature). This is about 17 % of the total change in radiative due to CO_2 and other greenhouse gases over this time period (other research studies have found similar percentages for the effect of methane over this period: e.g., Rodhe (1990) derived 15 %, while Hansen et al. (1989) and MacKay and Khalil (1991) got about 22 %). Inclusion of indirect effects on stratospheric water vapor, ozone, and carbon dioxide (see below) could increase this percentage appreciably; inclusion of the stratospheric water vapor effect in IPCC(1990) increases this percentage to 23 %.

Over the last decade, the change in atmospheric methane concentration is calculated to increase the radiative forcing by about 0.06 Wm^{-2}, about 11 % of the total increase in radiative forcing from greenhouse gases over this time period. Hansen et al. (1988) determined that methane was 12.2 % of the total change in radiative forcing over this period (note that these estimates ignore effects on radiative forcing due to changes in global ozone and aerosols over this time period).

Various studies have evaluated the potential effects on radiative forcing and surface temperature from a doubling of methane concentrations using radiative-convective models. For a doubling of methane concentration from 1.6 to 3.2 ppmv, effects on surface temperature range from 0.2 K to 0.3 K (Wang et al., 1976; Donner and Ramanathan, 1980; Lacis et al., 1981; Owens et al., 1985; Ramanathan et al., 1987; MacKay and Khalil, 1991), with the differences in model results primarily relating to uncertainties in the band strengths for methane infrared absorption. For a doubling from 1.7 to 3.4 ppmv, Owens et al. (1985) calculate a direct 0.34 K increase in surface temperature, along with an additional 0.26 K due to indirect effects from methane-induced effects on carbon dioxide and ozone. For a 25 % increase in methane concentrations, Ramanathan et al. (1985) determine a 0.08 K increase in surface temperature when overlap with the radiative absorption with other greenhouse gases is included, and a 0.19 K increase in surface temperature without overlap.

Other modeling studies have included increasing methane concentrations in studies evaluating scenarios for potential future changes in radiative forcing and global temperatures (Wang and Molnar, 1985; Ramanathan et al., 1985, 1987; WMO, 1985; Dickinson and Cicerone, 1986; Wang et al., 1986; Wigley, 1987; Hansen et al., 1988, 1989; IPCC, 1990). Table 1 shows the effect on radiative forcing due to assumed methane increases for several scenarios evaluated by IPCC (1990). For the IPCC Business-as-Usual scenario, the direct radiative effect of increasing methane accounts for 15.7 % of the increase in radiative forcing from 1765 to 2025 (14 % from 1990 to 2025). Similar effects to those in Table 1 were found in the other published scenario studies cited above. Table 2 show the corresponding effects for the change in radiative forcing (Wm^{-2}) due to increasing methane compared to total change due to all greenhouse gases as derived by IPCC (1996) for their high (IS92e), medium (IS92a), and low (IS92c) scenarios. These more recent scenarios generally show a somewhat increased role for methane in determining future radiative forcing. These studies do not consider the significant additional effects from chemical interactions as discussed in the next section.

Methane releases from northern gas fields and from gas hydrates may have been a significant contributor to the warming at the end of the last major glacial period (Nisbet, 1990a, b; MacDonald, 1990). Global warming could destabilize the storage of extensive amounts of methane in methane hydrates and clathrates, and lead to increased emissions of methane into the atmosphere, adding further to the greenhouse forcing (MacDonald, 1990). Climate change could also affect other natural sources of methane (Lashof, 1989).

Table 2. Change in radiative forcing (Wm^{-2}) due to increasing methane compared to total change due to all greenhouse gases as derived by IPCC (1996) for their high (IS92e), medium (IS92a), and low (IS92c) scenarios. The percentage effect of methane on the total change in radiative forcing is also shown.

Scenarios:

Time Period		1765-2025	1765-2050	1990-2025	1990-2050
IS92a					
	CH_4	0.66	0.83	0.19	0.36
	All	4.01	5.37	1.61	2.97
	%	16.4	15.4	11.8	12.1
IS92c					
	CH_4	0.59	0.65	0.12	0.18
	All	3.63	4.37	1.23	1.97
	%	16.2	14.9	9.7	9.1
IS92e					
	CH_4	0.69	0.89	0.22	0.42
	All	4.31	6.12	1.91	3.72
	%	16.0	14.5	11.5	11.3

4.2. Indirect Effects

There are several ways that methane, through its chemical interactions in the atmosphere, can indirectly influence climate. Oxidation of methane leads eventually to carbon dioxide, one of the most important greenhouse gases. About 450 Tg / yr of methane are destroyed by reaction with hydroxyl and converted to carbon dioxide, accounting for production of 340 Tg C / yr as carbon dioxide; in contrast, the production of carbon dioxide from anthropogenic fossil fuel use and cement manufacturing is about 6000 Tg C / yr. Other indirect effects on climate resulting from methane include: production of stratospheric water vapor, changes in tropospheric and stratospheric ozone, and changes in concentrations of tropospheric hydroxyl.

Even though the concentration of stratospheric water vapor is appreciably smaller than that in the troposphere, water vapor is such an important greenhouse gas that changes in the concentrations of stratospheric water vapor can influence global radiative forcing on climate. Therefore, an increase in stratospheric water vapor concentrations resulting from increasing methane concentrations further enhances the greenhouse effect, increasing the radiative forcing from the added methane. While earlier studies assuming a mixing ratio increase in methane suggested that the increase in water vapor could enhance the radiative forcing by as much as 30 % (IPCC, 1990; Wuebbles and Grant, 1991), more recent studies (Lelieveld and Crutzen, 1991, 1992; Brühl, 1993; Lelieveld et al., 1993; Hauglustaine et al., 1994; IPCC, 1995) using flux boundary condition changes in methane have shown the effect of the water vapor increase to be much smaller, 4-5% of the total direct and indirect radiative forcing from methane.

As discussed earlier, increasing water vapor from methane could be leading to an increased amount of polar stratospheric clouds. Ramanathan (1988b) notes that both water and ice clouds, when formed at cold lower stratospheric temperatures, are extremely efficient in enhancing the atmospheric greenhouse effect. He also notes that there is a distinct possibility that large increases in future methane may lead to a surface warming that increases nonlinearly with the methane concentration.

Changes in ozone associated with increasing methane can add 19 ± 12 % to the total radiative forcing from methane (IPCC, 1995; based on ours and other two-dimensional model studies). Determination of the effect due to OH changes is more difficult because of uncertainties associated with modeling the CH_4-CO-OH nonlinear relationships. The effect on the response time of methane to a perturbation can result in a much larger apparent lifetime for the perturbed methane than the overall atmospheric lifetime of methane (Prather, 1994; IPCC, 1995, 1996).

4.3 Global Warming Potentials

Global Warming Potentials (GWPs) have been developed as an analysis tool for policy makers in their evaluations of possible policy actions related to emissions of greenhouse gases. The GWP of a greenhouse gas is defined as the time-integrated commitment to climatic forcing from the instantaneous release of a kilogram of the gas relative to the climatic forcing from the release of 1 kg of carbon dioxide. Under this measure, the GWP for methane after a 100-year integration for the direct plus indirect methane effects is a value of 24.5 ± 7.5 (compared to 1.0 for carbon dioxide). As shown in IPCC (1990, 1995), indirect effects can more than double the direct methane GWP value (also see Fuglestvedt et al., 1996). Published GWPs are derived for integration periods from 20 to 500 years, with the 100-year values generally thought to provide a balanced representation of the various time horizons for climatic response. The GWPs for direct plus indirect effects from methane for the 20- and 500-year integrations are 62 ± 20 and 7.5 ± 2.5, respectively.

5. Summary

The growing concentrations of atmospheric methane require that the potential environmental effects from methane be understood. The necessity for furthering this understanding is made all the more urgent when one considers the variety of roles that methane plays in both the chemistry of the global atmosphere and in affecting the Earth's radiation balance. Research studies over the last few decades have shown that methane influences tropospheric ozone, hydroxyl radical, formaldehyde, and carbon monoxide concentrations, stratospheric chlorine and ozone chemistry, and, through its infrared absorption, the radiative forcing on climate. This chapter established that much has been learned over the last few decades about the effects of methane on atmospheric chemistry and climate. However, there are still remain significant uncertainties associated with determining the effects of methane on the global environment that require further study.

Acknowledgments. This study was supported in part through grants from the U. S. Environmental Protection Agency, from the National Aeronautics and Space Administration Atmospheric Chemistry Modeling and Analysis Program, and from the U. S. Department of Energy.

References

Bekki, S., K. S. Law, and J. A. Pyle, Effect of ozone depletion on atmospheric CH_4 and CO concentrations. *Nature, 371*, 595-597, 1994.

Blake, D. R., and F. S. Rowland, Continuing worldwide increase in tropospheric methane. 1978 to 1987, *Science, 239*, 1129-1131, 1988.

Blanchet, J. P. , The response of polar stratospheric clouds to increasing carbon dioxide. *Proceedings of the International Radiation Symposium*, Lille, France, August, 1988, A. Deepak Publishing, Hampton, Va, 1989.

Bojkov, R. J., Surface ozone during the second half of the nineteenth century. *J. Clim. Appl. Meteor., 25*, 343-352, 1986.

Brasseur, G., and M. H. Hitchman, Stratospheric response to trace gas perturbations: Changes in ozone and temperature distribution. *Science, 240*, 634-637, 1988.

Brenninkmeijer, C. A. M., D. C. Lowe, M. R. Manning, R. J. Sparks and P. F. J. van Velthoven, The ^{13}C, ^{14}C, and ^{18}O isotopic composition of CO, CH_4, and CO_2 in the higher southern latitudes lower stratosphere. *J. Geophys. Res., 100*, 26,163-26,172, 1995.

Browell, E. V., M. A. Fenn, and others, Large-scale air mass characteristics observed over western Pacific during summertime. *J. Geophys. Res., 101*, 1691-1712, 1996.

Brühl, C., The impact of the future scenarios for methane and other chemically active gases on the GWP of methane. *Chemosphere, 26*, 731-738, 1993.

Burnett, E. B., and C.R. Burnett, Enhanced production of stratospheric OH from methane oxidation at elevated reactive chlorine levels in northern midlatitudes. *J. Atmos. Chem., 21*, 13-41, 1995.

Cicerone, R. J., and R. S. Oremland, Biogeochemical aspects of atmospheric methane. *Global Biogeochemical Cycles, 2*, 299-327, 1988.

Craig, H., and C. C. Chou, Methane record in polar ice cores. *Geophys. Res. Lett., 9*, 1221-1224, 1982.

Crosley, D. R., Measurement of HO_x radicals in the atmosphere. *J. Atmos. Sci., 25*, 3297-3298, 1995.

Crutzen, P. J., Tropospheric ozone: An overview. in *Tropospheric Ozone: Regional and Global Scale Interactions*, edited by I.S.A. Isaksen, pp. 3-11, D. Reidel, Boston, 1988.

Crutzen, P. J., and P. H. Zimmermann, The changing photochemistry of the troposphere. *Tellus, 43AB*, 136-151, 1991.

Daum P H., L.I. Kleinman, L. Newman, W. T. Luke, J. Weinstein-Lloyd, C.M. Berkowitz and K. M. Busness, Chemical and physical properties of plumes of anthropogenic pollutants transported over the North Atlantic during the North Atlantic Regional Experiment. *J. Geophys. Res.*, *101*, 29,029-29,042, 1996.

DeLuisi, J. J., D. U. Longenecker, C. L. Mateer, and D. J. Wuebbles, An analysis of northern mid-latitude Umkehr measurements corrected for stratospheric aerosols for 1979-1986. *J. Geophys. Res.*, *94*, 9837-9845, 1989.

DeMore, W. B., S. P. Sander, D. M. Golden, R. F. Hampson, M. J. Kurylo, C. J. Howard, A. R. Ravishankara, C. E. Kolb, and M. J. Molina, *Chemical Kinetics and Photochemical Data for Use in Stratospheric Modeling*, NASA Jet Propulsion Laboratory, JPL Publication 97-4, 1997.

Dickinson, R. E., and R. J. Cicerone, Future global warming from atmospheric trace gases. *Nature*, *319*, 109-115, 1986.

Donner, L., and V. Ramanathan, Methane and nitrous oxide: Their effects on the terrestrial climate. *J. Atmos. Sci.*, *37*, 119-124, 1980.

Ehhalt, D.H., On the consequence of a tropospheric CH_4 increase to the exospheric density. *J. Geophys. Res.*, *91*, 2843, 1986.

Ehhalt, D. H., H.-P. Dorn, and D. Poppe, The chemistry of the hydroxyl radical in the troposphere. *Proc. Royal Soc. Edinburgh*, *97B*,17-34, 1991.

Ellsaesser, H. W., J. E. Harries, D. Kley, and R. Penndorf, Stratospheric H_2O. *Planet. Space Sci.*, *28*, 827-835, 1980.

Fishman, J., Ozone in the troposphere. in *Ozone in the free atmosphere* edited by R.C. Whitten and S.S. Prasad, pp. 161-194, Van Nostrand Reinhold, New York, 1985.

Fishman, J., S. Solomon, and P. J. Crutzen, Observational and theoretical evidence in support of a significant in-situ photochemical source of tropospheric ozone. *Tellus*, *31*, 432-446, 1979.

Fuglestvedt, J. S., I. S. A., Isaksen and W.-C. Wang, Estimates of indirect Global Warming Potentials for CH_4, CO and NO_x. *Climatic Change*, *34*, 405-437, 1996.

Guenther, A., et al., A global model of natural volatile organic compound emissions. *J. Geophys. Res.*, *100*, 8873-8892, 1995.

Hansen, A.R., and G.D. Robinson, Water vapor and methane in the upper stratosphere: An examination of some of the Nimbus 7 measurements. *J. Geophys. Res.*, *94*, 8474-8484, 1989.

Hansen, J., A. Lacis, and M. Prather, Greenhouse effect of chlorofluorocarbons and other trace gases. *J. Geophys. Res.*, *94*, 16417-16421, 1989.

Hansen, J., I. Fung, A. Lacis, D. Rind, S. Lebedeff, R. Ruedy, G. Russell, and P. Stone, Global climate changes as forecast by Goddard Institute for Space Studies three-dimensional model. *J. Geophys. Res.*, *93*, 9341-9364, 1988.

Hauglustaine, D.A., C. Granier, G. P. Brasseur and G. Megie, The importance of atmospheric chemistry in the cal;culation of radiative forcing on the climate system. *J. Geophys. Res.*, *99*, 1173-1186, 1994.

Heikes, B. G., Meehye Lee, and others, Hydrogen peroxide and methylhydroperoxide distributions related to ozone and odd hydrogen over the North Pacific in the fall of 1991. *J. Geophys. Res.*, *101*, ¡891-1906, 1996.

Hough, A. M., The development of a two-dimensional global tropospheric model: The model chemistry. *J. Geophys. Res.*, *96*, 7325-7362, 1991.

Hough, A. M.,, and R. G. Derwent, Changes in the global concentration of tropospheric ozone due to human activities. *Nature*, *344*, 645-648, 1990.

Intergovernmental Panel on Climate Change, *Climate Change: The IPCC Scientific Assessment*, Houghton, J. T., G. J. Jenkins, and J. J. Ephraums (eds.), Cambridge University Press, Cambridge, 1990.

Intergovernmental Panel on Climate Change (IPCC), *Climate Change 1994: Radiative Forcing of Climate Change and An Evaluation of the IPCC IS92 Emission Scenarios*, J. T. Houghton, Meira Filho, L.G., Lee, H., Callander, B.A., Haites, E., Harris, N., and Maskell, K. (Eds.). Cambridge University press, Cambridge, UK, 1995.

Intergovernmental Panel on Climate Change (IPCC), *Climate Change 1995: The Science of Climate Change.* J.T. Houghton, Meira Filho, L.G., Callander, B.A., Harris, N., Kattenberg, A., and Maskell, K. (Eds.). Cambridge University Press, Cambridge, UK, 1996.

Isaksen, I. S. A., Is the oxidizing capacity of the atmosphere changing, in *The Changing Atmosphere*, edited by F.S. Rowland and I.S.A. Isaksen, pp. 141-157. New York: John Wiley & Sons, 1988.

Isaksen, I. S. A., and O. Hov, Calculation of trends in the tropospheric concentration of O_3, OH, CO, CH_4, and NOx. *Tellus, 39B*, 271-285, 1987.

Isaksen, I. S. A., and F. Stordal, Ozone perturbations by enhanced levels of CFCs, N_2O, and CH_4: A two-dimensional diabatic circulation study including uncertainty estimates. *J. Geophys. Res., 91*, 5249-5263, 1986.

JAS (special issue), Measurement of HO_x radicals in the atmosphere, *J. Atmos. Science , 52*, 3297-3441, 1995.

JGR (special issue), 1993 Tropospheric OH photochemistry experiment. *J. Geophys. Res., 88*, 5,131-5,144, 1983.

Khalil, M. A. K., and R. A. Rasmussen, Sources, sinks, and seasonal cycles of atmospheric methane. *J. Geophys. Res.,102*, 6169-6510, 1983.

Khalil, M. A. K., and R. A. Rasmussen, Carbon monoxide in the earth's atmosphere. *Science, 224*, 54-56, 1984.

Khalil, M. A. K., and R. A. Rasmussen, Causes of increasing atmospheric methane: Depletion of hydroxyl radicals and the rise of emissions. *Atmospheric Environment, 13*, 397-407, 1985.

Khalil, M.A.K., and R. A. Rasmussen, Carbon monoxide in the earth's atmosphere; indications of a global increase. *Nature, 332*, 242-245, 1988.

Khalil, M. A. K., and R. A. Rasmussen, Temporal variations of trace gases in ice cores. In *The Environmental Record in Glaciers and Ice Sheets*, (H. Oeschger and C. C. Langway, Jr., eds.) John Wiley and Sons Limited, New York, p. 193-205, 1989.

Khalil, M.A.K., and R. A. Rasmussen, Atmospheric methane: recent global trends. *Environ. Sci. Tech., 24*, 549-553, 1990.

Khalil, M.A.K., and R. A. Rasmussen, Global decreases in atmospheric carbon monoxide concentration. *Nature, 370*, 639-641, 1994.

Khalil, M. A. K., R. A. Rasmussen, and M. J. Shearer, Trends of atmospheric methane during the 1960s and 1970s. 1 *J. Geophys. Res., 94*, 18,279-18,288, 1989.

Kleinman, L.I., P.H. Daum, Y-N. Lee, et al., Measurement of O_3 and related compounds over southern Nova Scotia 1. Vertical distributions. *J. Geophys. Res., 101*, 29,043-29,060, 1996.

Kley, D., P. J. Crutzen, H. G. J. Smit, H. Vomel, S.J. Oltmans, H. Grassl and V. Ramanathan, Observations of near-zero ozone concentrations over the convective Pacific: effects on air chemistry. *Science, 274*, 230-233, 1996.

Lacis, A., J. Hansen, P. Lee, T. Mitchell, and S. Lebedeff, Greenhouse effect of trace gases, 1970-1980. *Geophys. Res. Lett., 8*, 1035-1038, 1981.

Lacis, A.A., D. J. Wuebbles, and J. A. Logan, Radiative forcing of climate by changes in the vertical distribution of ozone. *J. Geophys. Res., 95*, 9971-9981, 1990.

Lashof, D. A., The dynamic greenhouse: feedback processes that may influence future concentrations of atmospheric trace gases and climatic change. *Climatic Change, 14*, 213-242, 1989.

Le Texier, L., S. Solomon, and R. R. Garcia, The role of molecular hydrogen and methane oxidation in the water vapor budget of the stratosphere. Q. J. Roy. Meteorol. Soc., 114, 281-296, 1988.

Lelieveld, J., and P. J. Crutzen, The role of clouds in tropospheric photochemistry. J. Atmos. Chem., 12, 229-267, 1991.

Lelieveld, J., and P. J. Crutzen, Indirect chemical effects of methane on global warming, Nature, 355, 339-342, 1992.

Lelieveld, J., P. J. Crutzen, and C. Brühl, Climate effects of atmospheric methane, Chemosphere, 26, 739-768, 1993.

Levine, J. S., C. P. Rinsland, and G.M. Tennille, The photochemistry of methane and carbon monoxide in the troposphere in 1950 and 1985. Nature, 318, 254, 1985.

Liu, S.C., M. McFarland, D. Kley, O. Zafiriou, and B.J. Huebert, Tropospheric NOx and O$_3$ budgets in the equatorial Pacific. J. Geophys. Res., 88, 1360-1368, 1983.

Liu, S.C., R.A. Cox, P. J. Crutzen, D.H. Ehhalt, R. Guicherit, A. Hofzumahaus, D. Kley, S.A. Penkett, L.F. Phillips, D. Poppe, and F.S. Rowland, Group report: Oxidizing capacity of the atmosphere. In The Changing Atmosphere, edited by F.S. Rowland and I.S.A. Isaksen, pp. 219-232. New York: John Wiley & Sons, 1988.

Logan, J. A., Tropospheric ozone: Seasonal behavior, trends, and anthropogenic influence. J. Geophys. Res., 90, 10463-10482, 1985.

Logan J. A., Trends in the vertical distribution of ozone: An analysis of ozonesonde data. J. Geophys. Res., 99, 25,553-25,585, 1994.

Logan, J. A., M. J. Prather, S.C. Wofsy, and M. B. McElroy, Tropospheric Chemistry: A Global Perspective. J. Geophys. Res., 86, 2210-7254, 1981.

Lowe, D.C. and U. Schmidt, Formaldehyde (HCHO) measurements in the nonurban atmosphere. J. Geophys. Res., 88, 10844-10858, 1983.

Lu, Y., and M. A. K. Khalil, Tropospheric OH: Model calculations of spatial, temporal, and secular variations. Chemosphere, 23, 397-444, 1991.

MacDonald, G. J., Role of methane clathrates in past and future climates. Climatic Change, 16, 247-281, 1990.

MacKay, R. M., and M. A. K. Khalil, Theory and development of a one dimensional time dependent radiative convective climate model. Chemosphere, 22, 383-417, 1991.

Miller, A. J., R. M. Nagatani, G. C. Tiao, X. F. Niu, G. C. Reinsel, D. Wuebbles, and K. Grant, Comparisons of observed ozone and temperature trends in the lower stratosphere. Geophys. Res. Lett., 19, 929-932, 1992.

Nisbet, E., Did the release of methane from hydrates accelerate the end of the last ice age? Can. J. Earth Sci., 27, 148-157, 1990a.

Nisbet, E., Climate change and methane. Nature, 347, 23, 1990b.

Novelli, P. C., K. A. Masarie, P. T. Tans, and P. M. Lang, Recent changes in atmospheric carbon monoxide. Science, 263, 1587-1590, 1994.

Oltmans, S. J., and H. Levy II, Surface ozone measurements from a global network. Atmos. Environ., 28, 9-24, 1994.

Owens, A. J., J. M. Steed, D. L. Filkin, C. Miller, and J. P. Jesson, The potential effects of increased methane on atmospheric ozone. Geophys. Res. Letters, 9, 1105-1108, 1982.

Owens, A. J., C. H. Hales, D. L. Filkin, C. Miller, J. M. Steed, and J. P. Jesson, A coupled one-dimensional radiative-convective, chemistry-transport model of the atmosphere: 1. Model structure and steady state perturbation calculations. J. Geophys. Res., 90, 2283-2311, 1985.

Parameswaran, K., and B. V. Krishna Murthy, Altitude profiles of tropospheric water vapor at low latitudes. J. Appl. Meteor., 29, 665-679, 1990.

Penkett, S. A., Indications and causes of ozone increase in the troposphere. In The Changing Atmosphere, edited by F.S. Rowland and I.S.A. Isaksen, pp. 91-103, John Wiley & Sons, New York, 1988.

Pinto, J. P., and M. A. K. Khalil, The stability of tropospheric OH during ice ages, inter-glacial epochs and modern times. *Tellus, 43B*, 347-352, 1991.

Prather, M. J., Lifetimes and eigenstates in atmospheric chemistry. *Geophys. Res. Lett., 21*, 801-804, 1994.

Prinn, R., D. Cunnold, P. Simmonds, F. Alyea, R. Boldi, A. Crawford, P. Fraser, D. Gutzler, D. Hartley, R. Rosen, and R. Rasmussen, Global averageconecntartion and trend for hydroxyl radicals deduced from ALE/GAGE trichloroethane (methyl chloroform) data for 1978-1990. *J. Geophys. Res., 97*, 2445-2461, 1992.

Prinn, R. G., R. F. Weiss, B. R. Miller, J. Huang, F. N. Alyea, D. M. Cunnold, P. J. Fraser, D. E. Hartley, and P.G. Simmonds, Atmospheric trends and lifetime of CH_3CCl_3 and global OH concentrations. *Science, 269*, 187-192, 1995.

Ramanathan, V., The greenhouse theory of climate change: A test by an inadvertent global experiment. *Science, 240*, 293-299, 1988a.

Ramanathan, V., The radiative and climatic consequences of the changing atmospheric composition of trace gases. In *The Changing Atmosphere*, edited by F. S. Rowland and I. S. A. Isaksen, pp. 159-186. New York: John Wiley & Sons, 1988b.

Ramanathan, V., R. J. Cicerone, H. B. Singh, and J. T. Kiehl, Trace gas trends and their potential role in climate change. *J. Geophys. Res., 90*, 5547-5566, 1985.

Ramanathan, V., L. Callis, R. Cess, J. Hansen, I. Isaksen, W. Kuhn, A. Lacis, F. Luther, J. Mahlman, R. Reck, and M. Schlesinger, Climate-chemical interactions and effects of changing atmospheric trace gases. *Rev. of Geophys., 25*, 1441-1482, 1987.

Rasmussen, R. A., and M. A. K. Khalil, Atmospheric methane (CH_4): trends and seasonal cycles. *J. Geophys. Res., 86*, 9,826-9,832, 1981.

Roble, R. G., and R. E. Dickinson, How will changes in carbon dioxide and methane modify the mean structure of the mesosphere and thermosphere? *Geophys. Res. Ltrs., 16*, 1441-1444, 1989.

Rodhe, H., A comparison of the contribution of various gases to the greenhouse effect. *Science, 248*, 1217-1219, 1990.

Schwab, J. J., E. M. Weinstock, J. B. Nee, and J. G. Anderson, In situ measurement of water vapor in the stratosphere with a cryogenically cooled Lyman-alpha hygrometer. *J. Geophys. Res., 95*, 13,781-13,796, 1990.

Shine, K. P., The greenhouse effect. In *Ozone Depletion: Health and Environmental Consequences*, edited by R. R. Jones and T. Wigley, pp. 71-83. John Wiley & Sons, New York, 1989.

Solomon, S., The mystery of the Antarctic ozone hole, *Rev. Geophys., 26,* 131-148, 1988.

Steele, L. P., P. J. Fraser, R. A. Rasmussen, M. A. K. Khalil, T. J. Conway, A. J. Crawford, R. H. Gammon, K. A. Masarie, and K. W. Thoning. The global distribution of methane in the troposphere, *J., Atmos. Chem., 5*, 125-171, 1987.

Stolarski, R. S., P. Bloomfield, R. D. McPeters, and J. R. Herman, Total ozone trends deduced from Nimbus 7 TOMS data. *Geophys. Res. Lett., 18*, 1015-1018, 1991.

Stordal, F., and I. S. A. Isaksen, Ozone perturbations due to increases in N_2O, CH_4, and chlorocarbons: two-dimensional time-dependent calculations. *Tellus, 39B*, 333-353, 1987.

Strand, A., and O. Hov, A two-dimensional zonally averaged transport model including convective motions and a new strategy for the numerical solution. *J. Geophys. Res., 98*, 9023-9027, 1993.

Thomas, G. E., J. J. Olivero, E. J. Jensen, W. Schroeder, and O. B. Toon, Relation between increasing methane and the presence of ice clouds at the mesopause. *Nature, 338*, 490-492, 1989.

Thompson, A. M., The oxidizing capacity of earths atmosphere: probable past and future changes. *Science, 256*, 1157-1165, 1992.

Thompson, A. M., and R. J. Cicerone, Possible perturbations to atmospheric CO, CH_4, and OH. *J. Geophys. Res., 91*, 10853-10864, 1986a.

Thompson, A. M., and R. J. Cicerone, Atmospheric CH_4, CO, and OH from 1860 to 1985. *Nature, 321*, 148-150, 1986b.

Thompson, A. M., and M. Kavanaugh, Tropospheric CH_4/CO/NOx: The next fifty years. in *Effects of Changes in Stratospheric Ozone and Global Climate*, vol. 2, United Nations Environmental Program Report, 1986.

Thompson, A. M., R. W. Stewart, M. A. Owens, and J. A. Herwehe. Sensitivity of tropospheric oxidants to global chemical and climate change. *Atmospheric Environment, 23*, 519-532, 1989.

Thompson, A. M., M. A. Huntley, and R. W. Stewart, Perturbations to tropospheric oxidants, 1985-2035: 1. Calculations of ozone and OH in chemically coherent regions, *J. Geophys. Res., 95*, 9829-9844, 1990.

Tiao, G. C., G. C. Reinsel, J. H. Pedrick, G. M. Allenby, C. L. Mateer, A. J. Miller, and J. J. DeLuisi, A statistical trend analysis of ozonesonde data. *J. Geophys. Res., 91*, 13121-13136, 1986.

Volz, A., and D. Kley, Evaluation of the Monteouris series of ozone measurements made in the nineteenth century. *Nature, 332*, 240-242, 1988.

Wallace, L., and W. Livingston, Spectroscopic observations of atmospheric trace gases over Kitt Peak: 1. Carbon dioxide and methane from 1979 to 1985. *J. Geophys. Res., 85*, 9,823-9,827, 1990.

Wang, W.C., Y. L. Yung, A.A. Lacis, T. Mo, and J. E. Hansen, Greenhouse effects due to man-made perturbations of trace gases. *Science, 194*, 685-690, 1976.

Wang, W.C., and G. Molnar, A model study of the greenhouse effects due to increasing atmospheric CH_4, N_2O, CF_2Cl_2, and $CFCl_3$. *J. Geophys. Res., 90*, 12971-12980, 1985.

Wang, W.C., D.J. Wuebbles, W. M. Washington, R. G. Isaacs, and G. Molnar, Trace gases and other potential perturbations to global climate. *Rev. of Geophysics, 24*, 110-140, 1986.

Wang, W.C., M.P. Dudek, X. Z. Liang, J. T. Kiehl, Inadequacy of effective CO_2 as a proxy in simulating the greenhouse effect of other radiatively active gases. *Nature, 350*, 573-577, 1991.

Wennberg, P.O., R. C. Cohen, R. M. Stimpfle, J. P. Koplow et al., Removal of stratospheric O_3 by radicals: In situ measurements of OH, HO_2, NO, NO_2, ClO, and BrO. *Science, 266*, 398-404, 1994.

Wigley, T. M. L., Relative contributions of different trace gases to the greenhouse effect. *Climate Monitor, 16*, 14-28, 1987.

World Meteorological Organization, Atmospheric Ozone 1985, WMO Global Ozone Res. and Monit. Proj., Report 16, Geneva: WMO, 1985.

World Meteorological Organization, Report of the International Ozone Trends Panel 1988, Global Ozone Res. and Monit. Proj., Report No. 18, Geneva: WMO, 1988.

World Meteorological Organization, Scientific Assessment of Stratospheric Ozone 1989, Global Ozone Res. and Monit. Proj., Report 20, Geneva: WMO, 1989.

World Meteorological Organization, Scientific Assessment of Ozone Depletion: 1991, Global Ozone Res. and Monit. Proj. Report 25, Geneva: WMO, 1991.

World Meteorological Organization, Scientific Assessment of Ozone Depletion: 1994, Global Ozone Research and Monitoring Project, Report No. 37, Geneva: WMO, 1995.

Wuebbles, D. J. and J. Edmonds, A primer on greenhouse gases. Lewis Publishers, Chelsea, MI, 1991.

Wuebbles, D. J., and K.E. Grant, Indirect effects on climatic forcing from stratospheric water vapor resulting from increased concentrations of CH_4 and H_2. Lawrence Livermore National Laboratory, 1991; results also described in WMO, 1991.

Wuebbles, D. J., and D. E. Kinnison, Predictions of future ozone changes. *Int. J. Environ. Studies, 51*, 269-283, 1996.

Wuebbles, D. J., and J. Tamaresis, The role of methane in the global environment., *Atmospheric Methane*, M.A.K. Khalil, editor, Springer-Verlag Publishers, 1993.

Wuebbles, D. J., F. M. Luther, and J. E. Penner, Effect of coupled anthropogenic perturbations on stratospheric ozone. *J. Geophys. Res.*, *88*, 1,444-1,456, 1983.

Wuebbles, D. J., K. E. Grant, P. S. Connell, and J. E. Penner, The role of atmospheric chemistry in climate change. *JAPCA*, *39*, 22-28, 1989.

Wuebbles, D. J., D. E. Kinnison, K. E. Grant, and J. Lean, The effect of solar flux variations and trace gas emissions on recent trends in stratospheric ozone and temperature. *J. Geomagnetism and Geoelectricity*, *43*, 709-718, 1991a.

Wuebbles, D. J., J. S. Tamaresis, and D. E. Kinnison, Effects of increasing methane on tropospheric and stratospheric chemistry, paper presented at the NATO Advanced Research Workshop on the Atmospheric Methane Cycle: Sources, Sinks, Distributions, and Role in Global Change, Portland, Oregon, 1991b.

Zander, R., H. Demoulin, D.H. Ehhalt, U. Schmidt, and C. P. Rinsland, Secular increase of the total vertical column abundance of carbon monoxide above central Europe since 1950. *J. Geophys. Res.*, *94*, 11,021-11,028, 1989.

Index

Printing: Druckhaus Beltz, Hemsbach
Binding: Buchbinderei Schäffer, Grünstadt